W0234849

Handbook of
Soil Chemistry

Handbook of
Soil Chemistry

Contributors :
**Rebecca Tirado-Corbalá,
Ray G. Anderson,** *et al.*

KOROS PRESS LIMITED
London, UK

Handbook of Soil Chemistry
Contributors : Rebecca Tirado-Corbalá *and* Ray G. Anderson, *et al.*

Published by Koros Press Limited

www.korospress.com

United Kingdom

Copyright 2016

Printed in 2017 for Sale in the Indian Subcontinent

Handbook of Soil Chemistry

ISBN: 978-1-78163-558-2

British Library Cataloguing in Publication Data
A CIP record for this book is available from the British Library

Exclusively distributed by CBS Publishers & Distributors Pvt. Ltd.

Sales & Distribution Rights only for India, Pakistan, Bangladesh, Sri Lanka, Nepal and Bhutan.This book is not to be sold outside these territories.

PREFACE

This book is intended as a reference for chemists and environmentalists who find that they need to analyze soil, interpret soil analysis, or develop analytical or instrumental analysis for soil. Soil scientists will also find it valuable when confronted by soil analyses that are not correct or appear to be incorrect or when an analysis does not work at all.

There are two themes in this work: (1) that all soil is complex and (2) that all soil contains water. The complexity of soil cannot be overemphasized. It contains inorganic and organic atoms, ions, and molecules in the solid, liquid, and gaseous phases. All these phases are both in quasi-equilibrium with each other and constantly changing. This means that the analysis of soil is subject to complex interferences that are not commonly encountered in standard analytical problems. The overlap of emission or absorption bands in spectroscopic analysis is only one example of the types of interferences likely to be encountered.

Soil is the most complicated of materials and is essential to life. It may be thought of as the loose material covering the dry surface of the earth, but it is much more than that. To become soil, this material must be acted on by the soil forming factors: time, biota, topography, climate, and parent material. These factors produce a series of horizons in the soil that make it distinct from simply ground-up rock. Simply observing a dark-colored surface layer overlying a reddish lower layer shows that changes in the original parent material have taken place. The many organisms growing in and on soil, including large, small, and microscopic plants, animals, and microorganisms also differentiate soil from ground-up rock.

There are other less obvious physical changes constantly taking place in soil. Soil temperature changes dramatically from day to night, week to week, and season to season. Even in climates where the air temperature is relatively constant, soil temperatures can vary by 20 degrees or more from day to night. Moisture levels can change from saturation to air dry, which can be as low as 1% moisture on a dry-weight basis. These changes have dramatic effects on the chemical reactions in the soil. Changes in soil water content alter the concentration of soil constituents and thus also their reaction rates.

Not only are soil's physical and observable characteristics different from those of ground-up rock; so also are its chemical characteristics. Soil is a mixture of inorganic and organic solids, liquids, and gases. In these phases can be found inorganic and organic molecules, cations, and anions. Inorganic and organic components can be preset as simple or complex ions. Crystalline material occur having different combinations of components from, for example, 1 : 1 and 2 : 1 clay minerals, leading to different structures with different physical and chemical characteristics with different surface functionalities and chemical reactivities.

CONTENTS

This page left intentionally blank.

LIST OF CONTRIBUTORS

Rebecca Tirado-Corbalá

USDA, Agricultural Research Service, San Joaquin Valley Agricultural Sciences Center, Water Management Research Unit, 9611 S. Riverbend Ave., Parlier, CA 93648-9757, USA; E-Mails: ray.anderson@ars.usda.gov (R.G.A.); dong.wang@ars.usda.gov (D.W.); james.ayars@ars.usda.gov (J.E.A.)

and

Department of Agro-environmental Sciences, University of Puerto Rico, Mayagüez, Puerto Rico, USA

Ray G. Anderson

USDA, Agricultural Research Service, San Joaquin Valley Agricultural Sciences Center, Water Management Research Unit, 9611 S. Riverbend Ave., Parlier, CA 93648-9757, USA; E-Mails: ray.anderson@ars.usda.gov (R.G.A.); dong.wang@ars.usda.gov (D.W.); james.ayars@ars.usda.gov (J.E.A.)

and

USDA-Agricultural Research Service, U.S. Salinity Laboratory Contaminant Fate and Transport Unit, 450 W. Big Springs Rd., Riverside, CA 92507-4617, USA

Dong Wang

USDA, Agricultural Research Service, San Joaquin Valley Agricultural Sciences Center, Water Management Research Unit, 9611 S. Riverbend Ave., Parlier, CA 93648-9757, USA; E-Mails: ray.anderson@ars.usda.gov (R.G.A.); dong.wang@ars.usda.gov (D.W.); james.ayars@ars.usda.gov (J.E.A.)

James E. Ayars

USDA, Agricultural Research Service, San Joaquin Valley Agricultural Sciences Center, Water Management Research Unit, 9611 S. Riverbend Ave., Parlier, CA 93648-9757, USA; E-Mails: ray.anderson@ars.usda.gov (R.G.A.); dong.wang@ars.usda.gov (D.W.); james.ayars@ars.usda.gov (J.E.A.)

This page left intentionally blank.

Chapter 1

INTRODUCTION TO SOIL CHEMISTRY

SOIL DESCRIPTION AND CLASSIFICATION

It is necessary to adopt a formal system of soil description and classification in order to describe the various materials found in ground investigation. Such a system must be comprehensive (covering all but the rarest of deposits), meaningful in an engineering context *(so that engineers will be able to understand and interpret)* and yet relatively concise. It is important to distinguish between description and classification :

Description of soil is a statement describing the physical nature and state of the soil. It can be a description of a sample, or a soil *in situ*. It is arrived at using visual examination, simple tests, observation of site conditions, geological history, etc.

Soil classification is the separation of soil into classes or groups each having similar characteristics and potentially similar behaviour. A classification for engineering purposes should be based mainly on mechanical properties, *e.g.* permeability, stiffness, strength. The class to which a soil belongs can be used in its description.

Basic Characteristics of Soils

Soils consist of grains (mineral grains, rock fragments, etc.) with water and air in the voids between grains. The water and air contents are readily changed by changes in conditions and location : soils can be perfectly dry (have no water content) or be fully saturated (have no air content) or be partly saturated (with both air and water present). Although the size and shape of the solid (granular) content rarely changes at a given point, they can vary considerably from point to point.

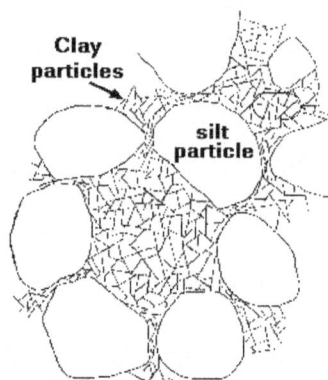

First of all, consider soil as a engineering material – it is not a coherent solid material like steel and concrete, but is a particulate material. It is important to understand the significance of particle size, shape and composition, and of a soil's internal structure or fabric.

Soil as an Engineering Material

The term "soil" means different things to different people : To a geologist it represents the products of past surface processes. To a pedologist it represents currently occurring physical and chemical processes. To an engineer it is a material that can be :

> **built on** : foundations to buildings, bridges.
> **built in** : tunnels, culverts, basements.
> **built with** : roads, runways, embankments, dams.
> **supported** : retaining walls, quays.

Soils may be **described** in different ways by different people for their different purposes. Engineers' descriptions give engineering terms that will convey some sense of a soil's current state and probable susceptibility to future changes (*e.g.* in loading, drainage, structure, surface level).

Engineers are primarily interested in a soil's mechanical properties : strength, stiffness, permeability. These depend primarily on the nature of the soil grains, the current stress, the water content and unit weight.

Size Range of Grains

clay	silt	sand	gravel	cobble	boulder

| | 0.006 | 0.02 | 6 | 20 | 6 | 20 | |
0.002mm   0.06mm   2mm   60mm   200mm

The range of particle sizes encountered in soil is very large : from boulders with a controlling dimension of over 200 mm down to clay particles less than 0.002 mm (2μm). Some clays contain particles less than 1 μμ in size which behave as colloids, *i.e.* do not settle in water due solely to gravity.

In the British Soil Classification System, soils are classified into named Basic Soil Type groups according to size, and the groups further divided into coarse, medium and fine sub-groups :

Very coarse soils	BOULDERS		> 200 mm
	COBBLES		60 - 200 mm
Coarse soils	G GRAVEL	coarse	20 - 60 mm
		medium	6 - 20 mm
		fine	2 - 6 mm
	S SAND	coarse	0.6 - 2.0 mm
		medium	0.2 - 0.6 mm
		fine	0.06 - 0.2 mm
Fine soils	M SILT	coarse	0.02 - 0.06 mm
		medium	0.006 - 0.02 mm
		fine	0.002 - 0.006 mm
	C CLAY		< 0.002 mm

Aids to Size Identification

Soils possess a number of physical characteristics which can be used as aids to size identification in the field. A handful of soil rubbed through the fingers can yield the following :

SAND (and coarser) particles are visible to the naked eye.

SILT particles become dusty when dry and are easily brushed off hands and boots.

CLAY particles are greasy and sticky when wet and hard when dry, and have to be scraped or washed off hands and boots.

Shape of Grains

- Shape characteristics of SAND grains
- Shape characteristics of CLAY grains
- Specific surface

The majority of soils may be regarded as either SANDS or CLAYS :

SANDS include gravelly sands and gravel-sands. Sand grains are generally broken rock particles that have been formed by physical weathering, or they are the resistant components of rocks broken down by chemical weathering. Sand grains generally have a **rotund** shape.

CLAYS include silty clays and clay-silts; there are few pure silts (*e.g.* areas formed by windblown Löess). Clay grains are usually the product of chemical weathering or rocks and soils. Clay particles have a **flaky** shape.

There are major differences in engineering behaviour between SANDS and CLAYS (*e.g.* in permeability, compressibility, shrinking/swelling potential). The shape and size of the soil grains has an important bearing on these differences.

Shape Characteristics of SAND Grains

SAND and larger-sised grains are **rotund**. Coarse soil grains (silt-sised, sand-sised and larger) have different shape characteristics and surface roughness depending on the amount of wear during transportation (by water, wind or ice), or after crushing in manufactured aggregates. They have a relatively low specific surface (surface area).

Rounded : Water- or air-worn; transported sediments

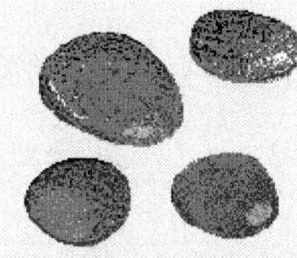

Irregular : Irregular shape with round edges; glacial sediments (sometimes sub-divided into 'sub-rounded' and 'sub-angular')

Angular : Flat faces and sharp edges; residual soils, grits

Flaky : Thickness small compared to length/breadth; clays

Elongated : Length larger than breadth/thickness; scree, broken flagsto

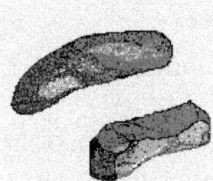

Flaky & Elongated : Length>Breadth>Thickness; broken schists and slates

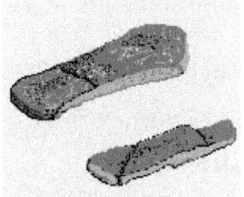

Shape Characteristics of CLAY Grains

CLAY particles are **flaky**. Their thickness is very small relative to their length & breadth, in some cases as thin as 1/100th of the length. They therefore have high to very high specific surface values. These surfaces carry a small negative electrical charge, that will attract the positive end of water molecules. This charge depends on the soil mineral and may be affected by an electrolite in the pore water. This causes some additional forces between the soil grains which are proportional to the specific surface. Thus a lot of water may be held asadsorbed water within a clay mass.

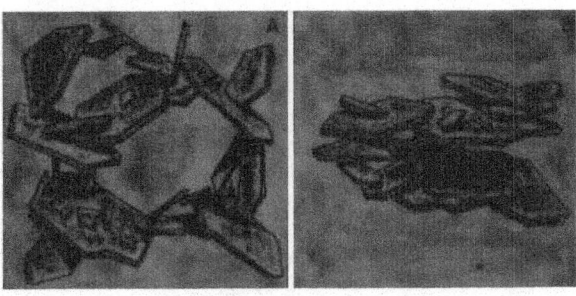

Anordnung der Tonteilchen nach der Sedimentation
A) in Meerwasser and B) in Susswasser.

Specific Surface

- Examples

Specific surface is the ratio of surface area per unit wight.

Surface forces are proportional to surface area (*i.e.* to d^2).

Self-weight forces are proportional to volume (*i.e.* to d^3).

Therefore $\qquad \dfrac{\text{Surface force}}{\text{self weight forces}} \propto \dfrac{1}{d}$

Also, specific surface = $\dfrac{\text{area}}{\rho * \text{volume}} \propto \dfrac{1}{d}$

Hence, specific surface is a measure of the relative contributions of surface forces and self-weight forces.

The specific surface of a 1mm cube of quartz ($\rho = 2.65 \, \text{gm}/\text{cm}^3$) is 0.00023 m^2/N

SAND grains (size 2.0 - 0.06 mm) are close to cubes or spheres in shape, and have specific surfaces near the minimum value.

CLAY particles are flaky and have much greater specific surface values.

Examples of Specific Surface

The more elongated or flaky a particle is the greater will be its specific surface.

Click on the following examples :

cubes

Cubes

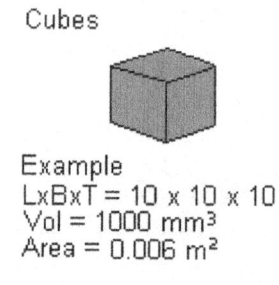

Example
LxBxT = 10 x 10 x 10
Vol = 1000 mm³
Area = 0.006 m²

Rods

Sheets

Example
LxBxT=1000x1000x0.001
Vol = 1000 mm³
Area = 20 m²

sheets

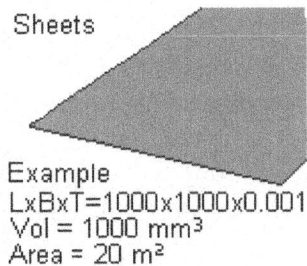

Sheets

Example
LxBxT=1000x1000x0.001
Vol = 1000 mm³
Area = 20 m²

Examples of mineral grain specific surfaces :

Mineral/Soil	Grain width d (μm)	Thickness	Specific Surface m²/N
Quartz grain	100	d	0.0023
Quartz sand	2.0 - 0.06	≈d	0.0001 - 0.004
Kaolinite	2.0 - 0.3	≈0.2d	2
Illite	2.0 - 0.2	≈0.1d	8
Montmorillonite	1.0 - 0.01	≈0.01d	80

Structure or Fabric

Natural soils are rarely the same from one point in the ground to another. The content and nature of grains varies, but more importantly, so does the arrangement of these.

The arrangement and organisation of particles and other features within a soil mass is termed its **structure** or **fabric**. This includes bedding orientation, stratification, layer thickness, the occurrence of joints and fissures, the occurrence of voids, artefacts, tree roots and nodules, the presence of cementing or bonding agents between grains.

Structural features can have a major influence on *in situ* properties.

- Vertical and horizontal permeabilities will be different in alternating layers of fine and coarse soils.
- The presence of fissures affects some aspects of strength.
- The presence of layers or lenses of different stiffness can affect stability.
- The presence of cementing or bonding influences strength and stiffness.

ORIGINS, FORMATION AND MINERALOGY

Soils are the results of geological events (except for the very small amount produced by man). The nature and structure of a given soil depends on the geological processes that formed it :

breakdown of parent rock : weathering, decomposition, erosion.

transportation to site of final deposition : gravity, flowing water, ice, wind.

environment of final deposition : flood plain, river terrace, glacial moraine, lacustrine or marine.

subsequent conditions of loading and drainage — little or no surcharge, heavy surcharge due to ice or overlying deposits, change from saline to freshwater, leaching, contamination.

Origins of Soils from Rocks

All soils originate, directly or indirectly, from solid rocks in the Earth's crust :

igneous rocks

crystalline bodies of cooled magma

e.g. granite, basalt, dolerite, gabbro, syenite, porphyry

sedimentary rocks

layers of consolidated and cemented sediments, mostly formed in bodies of water (seas, lakes, etc.)
e.g. limestone, sandstones, mudstone, shale, conglomerate

metamorphic rocks

formed by the alteration of existing rocks due to heat from igneous intrusions (*e.g.* marble, quartzite, hornfels) or pressure due to crustal movement (*e.g.* slate, schist, gneiss).

Weathering of Rocks

Physical Weathering

Physical or mechanical processes taking place on the Earth's surface, including the actions of water, frost, temperature changes, wind and ice; cause disintegration and wearing. The products are mainly coarse soils (silts, sands and gravels). Physical weathering produces Very Coarse soils and Gravels consisting of broken rock particles, but Sands and Silts will be mainly consists of mineral grains.

Chemical Weathering

Chemical weathering occurs in wet and warm conditions and consists of degradation by decomposition and/or alteration. The results of chemical weathering are generally fine soils with separate mineral grains, such as Clays and Clay-Silts. The type of clay mineral depends on the parent rock and on local drainage. Some minerals, such as quartz, are resistant to the chemical weathering and remain unchanged.

Quartz

A resistant and enduring mineral found in many rocks (*e.g.* granite, sandstone). It is the principal constituent of sands and silts, and the most abundant soil mineral. It occurs as equidimensional hard grains.

Haematite

A red iron (ferric) oxide : resistant to change, results from extreme weathering. It is responsible for the widespread red or pink colouration in rocks and soils. It can form a cement in rocks, or a duricrust in soils in arid climates.

Micas

Flaky minerals present in many igneous rocks. Some are resistant, *e.g.* muscovite; some are broken down, *e.g.* biotite.

Clay Minerals

These result mainly from the breakdown of feldspar minerals. They are very flaky and therefore have very large surface areas. They are major constituents of clay soils, although clay soil also contains silt sised particles.

Clay Minerals

Clay minerals are produced mainly from the chemical weathering and decomposition of feldspars, such as orthoclase and plagioclase, and some micas. They are small in size and very flaky in shape.

The key to some of the properties of clay soils, *e.g.* plasticity, compressibility, swelling/shrinkage potential, lies in the structure of clay minerals.

There are three main groups of clay minerals :

kaolinites

(include kaolinite, dickite and nacrite) formed by the decomposition of or-thoclase feldspar (*e.g.* in granite); kaolin is the principal constituent in china clay and ball clay.

illites

(include illite and glauconite) are the commonest clay minerals; formed by the decomposition of some micas and feldspars; predominant in marine clays and shales (*e.g.* London clay, Oxford clay).

montmorillonites

(also called smectites or fullers' earth minerals) (include calcium and sodium momtmorillonites, bentonite and vermiculite) formed by the alteration of basic igneous rocks containing silicates rich in Ca and Mg; weak linkage by cations (*e.g.* Na+, Ca++) results in high swelling/shrinking potential.

Transportation and Deposition

The effects of weathering and transportation largely determine the basic **nature** of the soil (*i.e.* the size, shape, composition and distribution of the grains). The environment into which deposition takes place, and subsequent geological events that take place there, largely determine the **state** of the soil, (*i.e.* density, moisture content) and the **structure** or fabric of the soil (*i.e.* bedding, stratification, occurrence of joints or fissures, tree roots, voids, etc.)

Transportation

Due to combinations of gravity, flowing water or air, and moving ice. In water or air : grains become sub-rounded or rounded, grain sizes are sorted, producing poorly-graded deposits. In moving ice : grinding and crushing occur, size distribu-tion becomes wider, deposits are well-graded, ranging from rock flour to boulders.

Deposition

In flowing water, larger particles are deposited as velocity drops, *e.g.* gravels in river terraces, sands in floodplains and estuaries, silts and clays in lakes and seas. In still water : horizontal layers of successive sediments are formed, which may change with time, even seasonally or daily.

- Deltaic & shelf deposits : often vary both horizontally and vertically.
- From glaciers, deposition varies from well-graded basal tills and boulder clays to poorly-graded deposits in moraines and outwash fans.
- In arid conditions : scree material is usually poorly-graded and lies on slopes.
- Wind-blown Löess is generally uniformly-graded and false-bedded.

Loading and Drainage History

The current state (*i.e.* density and consistency) of a soil will have been profoundly influenced by the history of loading and unloading since it was deposited. Changes in drainage conditions may also have occurred which may have brought about changes in water content.

Loading/Unloading History

Initial Loading

During deposition the load applied to a layer of soil increases as more layers are deposited over it; thus, it is compressed and water is squeezed out; as deposition continues, the soil becomes stiffer and stronger.

Unloading

The principal natural mechanism of unloading is erosion of overlying layers. Unloading can also occur as overlying ice-sheets and glaciers retreat, or due to large excavations made by man. Soil expands when it is unloaded, but not as much as it was initially compressed; thus it stays compressed — and is said to be over-consolidated. The degree of overconsolidation depends on the history of loading and unloading.

Drainage History

Chemical Changes

Some soils initially deposited loosely in saline water and then inundated with fresh water develop weak collapsing structure. In arid climates with intermittent rainy periods, cycles of wetting and drying can bring minerals to the surface to form a cemented soil.

Climate Changes

Some clays (*e.g.* montmorillonite clays) are prone to large volume changes due to wetting and drying; thus, seasonal changes in surface level occur, often causing foundation damage, especially after exceptionally dry summers. Trees extract water from soil in the process of evapotranspiration; The soil near to trees can therefore either shrink as trees grow larger, or expand following the removal of large trees.

Grading and Composition

The recommended standard for soil classification is the British Soil Classification System, and this is detailed in BS 5930 Site Investigation.

Coarse soils

- Particle size tests
- Typical grading curves
- Grading characteristics
- Sieve analysis example

Coarse soils are classified principally on the basis of particle size and grading.

Very coarse soils	BOULDERS		> 200 mm
	COBBLES		60 - 200 mm
Coarse soils	G GRAVEL	coarse	20 - 60 mm
		medium	6 - 20 mm
		fine	2 - 6 mm
	S SAND	coarse	0.6 - 2.0 mm
		medium	0.2 - 0.6 mm
		fine	0.06 - 0.2 mm

Particle Size Tests

The aim is to measure the distribution of particle sizes in the sample. When a wide range of sizes is present, the sample will be sub-divided, and separate tests carried out on each sub-sample. Full details of tests are given in BS 1377 : "Methods of test for soil for civil engineering purposes".

Particle-size Tests

Wet sieving to separate fine grains from coarse grains is carried out by washing the soil specimen on a 60μm sieve mesh.

Dry sieving analyses can only be carried out on particles > 60 μm. Samples (with fines removed) are dried and shaken through a nest of sieves of descending size. **Sedimentation** is used only for fine soils. Soil particles are allowed to settle from a suspension. The decreasing density of the suspension is measured at time intervals. Sizes are determined from the settling velocity and times recorded. Percentages between sizes are determined from density differences.

Particle-size Analysis

The cumulative percentage quantities finer than certain sizes (*e.g.* passing a given size sieve mesh) are determined by weighing. Points are then plotted of **per cent finer (passing)** against **log size**. A smooth S-shaped curve drawn through these points is called a **grading curve**. The position and shape of the grading curve determines the soil class. Geometrical **grading characteristics** can be determined also from the grading curve.

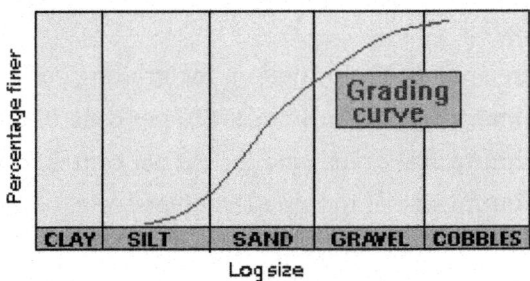

Typical grading curves

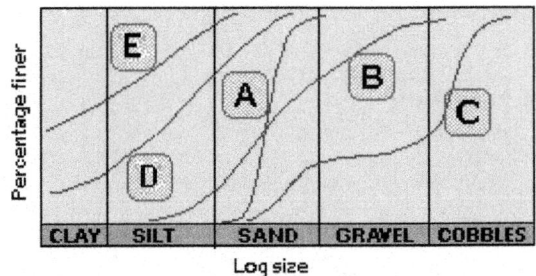

Both the position and the shape of the grading curve for a soil can aid its identity and description.

Some typical grading curves are shown in the figure :

A - a poorly-graded medium SAND (probably estuarine or flood-plain alluvium)

B - a well-graded GRAVEL-SAND (*i.e.* equal amounts of gravel and sand)

C - a gap-graded COBBLES-SAND

D - a sandy SILT (perhaps a deltaic or estuarine silt)

E - a typical silty CLAY (*e.g.* London clay, Oxford clay)

Grading Characteristics

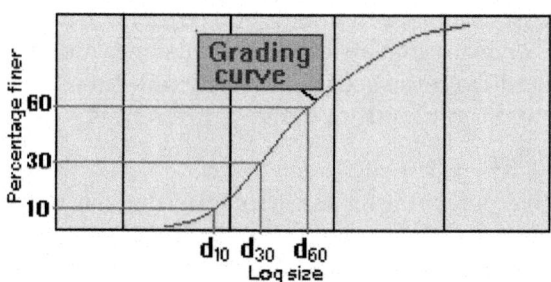

A grading curve is a useful aid to soil description. Grading curves are often included in ground investigation reports. Results of grading tests can be tabulated using geometric properties of the grading curve. These properties are called **grading characteristics**

First of all, three points are located on the grading curve :

d_{10} = the maximum size of the smallest 10 per cent of the sample

d_{30} = the maximum size of the smallest 30 per cent of the sample

d_{60} = the maximum size of the smallest 60 per cent of the sample

From these the grading characteristics are calculated :

Effective size

d_{10}

Uniformity coefficient

$C_u = d_{60} / d_{10}$

Coefficient of gradation

$C_k = d_{30}^2 / d_{60} d_{10}$

Both C_u and C_k will be 1 for a single-sised soil

Cu > 5 indicates a well-graded soil

Cu < 3 indicates a uniform soil

Ck between 0.5 and 2.0 indicates a well-graded soil

Ck < 0.1 indicates a possible gap-graded soil.

Sieve Analysis Example

The results of a dry-sieving test are given below, together with the grading analysis and grading curve. Note carefully how the tabulated results are set out and calculated. The grading curve has been plotted on special semi-logarithmic paper; you can also do this analysis using a spreadsheet.

Sieve mesh size (mm)	Mass retained (g)	Percentage retained	Percentage finer (passing)
14.0	0	0	100.0
10.0	3.5	1.2	98.8
6.3	7.6	2.6	86.2
5.0	7.0	2.4	93.8
3.35	14.3	4.9	88.9
2.0	21.1	7.2	81.7
1.18	56.7	19.4	62.3
0.600	73.4	25.1	37.2
0.425	22.2	7.6	29.6
0.300	26.9	9.2	20.4
0.212	18.4	6.3	14.1
0.150	15.2	5.2	8.9
0.063	17.5	6.0	2.9
Pan	8.5	2.9	
TOTAL	292.3	100.0	

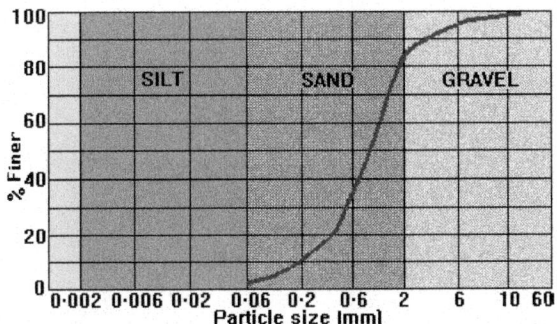

The soil comprises : 18 per cent gravel, 45 per cent coarse sand, 24 per cent medium sand, 10 per cent fine sand, 3 per cent silt, and is classified therefore as : a well-graded gravelly SAND.

Fine Soils

- Consistency limits and plasticity
- Plasticity index
- The plasticity chart and classification
- Activity.

In the case of fine soils (*e.g.* CLAYS and SILTS), it is the shape of the particles rather than their size that has the greater influence on engineering properties. Clay soils have flaky particles to which water adheres, thus imparting the property of plasticity.

Consistency Limits and Plasticity

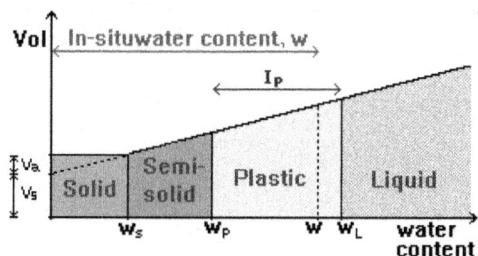

Consistency varies with the water content of the soil. The consistency of a soil can range from (dry) *solid* to *semi-solid* to *plastic* to *liquid* (wet). The water contents at which the consistency changes from one state to the next are called **consistency limits** (or **Atterberg limits**).

Two of these are utilised in the classification of fine soils :

Liquid limit (w_L) - change of consistency from plastic to liquid

Plastic limit (w_P) - change of consistency from brittle/crumbly to plastic

Measures of liquid and plastic limit values can be obtained from laboratory tests.

Plasticity Index

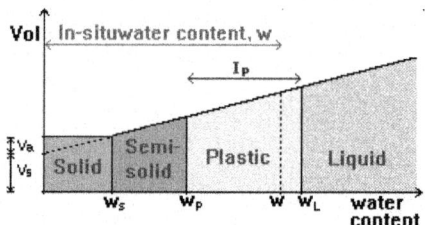

The consistency of most soils in the ground will be plastic or semi-solid. Soil strength and stiffness behaviour are related to the range of plastic consistency. The range of water content over which a soil has a plastic consistency is termed the **Plasticity Index** (I_P or PI).

$$I_P = \text{liquid limit - plastic limit}$$
$$= w_L - w_P$$

The Plasticity Chart and Classification

In the BSCS fine soils are divided into *ten* classes based on their measured plasticity index and liquid limit values : CLAYS are distinguished from SILTS, and five divisions of plasticity are defined :

Low plasticity	w_L = < 35 per cent
Intermediate plasticity	w_L = 35 - 50 per cent
High plasticity	w_L = 50 - 70 per cent
Very high plasticity	w_L = 70 - 90 per cent
Extremely high plasticity	w_L = > 90 per cent

A plasticity chart is provided to aid classification.

Activity

So-called 'clay' soils are not 100 per cent clay. The proportion of clay mineral flakes (< 2 μm size) in a fine soil affects its current state, particularly its tendency to swell and shrink with changes in water content. The degree of plasticity related to the clay content is called the **activity** of the soil.

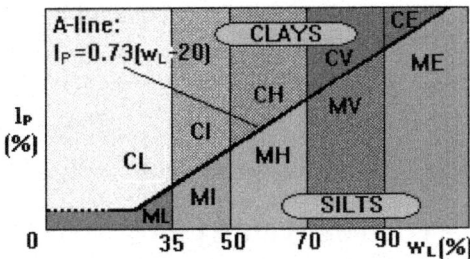

Activity = I_p / (per cent clay particles)

Some typical values are :

Mineral	Activity	Soil	Activity
Muscovite	0.25	Kaolin clay	0.4-0.5
Kaolinite	0.40	Glacial clay and loess	0.5-0.75
Illite	0.90	Most British clays	0.75-1.25
Montmorillonite	> 1.25	Organic estuarine clay	> 1.25

Specific Gravity

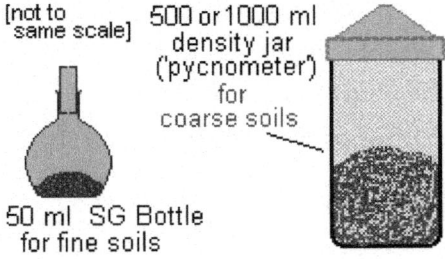

[not to same scale]

500 or 1000 ml density jar ('pycnometer') for coarse soils

50 ml SG Bottle for fine soils

Specific gravity (G_s) is a property of the mineral or rock material forming soil grains. It is defined as

$$G_S = \frac{\text{mass of a soil grain}}{\text{mass of an equal volume of water}}$$

Method of Measurement

For fine soils a 50 ml density bottle may be used; for coarse soils a 500 ml or 1000 ml jar. The jar is weighed empty (M_1). A quantity of dry soil is placed in the jar and the jar weighed (M_2). The jar is filled with water, air removed by stirring, and weighed again (M_3). The jar is emptied, cleaned and refilled with water - and weighed again (M_4).

$$G_S = \frac{\text{Mass of soil}}{\text{Mass of water displaced by soil}}$$

$$= \frac{M_2 - M_1}{\left(M_4 - M_1\right) - \left(M_3 - M_2\right)}$$

[The range of G_s for common soils is 2.64 to 2.72]

VOLUME-WEIGHT PROPERTIES

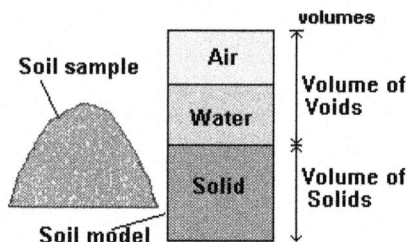

The volume-weight properties of a soil define its state. Measures of the amount of void space, amount of water and the weight of a unit volume of soil are required in engineering analysis and design.

Soil comprises three constituent phases :

Solid : rock fragments, mineral grains or flakes, organic matter.

Liquid : water, with some dissolved compounds (*e.g.* salts).

Gas : air or water vapour.

In natural soils the three phases are intermixed. To aid analysis it is convenient to consider a **soil model** in which the three phases are seen as separate, but still in their correct proportions.

Volumes of Solid, Water and Air : The Soil Model

- Degree of saturation
- Air-voids content

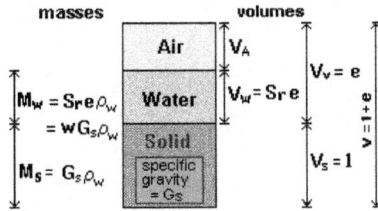

The soil model is given dimensional va ues for the solid, water and air components : Total volume, $V = V_s + V_w + V_a$

Since the amounts of both water and air are variable, the volume of solids present is taken as the reference quantity. Thus, the following relational volumetric quantities may be defined :

$$\text{Void ratio, } e = \frac{\text{Volume of voids}}{\text{Volume of solids}} = \frac{V_v}{V}$$

$$\text{Porosity, } n = \frac{\text{Volume of voids}}{\text{Total volume}} = \frac{V_v}{V}$$

$$\text{Specific volume, } v = 1 + e = \frac{V}{V_s}$$

Note also that :

$n = e / (1 + e)$

$e = n / (1 - n)$

$v = 1 / (1 - n)$

Typical void ratios might be 0.3 (*e.g.* for a dense, well graded granular soil) or 1.5 (*e.g.* for a soft clay).

Degree of Saturation

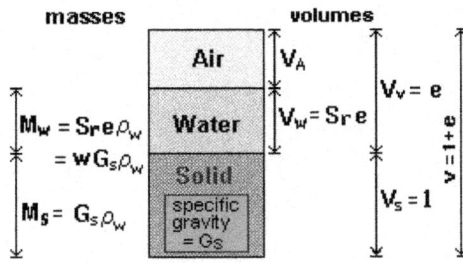

The volume of water in a soil can only vary between zero (*i.e.* a dry soil) and the volume of voids; this can be expressed as a ratio :

$$\text{Degree of saturation, } S_r = \frac{\text{Volume of water}}{\text{Volume of voids}} = \frac{V_w}{V_v}$$

For a perfectly *dry* soil :

$S_r = 0$

For a *saturated* soil :

$S_r = 1$

Note : In clay soils as the amount water increases the volume and therefore the volume of voids will also increase, and so the degree of saturation may remain at Sr = 1 while the actual volume of water is increasing.

Air-voids Content

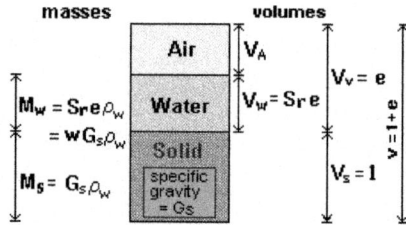

The **air-voids volume, V_a**, is that part of the void space not occupied bwater.

$V_a = V_v - V_w$

$= e - e.S_r$

$= e.(1 - S_r)$

Air-voids content, A_v

A_v = (air-voids volume) / (total volume)

$= Va / V$

$= e.(1 - S_r) / (1+e)$

$= n.(1 - S_r)$

For a perfectly dry soil :

$Av = n$

For a saturated soil :

$Av = 0$

Masses of Solid and Water : Water Content

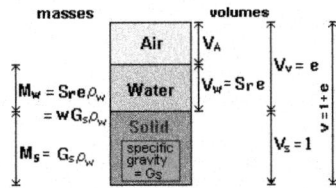

The mass of air may be ignored. The mass of solid particles is usually expressed in terms of their particle density or grain specific gravity.

Grain Specific Gravity

$$G_S = \frac{\text{mass of a soil grain}}{\text{mass of an equal volume of water}}$$

Hence the mass of solid particles in a soil

$M_s = V_s \cdot G_s \cdot \rho_w$

(ρ_w = density of water = 1.00 Mg/m³)

[Range of G_s for common soils : 2.64-2.72]

Particle Density

rs = mass per unit volume of particles

= Gs .rw

The ratio of the mass of water present to the mass of solid particles is called the *water* content, or sometimes the moisture content.

$$\text{Water content, } w = \frac{\text{Mass of water}}{\text{Mass of solids}} = \frac{M_w}{M_s}$$

From the soil model it can be seen that

$w = (S_r \cdot e \cdot \rho_w) / (G_s \cdot \rho_w)$

Giving the useful relationship :

w .Gs = Sr .e

Densities and Unit Weights

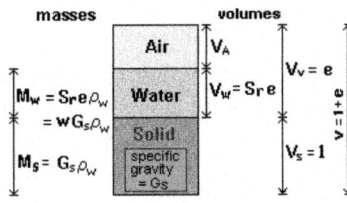

Density is a measure of the quantity of *mass* in a unit volume of material. **Unit weight** is a measure of the *weight* of a unit volume of material.

There are two basic measures of density or unit weight applied to soils : Dry density is a measure of the amount of *solid* particles per unit volume. Bulk density is a measure of the amount of *solid + water* per unit volume.

$$\text{Dry density, } \rho_d = \frac{\text{Mass of solids}}{\text{Total volume}} = \frac{M_S}{V} = \frac{G_s \rho_w}{1 + e}$$

$$\text{Bulk density, } \rho = \frac{\text{Total mass}}{\text{Total volume}} = \frac{M_S + M_w}{V} = \frac{G_s \rho_w + S_r e \rho_w}{1 + e}$$

The preferred units of density are :

Mg/m^3, kg/m^3 or g/ml.

The corresponding *unit weights* are :

$$\text{Dry unit weight, } \gamma_d = \frac{\text{Dry weight}}{\text{Total volume}} = \frac{W_s}{V} = \frac{G_s \gamma_w}{1 + e} = 9.81 \, \rho_d$$

$$\text{Unit weight, } \gamma = \frac{\text{Total weight}}{\text{Total volume}} = \frac{W_s + W_w}{V} = \frac{G_s \gamma_w + S_r e \gamma_w}{1 + e} = 9.81 \, \rho$$

Also, it can be shown that

$\rho = \rho_d (1 + w)$ and

$\gamma = \gamma_d (1 + w)$

Laboratory Measurements

- Water content
- Unit weight

It is important to quantify the state of a soil immediately it is received in the testing laboratory and just prior to commencing other tests (*e.g.* shear tests, compression tests, etc.).

The water content and unit weight are particularly important, since these could change during transportation and storage.

Some physical state properties are calculated following the practical measurement of others; *e.g.* void ratio from porosity, dry unit weight from unit weight & water content.

Water Content

The most usual method of determining the water content of soil is to weigh a small representative specimen, drying it to constant weight and then weighing it again. Drying can be carried out using an electric oven set at 104-105° Celsius or using a microwave oven.

Example : *A sample of soil was placed in a tin container and weighed, after which it was dried in an oven and then weighed again. Calculate the water content of the soil.*

Weight of tin empty	= 16.16 g
Weight of tin + moist soil	= 37.82 g
Weight of tin + dry soil	= 34.68 g
Water content, w	= (mass of water) / (mass of dry soil)
	= (37.82 - 34.68) / (34.68 - 16.16)
	= 0.169
Percentage water content	= 16.9 per cent

Unit Weight

Clay soils : Specimens are usually prepared in the form of regular geometric shapes, (*e.g.* prisms, cylinders) of which the volume is easily computed. **Sands and gravels** : Specimens have to be placed in a container to determine volume (*e.g.* a cylindrical can).

Example

A soil specimen had a volume of 89.13 ml, a mass before drying of 174.45 g and after drying of 158.73 g; the water content was 9.9 per cent . Determine the bulk and dry densities and unit weights.

Bulk density
ρ = (mass of specimen) / (volume of specimen)
= 174.45 / 89.13 g/ml
= 1.957 Mg/m³
[1 g/ml = 1 Mg/m³]

Unit weight
γ = 9.81m/s² x ρ Mg/m³
= 19.20 kN/m³

Dry density
ρ_d = (mass after drying) / (volume)
= 158.73 / 89.13
= 1.781 Mg/m³
ρ_d = ρ / (1 + w)
= 1.957 / (1+0.099)
= 1.781 Mg/m³

Dry unit weight
γ_d = γ / (1 + w)
= 19.20 / (1+0.099)
= 17.47 kN/m³

Field Measurements

Measurements taken in the field are mostly to determine density/unit weight. The most common application is the determination of the density of rolled and compacted fill, *e.g.* in road bases, embankments, etc.

Note : These methods are covered in detail by BS1377. You should understand the general principle that density is calculated from the mass and volume of a sample. How a sample of known volume is obtained depends on the nature of the soil. You are not expected to remember the details of each method.

The Core Cutter Method

This method is suitable for soft fine grained soils.

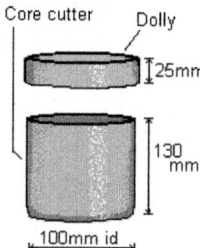

A steel cylinder is driven into the ground, dug out and the soil shaved off level. The mass of soil is found by weighing and deducting the mass of the cylinder. Small samples are taken from both ends and the water content determined.

The Sand-pouring Cylinder Method

This method is suitable for stony soils

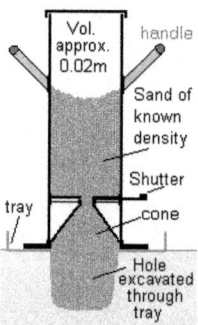

Using a special tray with a hole in the centre, a hole is formed in the soil and the mass of soil removed is weighed.

The volume of the hole is calculated from the mass of clean dry running sand required to fill the hole.

The sand-pouring cylinder is used to fill the hole in a controlled manner. The mass of sand required to fill the hole is equal to the difference in the weight of the cylinder before and after filling the hole, less an allowance for the sand left in the cone above the hole.

Bulk Density

ρ = (mass of soil) / (volume of core cutter or hole)

CURRENT STATE OF SOIL

The state of soil is essentially the closeness of packing of the grains in the range :

Closely packed	→ Loosely packed
Dense	→ Loose
Low water content	→ High water content
Strong and stiff	→ Weak and soft

The important indicators of the current state of a soil are :

current stresses : vertical and horizontal effective stresses

current water content : effecting strength and stiffness in fine soils

liquidity index : indicates state in fine soils

density index : indicates state of compaction in coarse soils

history of loading and unloading : degree of overconsolidation.

Engineering operations (*e.g.* excavation, loading, unloading, compaction, etc.) on soil bring about changes in its state. Its initial state is the result of processes of erosion and deposition. It is possible for the engineer to predict changes that could result from a proposed engineering operation : changes from the soil's current state to a new future state.

Soil History : Deposition and Erosion

Original Deposition

Most soils are formed in layers or lenses by deposition from moving water, ice or wind.

One-dimensional compression occurs as overlying layers are added. Vertical and horizontal stresses increase with deposition.

Erosion

Erosion causes unloading; stresses decrease; some vertical expansion occurs.

Plastic strain has occurred; the soil remains compressed, *i.e.* overconsolidated.

Subsequent Changes

Subsequent changes may occur in the depositional environment : further loading/unloading due to glaciation, land movement, engineering; and ageing processes.

Soil History : Ageing

The term *ageing* includes processes that occur with time, except loading and unloading. Ageing processes are independent of changes in loading.

Vibration and Compaction

Coarse soils can be made more dense by vibration or compaction at essentially constant effective stress

Creep

Fine soils creep and continue to compress and distort at constant effective stress after primary consolidation is complete.

Cementing and Bonding

Intergranular cementing and bonding occurs due to deposition of minerals from groundwater, *e.g.* calcium carbonate; disturbance due to excavation fractures the bonding and reduces strength.

Weathering

Physical and chemical changes take place in soils near the ground surface due to the influence of changes in rainfall and temperature.

Changes in Salinity

Changes in the salinity of groundwater are due to changes in relative sea and land levels, thus soil originally deposited in sea water may later have fresh water in its pores, such soils may be prone to sudden collapse.

Density Index (Relative Density)

The void ratio of coarse soils (sands and gravels) varies with the state of packing between the loosest practical state in which it can exist and the densest. Some engineering properties are affected by this,

Shear strength

Seepage :

The flow of water through soil.

seepage force (J) :

The force transmitted to a body of soil due to the seepage of groundwater.

$J = i \gamma_w V$

seepage pressure (j) :

The seepage force per unit volume.

$j = i \gamma_w$

seepage velocity (v_s) :

The average velocity at which groundwater flows through the pores; the ratio of the volume flow rate to the average area of voids in a cross-section. $v_s = q / A_v$

sensitivity (S_t) :

$$S_t = \frac{\text{undisturbed undrained strength}}{\text{remoulded undrained strength}}$$

A measure of the change in strength of clays upon disturbance : For ordinary clays S_t = 1 to 4, sensitive clays 4 to 8, 'quick' clays 16 - 100.

settlement (s, ρ) :

The downward movement of ground or ground surface; the downward movement of a foundation.

shape factors (s_c s_q s_g) :

Factors used in a general equation giving ultimate bearing capacity which provide adjustment relating to the shape (*e.g.* strip, square, circle)

shear modulus (G') :

The ratio of the change in shear stress to the resulting change in shear strain.

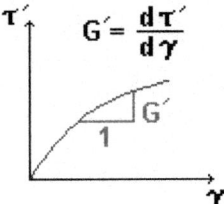

shear strain (γ) :

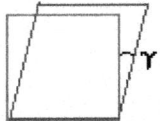

The angular distortion or change in shape of a body of material.

shear strength (γ_f) :

The maximum shear stress which a material can sustain under a given set of conditions. In soil mechanics it is necessary to refer shear strength to the strain at which the strength is measured.

critical shear strength

$\tau_c = c'_c + \sigma' \tan\phi'_c$

peak shear strength

$\tau_p = c'_p + \sigma' \tan\phi'_p$

undrained shear strength

$\tau_c = s_u$

residual shear strength

$\tau_r = c'_r + \sigma' \tan\phi'_r$

shear stress (τ) :

The force per unit area acting tangentially to a given plane or surface. Units : kPa.

short-term conditions :

Conditions in the ground when it is undrained; loading or unloading will cause changes in pore pressures, but not immediately in volume; these excess pore pressures dissipate with time due to consolidation.

shrinkage limit (w_s) :

(Also SL) The water content below which further reduction in water content causes no further reduction in volume.

skin friction stress (f_s) :

The shear stress on the shaft of a pile of a pile or caisson or cone penetrometer. Note : Although known as 'skin friction', the shear stress may not vary with normal stress.

soil suction :

Negative pore pressure created by capillary attraction in fine soils and in unsaturated soils.

specific gravity (G_s) :

The ratio of the mass of a body or a substance to the mass of an equal volume of water; the ratio of the density of a body or a substance to that of water.

$$G_s = \frac{\text{mass of body}}{\text{mass of the same volume of water}}$$

specific surface (S_s) :

The total surface area of all particles in a unit mass of soil. Units : m^2/g.

specific volume (v) :

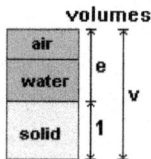

The total volume of a quantity of soil containing a unit volume of soil grains. $v = 1 + e$

state boundary surface :

The state boundary surface (SBS) is the boundary (usually in q' :p' :v space) to all possible stress-volume states of a soil. If a soil with a state on the state boundary surface is unloaded, the subsequent state will be **inside** the SBS, upon re-loading the subsequent state will move back **onto** (but not beyond) the SBS. In other words, stable states cannot exist outside the SBS.

state parameter (S_s, S_v) :

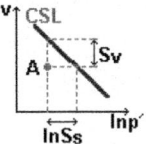

A measure of the distance between the current state (A) and the critical state; expressed as a ratio of stresses or as a difference of specific volumes

$$S_s = p'_A / p'_c$$
$$S_v = v_A - v_c$$

steady state pore pressure (u_o) :

The pore pressure at equilibrium when all excess pore pressures have fully dissipated.

Stiffness :

Susceptibility to distortion or volume change under load.

Train :

A measure of the change in size or shape of a body, relative to its original size or shape. (Direct strain is the ratio of change in length to original length; shear strain is the angle of distortion; volumetric strain is the ratio of change in volume to original volume.)

stream function (Ψ) :

A function introduced in the solution of the Laplace equations defining two-dimensional seepage flow; the flow-quantity inte val ($\Delta\theta$) between stream lines (flow lines) can be stated as $\Delta\Psi = \Delta\theta$.

See also potential function.

Stress :

The intensity of force per unit area; normal stress is applied perpendicularly to a surface or plane, **shear stress** is applied tangentially to a surface or plane.

stress history :

The past history of loading and unloading associated with a soil.

summation (Σ) :

The symbol Σ when placed in front of a quantity indicates that the quantity is a sum or total.

swelling / recompression index (C_s, κ) :

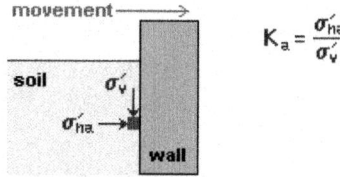

The slope of the swelling (unloading) and recompression (reloading) line. C_s = 2.303 κ

where κ = slope of v :ln σ´ swelling/recompression line.

Compressibility

capillary rise (h_c) :

The height to which water will rise above the **water table** due to capillarity.

clay :

Soil particles less than 0.002mm in size.

coarse-grained soils :

Soils with less than 35 per cent by mass of grains less than 0.06 mm in size. Coarse soil : grain size is predominantly 0.06-60 mm

Very coarse soil : grain size is predominantly greater than 60mm

coefficient of active earth pressure (K_a) :

movement ⟶

soil $σ'_v$

$σ'_{ha}$ ⟶

wall

$$K_a = \frac{σ'_{ha}}{σ'_v}$$

The ratio of the minimum horizontal effective stress to the vertical effective stress at a point in a soil mass retained by a surface as the surface moves away from the soil.

coefficient of consolidation (c_v and c_h) :

$$c_v \frac{\partial^2 \bar{u}}{\partial z^2} = \frac{\partial \bar{u}}{\partial t}$$

A measure of the rate of change of volume during primary consolidation; referred to either vertical drainage (c_v) or horizontal drainage (c_h). Units : m^2/s

coefficient of curvature of grading (C_z) :

(Also curvature coefficient) A measure of the shape of a grading curve :
$C_z = d_{30}^2 / (d_{60} \times d_{10})$.

coefficient of earth pressure at rest (K_o) :

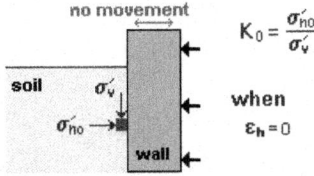

The ratio of horizontal effective stress to vertical effective stress at a point in a soil mass loaded in conditions of zero horizontal strain.

coefficient of friction (μ) :

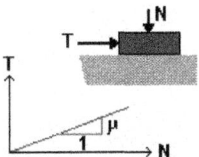

The ratio between the tangential force (T) required to cause a body to slide along a plane and the normal force (N) between the body and the plane :
$T = \mu N$.

Along a wall or foundation surface :

$\mu = \tan \delta$

where δ = angle of wall friction

coefficient of passive earth pressure (K_p) :

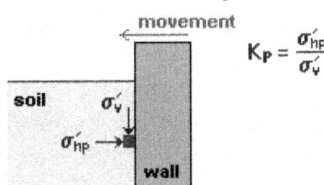

The ratio of the maximum horizontal effective stress to the vertical effective stress at a point in a soil mass retained by a surface as the surface moves towards from the soil.

coefficient of permeability (k) :

The constant average discharge velocity (v) of water passing through soil when the hydraulic gradient (i) is 1.0; defined by Darcy's law : $v = k.i$

coefficient of secondary consolidation (C_a) :

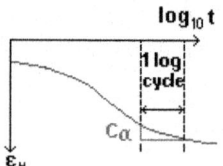

The change in volumetric strain per $\log_{10}$ cycle of time after primary consolidation is complete.

coefficient of uniformity of grading (C_u) :

(Also uniformity coefficient) A measure of the slope of a grading curve and therefore the uniformity of the soil in particle size analyses :
$C_u = d_{60} / d_{10}$

coefficient of volume compressibility (m_v) :

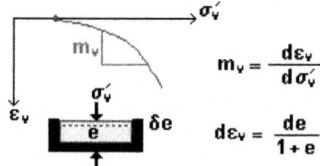

The change in volumetric strain per unit volume per unit change in effective stress in one-dimensional compression. (Units m²/MN) Compressibility is the reciprocal of stiffness.

cohesion (c') :

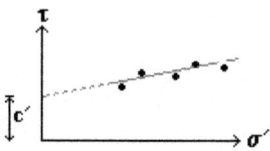

Apparent cohesion (c′) is the intercept on the shear stress axis of a straight-line Mohr-Coulomb envelope. For critical states and residual states, $c'_c = c'_r = 0$ in most cases. For peak states, a curved strength envelope also passes through the origin, but a straight-line fitted to a small number of results is often extrapolated to give an intercept $c'_p > 0$. (In physics, cohesion is described as 'the force that holds together molecules or like particles within a substance'.) See also adhesion and true cohesion.

compaction :

Volume change in soil in which air is expelled, but with the water content remaining constant. Compaction may occur due to vibration in loose sands and gravels, and in fill due to self-weight. In soil constructions, compaction is achieved by rolling, tamping or vibrating.

compatibility :

The relationship between the strains in a deforming body so that no holes appear and no material is destroyed.

compressibility of pore fluid (C_v) :

$$C_v = \frac{d\varepsilon_v}{d\sigma}$$

The ratio of the change in volumetric strain to the change in isotropic stress. For a saturated soil, C_v = ¥

compression index (C_c) :

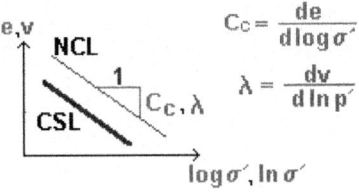

The slope of the normal compression line (NCL) and critical state line (CSL).

cone resistance (q_c) :

The resistance force divided by the end area of the cone tip, measured during the cone penetration test.

cone penetration test :

A penetration test in which a sleeved cone (diameter 35.7 mm, end area 1000 mm², apex angle 60°) is pushed into the ground at a rate of 20 mm/min and the force required measured. The magnitude of this force divided by the end area is called the *cone penetration resistance* (q_c). The cone and the sleeve can be advanced separately and so the frictional resistance along the sleeve (1000 mm² area) or *local side friction* (f_s) can also be measured.

confined aquifer :

A *confined* aquifer is contained between two strata of low permeability; in flow analyses, a confined aquifer is often assumed to be saturated throughout its depth.

consistency index (I_c) :

A measure of the relationship between the current water content and the consistency limits, similar in form to the density index, but expressed in terms of water content.

$I_c = (w_L - w)/I_p$

At liquid limit, $I_c = 0$

At plastic limit, $I_c = 1.0$

consistency limits :

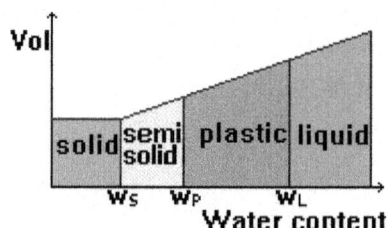

(Also called Atterberg limits) Measures of water content corresponding to changes in physical state of a soil :

liquid limit (w_L) = w/c at the change from liquid to plastic

plastic limit (w_P) = w/c at the change from plastic to semi-solid

shrinkage limit (w_s) = w/c below which no further shrinkage upon drying occurs

consolidation :

Volume change due to dissipation of excess pore pressure, usually with constant total stress.

consolidation settlement (s_c, ρ_c) :

The settlement of a foundation due to consolidation.

Coulomb's equation :

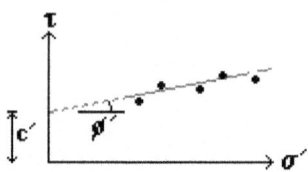

(After Charles Augustin Coulomb, 1736-1806) An equation relating the shear strength of soil to the normal effective stress on a failure plane.

$$\tau_f = c' + \sigma' \tan \phi'$$
$$= c' + (\sigma - u) \tan\phi'$$

creep :

Deformation or volume change which occurs in soil at constant effective stress progressing with time.

critical circle :

In slope stability analyses the slip circle corresponding to the lowest factor of safety.

critical ground slope angle (i_c) :

The ground slope angle that corresponds to a slope-stability factor of safety of 1.0.

critical height (H_c) :

The height of a slope (*e.g.* embankment, cutting, trench) for which the factor of safety against collapse is 1.0.

critical hydraulic gradient (i_c) :

The hydraulic gradient at which effective stresses becomes zero; with upward seepage, sand may become quicksand.

critical shear strength (τ_c) :

The shear stress developed along a slip surface during shearing at constant volume; also known as ultimate shear strength.

critical state :

The state of a soil in which it strains; critical states occur on the <u>critical state line.</u> Note : The critical state is not the same as the residual state; at the critical state, soil particles continue to rotate and the flow is turbulent; at the residual state, in clay soils, flat particles become aligned to the slip plane and the flow is laminar.

critical state line (CSL) :

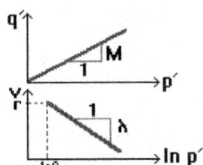

{bmc g17crsle.bmp}

The unique relationship at failure between deviator stress, average normal stress and volume (or shear stress, normal stress and void ratio) is defined by the critical state line.

$q' = Mp'$

$v = \Gamma - \lambda \ln p'$

$\tau' = \sigma'_n \tan\phi'$

$e = e_G - C_c \log \sigma'_n$

critical void ratio (e_c) :

The void ratio of a soil at which its volume remains constant during shearing. Note : critical void ratio depends on the mean stress.

current state of soil :

The current state of a soil is described by its voids ratio (e) or specific volume (v), the current stress (σ' or p') and the over-consolidation ratio or yield stress ratio (R_y).

Ppermeability

The property which allows the flow of water through a soil.

pF index :

A measure of soil suction : $pF = \log_{10}$ (suction head in cm), *i.e.* for a suction of 100 kPa, pF = 2.0

[Range for soils is pF = 0 to 7]

pH value :

A measure of acidity or alkalinity of groundwater or soil water extract based on the hydrogen ion content :

$pH = -\log_{10}$(hydrogen ion content)

pH < 7.0 indicates acidity.

pH > 7.0 indicates alkalinity.

piezometer :

An instrument used to measure *in situ* pore pressures; may be an open stand-pipe or an enclosed electronic pressure transducer.

piezometric surface :

An imaginary surface corresponding to the hydrostatic water level of a confined body of groundwater; the notional level to which artesian pressure would raise water in a (real or imaginary) standpipe.

pile spacing (s) :

The distance from centre to centre of piles in a group.

plane strain :

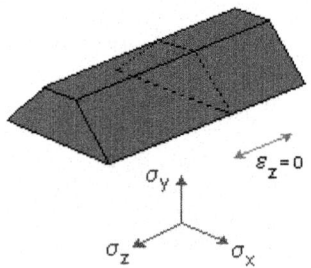

A two-dimensional stress state, where the out-of-plane strain (*i.e.* the strain normal to the plane being considered, ε_z) is zero. An example of a plane strain situation would be on a cross-section through a long structure being loaded in the x-y plane, such as an embankment dam.

plastic deformation :

The flow or distortion resulting in a permanent and irrecoverable change in shape or volume.

plastic limit (w_p) :

(**Also PL**) The moisture content above which a soil will have a plastic consistency, but below which it crumbles.

plastic strain :

Deformation or strain that is not recovered upon unloading.

plasticity :

1. The property of a soil (or other material) which allows it to deform continuously. 2. Plasticity theory is used to calculate plastic (irreversible) deformations.

plasticity index (I_p) :

(**Also PI**) The difference between the liquid limit and plastic limit.

$I_p = w_L - w_P$

Poisson's ratio (v, v') :

(After Simeon Poisson 1781-1840)

The ratio of the change in strain perpendicular to the direction of loading to the change in strain caused in the same direction.

$v' = - d\varepsilon_y / d\varepsilon_x$ *(for to loading or unloading in the x-direction)*.

For undrained loading of saturated soil,

$v_u = 0.5$

For drained loading or unsaturated soil,

$v' = 0.2 - 0.5$

pore air pressure (u_a) :

The pressure of air in a partially saturated soil; not necessarily the same as pore water pressure due to the surface tension on air-water interfaces within the voids.

pore pressure (u) :

The pressure exerted by the fluid within the pores or voids in a porous material; in saturated soil the pore pressure is the pore water pressure.

pore pressure coefficient (A) :

The ratio of the change in pore pressure to the change in deviator stress, *e.g.* in an undrained triaxial test; the value of A varies with strain and the overconsolidation ratio.

pore pressure coefficient (B) :

The ratio of the change in pore pressure to the change in isotropic stress in undrained loading.

$$B = \frac{\Delta u}{\Delta p}$$

For **saturated** soils, B = 1.

pore pressure force (U) :

The resultant force due to pore pressure acting on a given area.

pore pressure ratio (r_u) :

$$r_u = \frac{u}{\sigma_v}$$

At a given point in a body of soil, the ratio of the porewater pressure to the vertical overburden pressure.

pore water pressure (u_w) :

(See also pore pressure) In partially saturated soils the pressure exerted by the water in the voids may not be the same as the pore air pressure, due to the surface tension on air-water interfaces.

porosity (n) :

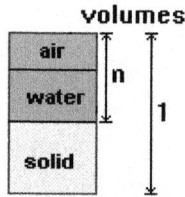

The ratio of void volume to total volume :

$n = V_v / V$

where

V_v = volume of voids

V = total volume

potential function (Φ) :

A function introduced in the solution of the Laplace equations defining two-dimensional seepage flow.

pressure head (h_w) :

(Also head) The height of a column of water required to develop a given pressure u at a given point.

$h_w = u / \gamma_w$

pressure in tension crack (p_w) :

The horizontal pressure exerted in a slope or against a retaining wall due to hydrostatic water pressure in tension cracks.

principal axes :

A set of orthogonal axes perpendicular to which the shear stresses and shear strains are zero and normal stresses and strains are referred to as principal stresses and principal strains.

principal strains (ε_1, ε_2, ε_3) :

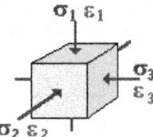

The strains occurring in the directions of the principal axes of strain. Note : the principal axes of stress and strain may not coincide.

principal stresses (σ_1, σ_2, σ_3) :

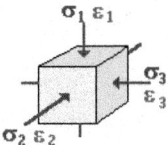

Normal stresses acting in the direction of principal axes of stress. Note : the principal axes of stress and strain may not coincide.

It is therefore useful to measure the *in situ* state and this can be done by comparing the *in situ* void ratio (e) with the minimum and maximum practical values (e_{min} and e_{max}) to give a **density index** I_D

$$I_D = \frac{e_{max} - e}{e_{max} - e_{min}}$$

e_{min} is determined with soil compacted densely in a metal mould
e_{max} is determined with soil poured loosely into a metal mould

Density index is also known as relative density

Relative states of compaction are defined :

Density index	State of compaction
0-15 per cent	Very loose
15-35	Loose
35-65	Medium
65-85	Dense
85-100 per cent	Very dense

Liquidity Index

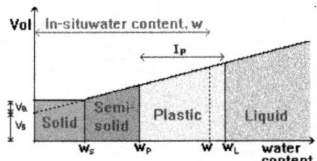

In fine soils, especially clays, the current state is dependent on the water content with respect to the consistency limits (or Atterberg limits). The **liquidity index** (I_L **or LI**) provides a quantitative measure of the current state :

$$I_L = \frac{w - w_P}{I_p} = \frac{w - w_P}{w_L - w_P}$$

where :

w_P = plastic limit and

w_L = liquid limit

Significant values of I_L indicating the consistency of the soil are :

$I_L < 0 \Rightarrow$ semi-plastic solid or solid

$0 < I_L < 1 \Rightarrow$ plastic

$1 < I_L \Rightarrow$ liquid.

Predicting Stiffness and Strength from Index Properties

Preliminary estimates of strength and stiffness can provide a useful basis for early design and feasibility studies, and also the planning of more detailed testing programmes. The following suggestions have been made; they are simple, but not necessarily reliable, and should be not be used in final design calculations.

Undrained Shear Strength

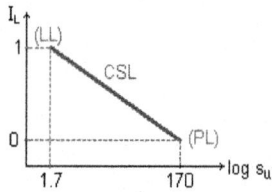

$s_u = 170 \exp(-4.6\ I_L)\ kN/m^2$

[Schofield and Wroth (1968)]

$s_u = (0.11 + 0.37\ I_P)\ \sigma'_{vo}\ kN/m^2$

where σ'_{vo} = vertical effective stress *in situ*

[Skempton and Bjerrum (1957)]

Stiffness

The slope of the critical state line may be estimated from :

$\lambda = I_P\ .G_s\ /\ 461$

[After Skempton and Northey (1953)]

The compressibility index may be estimated from :

$C_c = 1 \ln 10 = I_P\ G_s\ /\ 200$

(where I_P is in percentage units)

BS SYSTEM FOR DESCRIPTION AND CLASSIFICATION

BS 5930 Site Investigation recommends the terminology and a system for describing and classifying soils for engineering purposes. Without the use of a satisfactory system of description and classification, the description of materials found on a site would be meaningless or even misleading, and it would be difficult to apply experience to future projects.

BS Description System

A recommended protocol for describing a soil deposit uses **nine**characteristics; these should be written in the following order :

compactness

e.g. loose, dense, slightly cemented

bedding structure

e.g. homogeneous or stratified; dip, orientation

discontinuities

spacing of beds, joints, fissures

weathered state

degree of weathering

colour

main body colour, mottling

grading or consistency

e.g. well-graded, poorly-graded; soft, firm, hard

SOIL NAME

e.g. GRAVEL, SAND, SILT, CLAY; (upper case letters) plus silty-, gravlly-, with-fines, etc. as appropriate

soil class

(BSCS) designation (for roads & airfields) *e.g.* SW = well-graded sand

geological stratigraphic name

(when known) *e.g.* London clay

Not all characteristics are necessarily applicable in every case.

Example :

(i) Loose homogeneous reddish-yellow poorly-graded medium SAND (SP), Flood plain alluvium

(ii) Dense fissured unweathered greyish-blue firm CLAY. Oxford clay.

Definitions of Terms Used in Description

A table is given in BS 5930 Site Investigation setting out a recommended field indentification and description system. The following are some of the terms listed for use in soil descriptions :

- **Particle shape :** angular, sub-angular, sub-rounded, rounded, flat, elongate
- Compactness : loose, medium dense, dense (use a pick or driven peg, or density index)
- Bedding structure : homogeneous, stratified, inter-stratified
- Bedding spacing : massive(>2m), thickly bedded (2000-600 mm), medium bedded (600-200 mm), thinly bedded (200-60 mm), very thinly bedded (60-20 mm), laminated (20-6 mm), thinly laminated (<6 mm).
- Discontinuities : *i.e.* spacing of joints and fissure : very widely spaced(>2m), widely spaced (2000-600 mm), medium spaced (600-200 mm), closely spaced (200-60 mm), very closely spaced (60-20 mm), extremely closely spaced (<20 mm).
- Colours : red, pink, yellow, brown, olive, green, blue, white, grey, black
- Consistency : very soft (exudes between fingers), soft (easily mouldable), firm (strong finger pressure required), stiff (can be indented with fingers, but not moulded) very stiff (indented by sharp object), hard (difficult to indent).
- Grading : well graded (wide size range), uniform (very narrow size range), poorly graded (narrow or uneven size range).
- Composite soils : **In SANDS and GRAVELS : slightly clayey or silty (<5 per cent), clayey or silty (5-15 per cent), very clayey or silty(>15 per cent) In CLAYS and SILTS : sandy or gravelly (35-65 per cent)**

British Soil Classification System

The recommended standard for soil classification is the **British Soil Classification System**, and this is detailed in BS 5930 Site Investigation. Its essential structure is as follows :

Soil group		*Symbol*		*Recommended name*
Coarse soils			Fines per cent	
GRAVEL	G	GW	0 - 5	Well-graded GRAVEL
		GPu/GPg	0 - 5	Uniform/poorly-graded GRAVEL
	G-F	GWM/GWC	5 - 15	Well-graded silty/clayey GRAVEL
		GPM/GPC	5 - 15	Poorly graded silty/clayey GRAVEL
	GF	GML, GMI...	15 - 35	Very silty GRAVEL [plasticity sub-group...]
		GCL, GCI...	15 - 35	Very clayey GRAVEL [..symbols as below]

	S	SW	0 - 5	Well-graded SAND
SAND		SPu/SPg	0 - 5	Uniform/poorly-graded SAND
	S-F	SWM/SWC	5 - 15	Well-graded silty/clayey SAND
		GPM/GPC	5 - 15	Poorly graded silty/clayey SAND
	SF	SML, SMI...	15 - 35	Very silty SAND [plasticity sub-group...]
		SCL, SCI...	15 - 35	Very clayey SAND [..symbols as below]
Fine soils		**>35 per cent fines**	**Liquid limit per cent**	
SILT	M	MG		Gravelly SILT
		MS		Sandy SILT
		ML, MI...		[Plasticity sub-divisions as for CLAY]
CLAY	C	CG		Gravelly CLAY
		CS		Sandy CLAY
		CL	<35	CLAY of low plasticity
		CI	35 - 50	CLAY of intermediate plasticity
		CH	50 - 70	CLAY of high plasticity
		CV	70 - 90	CLAY of very high plasticity
		CE	>90	CLAY of extremely high plasticity
Organic soils	O			[Add letter 'O' to group symbol]
Peat	Pt			[Soil predominantly fibrous and organic]

Chapter 2

BASICS OF SOIL

Soils are vital, fragile, finite natural resources that are essential for the sustained production of food and fibre. Soils, however, are subject to degradation and erosion when mismanaged. Between 1950 and 1993, grain area per person worldwide decreased from 0.58 to 0.33 acres (0.23 to 0.13 hectares). As human populations increase, soil resources are used more intensively, with increasing probability that many practices will lead to deterioration of the resource. Competition between agricultural uses and non-agricultural uses of land, such as support of structures, disposal of wastes, and growing plants for recreational and aesthetic purposes will increase.

In ecosystems, soils, water, air, plants, animals and people have interdependent relationships. Soils are dynamic, living systems whose productivity, through management that often includes additions of nutrients, organic materials and water, can be sustained indefinitely. Soils exhibit unique physical and chemical sorptive qualities and dynamics reflective of their inorganic and organic composition. Cycling of carbon, nitrogen and other nutrient elements in nature involves transformations in soils.

Great diversity occurs among soils, sometimes in very small geographical areas, such as building lots in urban areas. The rise and fall of civilizations sometimes has been related to the wise use and misuse of natural resources including soil and water.

Definitions of soil vary, but one view is that the unconsolidated material at the earth's surface becomes soil when biological activity results in a noticeable accumulation of organic matter as revealed by a dark surface colour. Soils that form in loose, fine-grained material weathered from the rock immediately below them are called residual soils. More frequently, soils are formed in materials that have been transported away from the source rock. Examples of such materials are alluvial materials, which have been deposited from running water as in flood plains or deltas; lacustrine material which is deposited in lakes; glacial material which has been moved by ice, and aeolian material which has been transported and deposited by wind.

Many of the intensively used soils in the world are formed in transported materials. Their usefulness often is associated with topography, or with physical and chemical properties inherited directly from the transported material.

SOIL FORMATION

Soils are porous natural bodies composed of inorganic and organic matter. They form by interaction of the earth's crust with atmospheric and biological influences. They are dynamic bodies having properties that reflect the integrated effects of climate (atmosphere) and biotic activity (micro-organisms, insects, worms, burrowing animals, plants, etc.) on the unconsolidated remnants of rock at the earth's surface (parent material). These effects are modified by the topography of the landscape and of course continue to take place with the passage of time. Soils formed in parent materials over decades, centuries, or millennia may be lost due to accelerated erosion over a period of years or a few decades.

The exposed surfaces of soils are a common sight on almost any landscape not dominated by rock. The surface of a soil reveals very little about the depth of the soil or its subsurface characteristics. A vertical cross-sectional view of a soil is called a soil profile. Each of the horizontal layers which can be seen in the vertical section is called a soil horizon. Horizons are formed because of the integrated effects of climate and biosphere change and generally become less pronounced with depth. The depth of soils, usually 0.6 to 1.8 m, is determined by the depth to which the mantle material has been altered in a significant way. That part of the three-dimensional soil body in which the effects of climate and biological activity are most pronounced is the soil solum. In succeeding pages the nature and properties of soils, their management, and environmental public policy issues will be discussed.

SOIL COMPOSITION

The proportions of these components may vary between horizons in a soil or between similar horizons in different soils. The ratio of soil water to soil air depends upon whether the soil is wet or dry. The mineral matter, composed of particles ranging in size from the sub-microscopic to gravel or even rocks in some cases, accounts for the bulk of the dry weight of the soil and occupies some 40 to 60 per cent of the soil volume. Organic matter, derived from the waste products and remains of plants and animals, occurs in largest amounts in the surface soil, but even here seldom accounts for more than 10 per cent of the dry weight of the soil.

Soils are very porous bodies. Some 40 to 60 per cent of the volume is interparticle space, or pore space. The pores, highly irregular in shape and size but almost all interconnected by passages, contain soil water, soil air, or both of these. The soil water reacts chemically with the soil solids and usually contains dissolved substances and perhaps suspended particles. The soil air approaches equilibrium with atmospheric air through movement of individual gases.

Bedrock is the ultimate source of the inorganic component in soils. When rock is exposed at the surface of the earth's crust, it is broken down into smaller and smaller fragments by physical forces. The fragments may be altered or decomposed by chemical reaction of mineral matter with water and air. Hundreds, thousands, or even millions of years may be required for the weathering or physical and chemical alteration of rock to produce the ultimate end products in soils. Once particles reach a sufficiently small size they can be moved by wind, water or ice when exposed at the surface. It is common, therefore, for small particles to be moved from one location to another. A single particle might occur in several different soils over a period of 100,000 years. Eventually, these particles or their decomposition products reach the ocean where they are redeposited as marine sediments.

The silicate group of minerals is dominant in soil systems. The terms, clay mineral and layer silicate, are used almost interchangeably. The dominant chemical elements in silicate clays are oxygen, silicon, aluminum and iron. Important constituents in relatively small amounts are potassium, calcium, magnesium and sodium. Other elements occur in very small amounts in silicates. Carbonates, oxides, phosphates and sulfates are other mineral groups that occur commonly in parent materials.

SOIL PHYSICAL PROPERTIES

Physical Properties of Soil

Soils are porous and open bodies, yet they retain water. They contain mineral particles of many shapes and sizes and organic material which is colloidal (particles so small they remain suspended in water) in character. The solid particles lie in contact one with the other, but they are seldom packed as closely together as possible.

Texture

The size distribution of primary mineral particles, called soil texture, has a strong influence on the properties of a soil. Particles larger than 2 mm in diameter are considered inert. Little attention is paid to them unless they are boulders that interfere with manipulation of the surface soil. Particles smaller than 2 mm in diameter are divided into three broad categories based on size. Particles of 2 to 0.05 mm diameter are called sand; those of 0.05 to 0.002 mm diameter are silt; and the <0.002 mm particles are clay. The texture of soils is usually expressed in terms of the percentages of sand, silt, and clay. To avoid quoting exact percentages, 12 textural classes have been defined. Each class, named to identify the size separate or separates having the dominant impact on properties, includes a range in size distribution that is consistent with a rather narrow range in soil behaviour. The loam textural class contains soils whose properties are controlled equally by clay, silt and sand separates. Such soils tend to exhibit good balance between large and small pores; thus, movement of water, air and roots is easy and water retention is adequate. Soil texture, a stable and an easily determined soil characteristic, can be estimated by feeling and manipulating a moist sample, or it can be determined accurately by laboratory analysis. Soil horizons are sometimes separated on the basis of differences in texture.

Structure

Anyone who has ever made a mud ball knows that soil particles have a tendency to stick together. Attempts to make mud balls out of pure sand can be frustrating experiences because sand particles do not cohere (stick together) as do the finer clay particles. The nature of the arrangement of primary particles into naturally formed secondary particles, called aggregates, is soil structure. A sandy soil may be structureless because each sand grain behaves independently of all others. A compacted clay soil may be structureless because the particles are clumped together in huge massive chunks. In between these extremes, there is the granular structure of surface soils and the blocky structure of sub-soils. In some cases sub-soils may have platy or columnar types of structure. Structure may be further described in terms of the size and stability of aggregates. Structural class is based on aggregate size, while structural grade is based on aggregate strength. Soil horizons can be differentiated on the basis of structural type, class, or grade.

What causes aggregates to form and what holds them together? Clay particles cohere to each other and adhere to larger particles under the conditions that prevail in most soils. Wetting and drying, freesing and thawing, root and animal activity, and mechanical agitation are involved in the rearranging of particles in soils — including destruction of some aggregates and the bringing together of particles into new aggregate groupings. Organic materials, especially microbial cells and waste products, act to cement aggregates and thus to increase their strength. On the other hand, aggregates may be destroyed by poor tillage practices, compaction, and depletion of soil organic matter. The structure of a soil, therefore, is not stable in the sense that the texture of a soil is stable. Good structure, particularly in fine textured soils, increases total porosity because large pores occur between aggregates, allowing penetration of roots and movement of water and air.

Consistence

Consistence is a description of a soil's physical condition at various moisture contents as evidenced by the behaviour of the soil to mechanical stress or manipulation. Descriptive adjectives such as hard, loose, friable, firm, plastic, and sticky are used for consistence. Soil consistence is of fundamental importance to the engineer who must move the material or compact it efficiently. The consistence of a soil is determined to a large extent by the texture of the soil, but is related also to other properties such as content of organic matter and type of clay minerals.

Colour

The colour of objects, including soils, can be determined by minor components. Generally, moist soils are darker than dry ones and the organic component also makes soils darker. Thus, surface soils tend to be darker than sub-soils. Red, yellow and gray hues of sub-soils reflect the oxidation and hydration states or iron oxides, which are reflective of predominant aeration and drainage characteristics in subsoil. Red and yellow hues are indicative of good drainage and aeration, critical for activity of aerobic organisms in soils. Mottled zones, splotches of one or more colours in a matrix of different colour, often are indicative of a transition between well drained, aerated zones and poorly drained, poorly aerated ones. Gray hues indicate poor aeration. Soil colour charts have been developed for the quantitative evaluation of colours.

Soil Chemical Properties

Major Elements

Eight chemical elements comprise the majority of the mineral matter in soils. Of these eight elements, oxygen, a negatively-charged ion (anion) in crystal structures, is the most prevalent on both a weight and volume basis. The next most common elements, all positively-charged ions (cations), in decreasing order are silicon, aluminum, iron, magnesium, calcium, sodium, and potassium. Ions of these elements combine in various ratios to form different minerals. More than eighty other elements also occur in soils and the earth's crust, but in much smaller quantities.

Soils are chemically different from the rocks and minerals from which they are formed in that soils contain less of the water soluble weathering products, calcium, magnesium, sodium, and potassium, and more of the relatively insoluble elements such as iron and aluminum. Old, highly weathered soils normally have high concentrations of aluminum and iron oxides.

The organic fraction of a soil, although usually representing much less than 10 per cent of the soil mass by weight, has a great influence on soil chemical properties. Soil organic matter is composed chiefly of carbon, hydrogen, oxygen, nitrogen and smaller quantities of sulfur and other elements. The organic fraction serves as a reservoir for the plant essential nutrients, nitrogen, phosphorus, and sulfur, increases soil water holding and cation exchange capacities, and enhances soil aggregation and structure.

The most chemically active fraction of soils consists of colloidal clays and organic matter. Colloidal particles are so small (<0.0002 mm) that they remain suspended in water and exhibit a very large surface area per unit weight. These materials also generally exhibit net negative charge and high adsorptive capacity. Several different silicate clay minerals exist in soils, but all have a layered structure. Montmorillonite, vermiculite, and micaceous clays are examples of 2 : 1 clays, while kaolinite is a 1 : 1 clay mineral. Clays having a layer of aluminum oxide (octahedral sheet) sandwiched between two layers of silicon oxide (tetrahedral sheets) are called 2 : 1 clays. Clays having one tetrahedral sheet bonded to one octahedral sheet are termed 1 : 1 clays.

Cation Exchange

Silicate clays and organic matter typically possess net negative charge because of cation substitutions in the crystalline structures of clay and the loss of hydrogen cations from functional groups of organic matter. Positively-charged cations are attracted to these negatively-charged particles, just as opposite poles of magnets attract one another. Cation exchange is the ability of soil clays and organic matter to adsorb and exchange cations with those in soil solution (water in soil pore space). A dynamic equilibrium exists between adsorbed cations and those in soil solution. Cation adsorption is reversible if other cations in soil solution are sufficiently concentrated to displace those attracted to the negative charge on clay and organic matter surfaces. The quantity of cation exchange is measured per unit of soil weight and is termed cation exchange capacity. Organic colloids exhibit much greater cation exchange capacity than silicate clays. Various clays also exhibit different exchange capacities. Thus, cation exchange capacity of soils is dependent upon both organic matter content and content and type of silicate clays.

Cation exchange capacity is an important phenomenon for two reasons :

1. Exchangeable cations such as calcium, magnesium, and potassium are readily available for plant uptake and

2. Cations adsorbed to exchange sites are more resistant to leaching, or downward movement in soils with water.

Movement of cations below the rooting depth of plants is associated with weathering of soils. Greater cation exchange capacities help decrease these losses. Pesticides or organics with positively charged functional groups are also attracted to cation exchange sites and may be removed from the soil solution, making them less subject to loss and potential pollution.

Calcium (Ca^{++}) is normally the predominant exchangeable cation in soils, even in acid, weathered soils. In highly weathered soils, such as oxisols, aluminum (Al^{3+}) may become the dominant exchangeable cation.

The energy of retention of cations on negatively charged exchange sites varies with the particular cation. The order of retention is : aluminum > calcium > magnesium > potassium > sodium > hydrogen. Cations with increasing positive charge and decreasing hydrated size are most tightly held.

Calcium ions, for example, can rather easily replace sodium ions from exchange sites. This difference in replaceability is the basis for the application of gypsum ($CaSO_4$) to reclaim sodic soils (those with >15 per cent of the cation exchange capacity occupied by sodium ions). Sodic soils exhibit poor structural characteristics and low infiltration of water.

The cations of calcium, magnesium, potassium, and sodium produce an alkaline reaction in water and are termed bases or basic cations. Aluminum and hydrogen ions produce acidity in water and are called acidic cations. The percentage of the cation exchange capacity occupied by basic cations is called percent base saturation. The greater the percent base saturation, the higher the soil pH.

Soil pH

Soil pH is probably the most commonly measured soil chemical property and is also one of the more informative. Like the temperature of the human body, soil pH implies certain characteristics that might be associated with a soil. Since pH (the negative log of the hydrogen ion activity in solution) is an inverse, or negative, function, soil pH decreases as hydrogen ion, or acidity, increases in soil solution. Soil pH increases as acidity decreases.

A soil pH of 7 is considered neutral. Soil pH values greater than 7 signify alkaline conditions, whereas those with values less than 7 indicate acidic conditions. Soil pH typically ranges from 4 to 8.5, but can be as low as 2 in materials associated with pyrite oxidation and acid mine drainage. In comparison, the pH of a typical cola soft drink is about 3.

Soil pH has a profound influence on plant growth. Soil pH affects the quantity, activity, and types of micro-organisms in soils which in turn influence decomposition of crop residues, manures, sludges and other organics. It also affects other nutrient transformations and the solubility, or plant availability, of many plant essential nutrients. Phosphorus, for example, is most available in slightly acid to slightly alkaline soils, while all essential micro-nutrients, except molybdenum, become more available with decreasing pH. Aluminum, manganese, and even iron can become sufficiently soluble at pH < 5.5 to become toxic to plants. Bacteria which are important mediators of numerous nutrient transformation mechanisms in soils generally tend to be most active in slightly acid to alkaline conditions.

SOIL CLASSIFICATION

Soils, like other naturally occurring things, come in great variety and exhibit great ranges of properties. Using measurable and observable properties, such as the kind and arrangement of soil horizons, soils can be characterised and named. The soil series is the lowest category within soil taxonomy (classification system). All soils within a single series have uniform differentiating characteristics and arrangement of horizons. This does not mean that all soils within a series are identical; it does mean that they have the same horizonation, but the horizons may be of different thickness, colour, structure, etc., within prescribed limits. Some 15,000 soil series have been described and named in the United States. Most series names are taken from the name of a town, city, county, river or other constructed or natural feature near the location where the soil is first described and named.

All of the soils within a series will have developed in the same kind of parent material with comparable drainage characteristics and will be of similar age. The effects of climate and biological activity will have been very similar. Consequently, the soils within a series exhibit like properties and respond in like fashion to usage or manipulation.

Higher levels of classification are the family, sub-group, great group, sub-order and order. All these categories are given generic names which convey as much information as possible about the soil series to be classified within the group. Eleven soil orders form the highest level of classification. Soils classified within each order show only small differences in the kinds and relative strengths of processes that tend to develop soil horizons.

Considerable effort by soil scientists has been expended, and continues to be expended, in surveying soil resources. These surveys, when done in a detailed manner, require a soil scientist to traverse the landscape at frequent intervals, stopping periodically to auger or dig into the soil. The soil surveyor plots the occurrence of soils on a map which is subsequently formalised and eventually published in a soil survey report which includes not only the soil maps for a given area, usually a county, but all types of information about the county as well as descriptions of the properties of the soils in the area, their present and potential uses, and potential problems associated with the utilisation of the soils for both agricultural and engineering purposes. Such reports are very expensive in terms of labour and money, but they contain information that is of great value to those who utilise soils for any purpose. Unfortunately, many who could use the information in soil survey reports to great advantage do not know that the reports are available. With increased emphasis on planning, which includes land use, on county, state and regional bases, more utilisation of soil survey information is being made and even more detailed and more modern soil surveys are being urgently requested in many areas. Much of this information is being digitised for electronic transmission.

THE SOIL ENVIRONMENT

The principal environmental variables affecting life in soils include moisture, temperature, pH, aeration (*i.e.* presence or absence of sufficient oxygen), organic matter, and inorganic nutrients such as nitrogen and phosphorus. The balance of these factors controls the abundance and activities of the microbes and larger animals in soils which in turn have a marked influence on the critical processes of soil aggregation and degradation of plant and animal residues and the nutrient cycling that accompanies this latter process. To understand life in the soil, the soil is best viewed as an ex-tremely heterogeneous collection of microhabitats. The soil is a matrix of solids including sand, silt, clay, and organic matter particles as well as aggregates of various sizes formed from them, and pore space, which may be filled with air or water. Thus, depending on the mix of environmental variables, prevailing conditions can vary quite markedly over distances on the order of a few millimeters (1 inch = 25.4 millimeters).

Soil Water

Soil water is usually derived from rainfall or some other form of overland flow. The amount of water that enters the soil is a function of soil structure which in turn is a function of texture and higher-order structure *i.e.* aggregation (*e.g.* well-aggregated, porous soils allow greater infiltration). Larger pores created by root channels and animal burrows (ant tunnels, crayfish tunnels and worm holes) as well as other types of macropores also greatly facilitate movement of water into and through the soil profile. This is why these soil animals are usually regarded as highly beneficial in soil and why some scientists often use the abundance of earthworms as an indicator of a "healthy soil", though it should be remembered that highly fertile, productive soils do not always contain large numbers of earthworms and other soil animals.

Water moves through soil by mass flow (*i.e.* fluid movement through larger channels) and by capillary action, the slower movement through the highly tortuous network of very small channels and pores. Most soil water is held in this system of capillary pores and channels and can be held quite firmly through physical interactions with the solid components of the soil matrix. Matric water is held in small (usually microscopic) pores and is often adsorbed onto particle surfaces that can be held rather tightly. Because these interactions with the soil matrix essentially bind the soil water, energy on the part of plants and microbes is required to extract water. These forces are usually given as a matric potential or pressure and they indicate how tightly the water is held. These values are usually expressed as a negative value because they are a measure of the amount of suction required to extract water from the soil. It is this adsorption of water to soil solids that explains why plants can wilt in a soil having substantial clay content, even when some matric water is still present. Beyond a certain point, the water is held too tightly for the plants to extract and the so-called wilting point is reached. An often-stated value for the matric potential at the permanent wilting point is approximately -15 bars (-1.5 mPa). Plants exposed to lower matric (more negative) potentials may wilt and not recover. It should be stressed that the matric potential is a function of soil texture and that fine-textured soils (silts and clays) can contain moisture at very low matric potentials because of the much greater numbers of small particles (greater surface area) and the finer pore network which adsorbs and retains water. Thus in clayey soils, water is not as biologically available as it is in coarser-textured soils. A corollary, however, is that coarse-textured soils do not retain as much water because the larger particles do not adsorb water as well and the larger pores drain more rapidly. Thus, nutrients and other chemicals may be retained in a clay soil whereas they may leach, *i.e.* move downward, with water more rapidly through sandy soils. These textural concerns are of extreme importance to our understanding of the movement of materials into ground water.

Water occupying the pore network in soil is often called the soil solution. It is never pure water but rather it is a solution containing dissolved salts and gases. This is the solution which is most available to plants and microbes and provides some of the nutrients needed for growth. Many nutrients actually reach the roots of plants through mass flow of the water to the vicinity of the roots. Too much water in the soil disrupts the ideal balance between air-filled and water-filled pores, *i.e.* soil aeration. Salts dissolved in the soil solution give rise to an osmotic pressure or potential which can be measured. The osmotic potential can influence the movement of water into and out of cells and therefore it is important that the soil solution not be allowed to become too saline (salty) as it might in a soil experiencing increased salinity due to prolonged irrigation and poor drainage.

Under low osmotic (high negative) potentials (salty or sugary solutions), water is actually drawn out of cells resulting in physiological impairment, that is, being dehydrated beyond some critical point. This is the basis of preserving foods by adding high concentrations of salt.

Osmotic and matric potentials are the two major factors controlling the availability of water to soil life and of these two, it is the matric potential which most affects soil microbes. As a soil dries, life in the soil must work harder to extract water against ever increasing matric and osmotic potentials. For these reasons, adequate soil moisture is required for maximum rates of plant growth and to maintain the vital soil microbial processes essential for life on earth.

Soil Atmosphere

The soil atmosphere is the mixture of gases occupying that portion of the pore network that is not filled with water. It is derived from the overlying air but differs in a few prominent ways. Most notably, the soil atmosphere contains less oxygen than the overlying air and typically has 10- to 100-fold more carbon dioxide. You might surmise, and you would be correct, that these changes are driven by the aerobic respiration of the organisms inhabiting the soil. These changes are also a result of the physical characteristics of the soil environment. Generally, the rate of movement of oxygen back into the soil is not sufficient to keep up with all the respiratory demand because movement of gases is restricted by the highly elaborate network of fine pores and capillaries. Filling pores with water restricts the movement of gases even further, and it is for this reason that waterlogged soils often become anaerobic (*i.e.* depleted of molecular O_2). This condition should be avoided because it gives rise to a number of negative consequences, not the least of which is that plant roots are injured or killed, not just due to lack of O_2, but to the accumulation of toxic products such as organic acids (*e.g.* acetic acid) or hydrogen sulfide (rotten egg gas) arising from anaerobic activities of soil bacteria.

Most soil organisms are aerobic, *i.e.* they require O_2 to grow. However, some soil bacteria do not require O_2 and in fact may be killed or inhibited by it. These bacteria are called anaerobes and they function when there are low concentrations of soil O_2, for example, during periods of prolonged waterlogging. Still other bacteria, called facultative anaerobes, can grow in the presence or absence of O_2.

This discussion of soil water and the soil atmosphere reveals why it is so desirable to maintain a soil in a highly aggregated state. Aggregation is indispensable for maintaining the balance between air- and water-filled pores that is essential for the maintenance of aerobic, beneficial microbes and soil animals.

Soil Temperature

Soil temperature is controlled by a number of factors not the least of which is the amount of sunlight incident upon it and the amount of water it contains. Soils are warmed by solar radiation. For this reason, dark-coloured soils tend to absorb more heat and warm earlier in the spring than light-coloured soils do. Obviously, vegetative cover plays an important role in controlling the amount of solar radiation that reaches the soil surface. We all know the value of a good shade tree on a hot summer afternoon. Because the soil can become quite warm in the surface few centimeters, particularly when free of vegetation, it is not surprising that numbers of microbes and soil animals are lower near the imme-diate surface. It is frequently just too hot and too dry in this narrow band of soil. Deeper in the solum the soil warms according to its moisture content and the amount of irradiance received on the surface.

Temperature affects soil life in several ways. First, all organisms exhibit preferences for certain temperature ranges; the so-called minimum, maximum and optimum temperatures. Soil microbes can grow across a very wide range of temperatures so that some portion of the population may be active nearly year round. At the lower extremes of temperature (near freezing) are microbes called cryophiles which actually require or prefer these cold temperatures. At the other extreme are thermophiles, organisms which thrive in high-temperature environments like the boiling geothermal hot springs of Yellowstone National Park at temperatures near 94 °C (water boils below 100 °C because of the lower atmospheric pressure at the higher altitude). Thermophiles are also critical components of composting systems designed to accelerate the conversion of municipal refuse, paper etc., into compost for use in gardens and for disposal onto land. In the middle temperature range (15 to 40 °C) we find the mesophiles and, not surprisingly, ourselves (humans) and most other life forms. Most soil microbes are mesophiles. Most of life's important biological processes operate best in the mesophilic range and we find special adaptations in those organisms that live in colder or hotter environments.

Temperature affects soil microbes and microbial processes in another fundamental way. Namely, it governs the rate of chemical and biological reactions. Thus, biological reactions and processes tend to occur more rapidly as temperature rises to some point where some vital function (perhaps an important enzyme) of the cell is impaired or destroyed and the organism is killed. At colder temperatures, reaction rates slow down and may almost completely stop. This is the well known basis for preserving food from spoilage by keeping it cold in the freezer or refrigerator. We mentioned above that temperature controls biological processes in soil. One can easily see the value of early warming of soils in the spring for germination of seed and an early start of a crop growth cycle. One can also look at maps of soil organic matter content and easily ascertain the combined effects of soil moisture and temperature on maintenance of organic matter in the soil. In warm, moist climates, soil organic matter is difficult to maintain because the conditions for growth of soil microbes are favourable for longer periods of the year. In colder climates, soil organic matter can accumulate because of prolonged periods when the soil is very cold or even frozen in the near surface layers where most microbes reside.

SOIL BIOLOGICAL PROPERTIES

Soil Biota

The soil contains a vast array of life forms ranging from sub-microscopic (the viruses), to earthworms, to large burrowing animals such as gophers and ground squirrels. Microscopic life forms in the soil are generally called the "soil micro-flora" (though strictly speaking, not all are plants in the true sense of the word) and the larger animals are called macro-fauna.

Soil animals, especially, the earthworms and some insects tend to affect the soil favourably through their burrowing and feeding activities which tend to improve aeration and drainage through structural modifications of the soil solum. In general, they affect soil chemical properties to a lesser extent though their actions indirectly enhance microbial activities due to creation of a more favourable soil environment.

Soil Micro-organisms

Soil micro-organisms occur in huge numbers and display an enormous diversity of forms and functions. Major microbial groups in soil are bacteria (including actinomycetes), fungi, algae (including cyanobacteria) and protozoa.

Because of their extremely small cell size (one to several micrometers), enormous numbers of soil microbes can occupy a relatively small volume, hence space is rarely a constraint on soil microbes. Soil microbes can occur in numbers ranging up to several million or more in a gram of fertile soil (a volume approximately that of a red kidney bean). Note that the bacteria are clearly the most numerous of the soil microbes. Perhaps more important than the numbers of the various soils microbes is the microbial biomass contributed by the respective groups. It is the soil fungi which tend to contribute the most biomass among the microbial groups. In fact, it is because of their large contribution to the biomass that they are generally regarded as being the dominant decomposer microbes in the soil. You might find it surprising that there are literally "tons" of microbes beneath your feet as you walk across a grassland in Africa or Australia or through a cornfield in the American Midwest. Interesting-ly, a fungus discovered in the state of Michigan may be one of the largest living organ-isms on the planet.

A fungus, *Armillaria bulbosa*, discovered in the U.S. in the state of Michigan, could turn out to be earth's largest creature or at least among the largest. Scientists discovered the fungus growing among the roots of hardwood trees in a forest. The microscopic, branched filaments (called hyphae) of the fungus occupy a 14.8 ha (37-acre) area of land. Careful genetic analysis has shown the filaments constitute a single organism. Fungi generally radiate outward in a circular pattern as they grow through the soil. In fact, the fairy rings of mushrooms (named because ancient peoples thought they represented the paths of fairies dancing in the night) often seen in lawns or on golf courses actually represent the outer boundary of a developing fungus. Scientists estimate that the portion of the Michigan fungus they have been able to identify may weigh as much as 100 tons, slightly less than a blue whale. Imagine the bio-chemical capacity of a soil microorganism this large !

The significance of these large amounts of microbial biomass in the soil lies not only in their large bio-chemical capacity, but in the phenomenal diversity of bio-chemical reactions attributed to the soil microbial population. It is worth remembering that soil microbes not only interact with other members of their own group, they also interact with other microbial groups. It is quite common to find, for example, that degradation of plant materials occurs much more quickly in the presence of the mixed soil population than it does when one or more groups of soil microbes have been eliminated from the system.

Soil life can be divided into trophic (*i.e.* feeding) levels. At the base of the trophic levels lies the soil microbial population which degrades plant, animal and microbial bodies, and also serves as the food source for some of the levels above it. For example, soil protozoa consume enormous numbers of bacteria and even some fungal spores. These in turn are consumed by still larger soil animals (nematodes, mites, etc.) which in turn are eaten by still larger animals (*e.g.* worms and insects). Thus, nutrients flow through this microbial food web which lies at the heart of controlling soil fertility and plant productivity in the absence of external inputs such as fertilizers. In fact, the role of soil microbes in degrading organic materials and thereby regenerating a supply of carbon dioxide for plants is perhaps their most vital global function.

Nutrient Cycling by Soil Microbes

Soil microbes exert much influence in controlling the quantities and forms of various chemical elements found in soil. Most notable are the cycles for carbon, nitrogen, sulfur and phosphorus, all of which are elements important in soil fertility, and as we know today, may be involved in global environmental phenomena. The mineralisation (*i.e.*, the conversion of organic forms of the elements to their inorganic forms) of organic materials by soil microbes liberates carbon dioxide, ammonium (which is rapidly converted to nitrate by soil microbes), sulfate, phosphate and inorganic forms of other elements. This is the basis of nutrient cycling in all major ecosystems of the world. John Burroughs once said, *"Without death and decay, how could life go on?"* No doubt, he was referring to the mineralisation of nutrients from dead animals and plants. We now know that soil microbes accomplish this task with remarkable zeal and that in the process a substantial part (perhaps as much as one third) of the decomposing materials are converted to the bodies of soil microbes. This pool of microbial biomass constitutes a portion of the soil organic matter which turns over (cycles) fairly quickly and therefore represents a "fertility buffer" in the soil. Don't forget that the liberation of carbon dioxide through microbial respiration makes possible the continued photo-synthesis (*i.e.* carbon dioxide fixation) by algae and green plants which in turn produce more organic materials which may ultimately reach the soil, thereby completing the cycle.

In the world's agricultural soils, the source of our food supply, mineralisation of nitrogen by soil microbes is a most important process. In those soils not receiving external inputs of fertilizer nitrogen (*e.g.* most forested lands and many grasslands) the liberation of ammonium from organic debris makes possible the continued growth of new plant matter. Therefore, it is the soil microbial population which controls the productivity of these soils if other environmental factors (moisture, temperature) are suitable. In fact, fertilization of a soil represents our attempt to balance the competition between plants and soil microbes for available soil nitrogen. Nitrogen tied-up (assimilated into cell constituents) in microbial cells is not available for plants or other microbes until that tissue has been decomposed by other microbes. In other words, nitrogen contained in tissues is said to be immobilised. Microbes are the keys for the remobilisation of these nutrients. These mineralisation/immobilisation phenomena are common to all the elements but typically they are only agriculturally important for the macronutrients such as nitrogen, phosphorus and sulfur.

Aside from their role in controlling the rates of production of inorganic forms of nitrogen and sulfur, soil microbes, in particular soil bacteria, can control the forms of the ions in which these nutrients occur. For example, ammonium (NH_4^+) in the soil is usually rapidly oxidised by bacteria first to nitrite (NO_2^-) and then to nitrate (NO_3^-) which may readily leach through soil. Ammonium is oxidised to nitrite and then to nitrate by the bacteria *Nitrosomonas* and *Nitrobacter*, respectively. Thus, bacteria can influence the form and, thereby, the retention of nitrogen in the soil. Similarly, reduced sulfur compounds such as thiosulfate, elemental sulfur and even iron pyrite (FeS_2, "Fool's Gold") can be oxidised to sulfuric acid by soil bacteria. The bacteria which accomplish the oxidation of reduced nitrogen and sulfur compounds use these materials as energy sources to drive their metabolism. Unlike the decomposer microbes which use organic carbon compounds from organic matter for energy and to make cell matter (*e.g.* they are called heterotrophs), these specialised bacteria called chemoautotrophs obtain their carbon for cell synthesis from carbon dioxide or from dissolved carbonate.

There are many genera of bacteria that can oxidise reduced sulfur compounds. However, much of this activity, especially the oxidation of sulfur and pyrite, can be attributed to bacteria of the genus *Thiobacillus* (thio = sulfur; bacillus = rod-shaped bacterium). *Thiobacillus thiooxidans* can oxidise elemental sulfur to sulfuric acid. Sulfur, therefore, can be used to decrease the pH of an alkaline soil. *Thiobacillus ferrooxidans* attacks both the iron and sulfur in iron pyrite, generating sulfuric acid and dissolved iron in the process. This is also the basis of acid mine drainage associated with the mining of coal throughout the world.

The long-term application of ammonium-based fertilizers can likewise result in the acidification of agricultural soils through bacterial nitrification (the conversion of ammonium to nitrate with the concurrent production of acidity). Thus, we see that certain environmental problems can arise from the activities of these chemoautotrophic soil bacteria.

Another important aspect of nutrient cycling is that under certain circumstances nitrogen and sulfur may be converted to gaseous forms (volatilised) and lost to the atmosphere. Nitrogen in the form of nitrate can be converted to gases such as nitrous oxide (N_2O) and dinitrogen (N_2) through the process of denitrification (the bacterial reduction of NO_3^- to N_2O or N_2) by soil bacteria under anaerobic conditions. A consequence of denitrification is that nitrogen, a precious nutrient for plants, is lost from the soil. On the other hand, this process is a useful way to remove excess nitrate from wastewater.

Sulfur in the form of sulfate (SO_4^{-2}) is used by anaerobic bacteria like the genus *Desulfovibrio* which convert it to hydrogen sulfide gas (H_2S). Hydrogen sulfide reacts with metal ions and forms very insoluble metallic sulfides like pyrite (Fe_2S). In fact, it is probable that the pyrites associated with coal seams were deposited by the action of these bacteria eons ago. The black colour of salt marsh soils and the rotten egg smell associated with them are a result of the activities of the sulfate-reducing bacteria in these habitats. They attest to the occurrence of anaerobic conditions. Sulfur volatilisation from soil represents loss of a plant nutrient as well as a contribution of atmospheric sulfur which may contribute to the phenomenon of acid precipitation.

We mentioned above that nitrogen can be lost from agricultural soils as well as from other ecosystems. Fortunately, this "leak" in the terrestrial nitrogen cycle can be at least partially replaced through another important biological process called biological nitrogen fixation. In this process, which is unique to bacteria and a few other microbes, notably the cyanobacteria (blue-green algae), atmospheric dinitrogen (N_2) is captured and converted to plant-available forms. Biological nitrogen fixation is carried out by free-living bacteria and cyanobacteria and by symbiotic micro-organisms in a wide variety of mutualistically symbiotic associations with higher plants.

The most useful and probably the most widely recognised example of symbiotic nitrogen fixation is that of the *Rhizobium* — legume root-nodule symbiosis. Soil bacteria belonging to the genera *Rhizobium* and *Bradyrhizobium* (and a few others) are capable of inducing the formation of nodules on roots of specific legumes (plants like peas, beans, peanuts, soybeans, alfalfa etc.) and fixing large quantities of nitrogen in these structures. In the nodule, the bacteria are supplied with carbon sources (photosynthate from the plant) that they need in order to fix nitrogen. In return for this carbon, the bacteria fix atmospheric nitrogen which is converted to amino acids used by the plant for growth. The result of this unique plant-microbe partnership is that many legumes are self-sufficient for nitrogen, that is, they are nearly independent of a supply of nitrogen from the soil. It is no wonder that these plants are cultivated all over the world as sources of food, fibre and forage. Nearly two-thirds of the world's nitrogen supply is from biological nitrogen fixation. Legumes have been used since the beginning of recorded history as "soil improving" crops known as "green manures". Green manuring is the practice of growing a legume species for the sole purpose of returning it to the soil to serve as a source of nitrogen for an ensuing crop.

Soil Microbes and Bio-remediation

We have touched on the remarkable metabolic diversity and capacity of the soil micro-flora. This capacity is increasingly being harnessed and put to good use by humans. A most beneficial spin-off from our understanding of the metabolism of soil microbes has been the development of methods for the bio-remediation of soils contaminated by hazardous wastes or spilled petroleum products both on land and sea. Bio-remediation may be defined as the controlled use of micro-organisms for the destruction of chemical pollutants. A large number of processes have been developed to handle various wastes and for the cleanup of spilled organic materials. At the heart of all of these processes lies the premise that the metabolic activities of bacteria or fungi can be used to degrade many of the organic chemicals of commerce (solvents, pesticides, hydrocarbon fuels, etc.).

Either of two forms of bio-remediation is commonly employed. In bio-stimulation the environment into which the material has been spilled or otherwise introduced is made favourable for the rapid development of microbes. Typically, this process involves adding sufficient nitrogen and phosphorus fertilizer to overcome nutrient limitations to microbial growth and providing some mechanism for increased aeration of the system. These practices encourage development of the indigenous microbial population which usually contains microbes able to degrade the compounds of interest. In the practice of bio-augmentation, an external microbial population is added in order to speed up the degradation process. Numerous microbes have been developed for such purposes. However, the full measure of the usefulness of such microbial products is not yet known. Some inoculants have reportedly enhanced the remediation process and others have had little or no effect on the process. It is probable that in due time useful microbial products or processes will be developed for use in the clean-up of oil or other chemical spills. What is certain is that successful bio-remediation will require detailed knowledge of the factors which make some microbes more competitive than others in a given environment. Only when these details are established will we know how to use sound ecological principles to add microbes to these complex environments to insure their establishment and function in the clean-up process.

In March 1989, the Exxon Valdez oil tanker hit a reef in Prince William Sound, Alaska (USA) and released over 40 million liters of crude oil into the Sound within a 5-hour period. Over 1500 km (932 miles) of shoreline in the Sound and the Gulf of Alaska were contaminated to varying degrees by crude oil. The Exxon Valdez oil spill was a historic event because of the magnitude of the spill, the vastness and isolation of the area to be treated, and the large number of personnel and vehicles ultimately involved. The success of bio-remediation, particularly in a climate as cold as Alaska's, prompted regulatory agencies in the United States to view bio-remediation much more favourably over previous strategies of physical or chemical "entombment" (storage in cement tombs).

Because oil is inherently high in carbon and low in nitrogen and phosphorus, a portion of the shoreline was selected for bio-stimulation. After several potential fertilizer candidates were evaluated, a micro-emulsion, Inipol EAP22™ (henceforth, Inipol), was selected. Inipol (an "oleophilic" fertilizer) is a stable water-in-oil formulation that yields an N-P-K ratio of 7.3 : 0.8 : 0. The nitrogen source is urea and the phosphorus source is trilaureth (4)-phosphate. At room temperature, Inipol has the consistency and appearance of honey, and it must be heated to 90 °C (194 °F) before it can be sprayed on the soil. Inipol was applied as a thin coat to the shore at a rate of 306 ml m^{-2} (0.27 quart per square yard). As the microemulsion mixed with the weathered crude oil, the crude oil destabilised Inipol to release its urea-N. In addition, a surface-active organic material (oleic acid) in Inipol served as a readily degradable carbon and energy source to increase the activity and number of indigenous hydrocarbon-degrading bacteria. When the oleic acid was depleted, the increased biomass of hydrocarbon-degrading bacteria supported enhanced biodegradation of the petroleum. Visual observations and chemical assays showed dramatic evidence that bio-stimulation contributed to the remediation of the site. Although passive bio-remediation also undoubtedly occurred in the absence of the fertilizer nitrogen and phosphorus, the accelerated rate of biodegradation observed with Inipol was critical to a successful bio-remediation effort.

PLANT ESSENTIAL NUTRIENTS

Sixteen chemical elements are recognised as being essential for the growth of all plants. Five others, silicon, sodium, cobalt, vanadium, and nickel, have been recognised as necessary for the growth of some plant species. Although certain essential elements can exist in nature in a number of ionic orms, plants can use only specific ones.

Nitrogen

Other than carbon, hydrogen, and oxygen, nitrogen is the nutrient required by plants in the greatest quantity. The nitrogen concentration of plants ranges from about 0.5 to 5 per cent on a dry weight basis. Since most plants have a rather high nitrogen requirement and most soils can't supply sufficient nitrogen to meet this demand, nitrogen normally must be supplemented from organic or inorganic fertilizer sources.

The ultimate source of all nitrogen in soils is the atmosphere, which is approximately 78 per cent N_2. Although this quantity represents an almost unlimited supply, nitrogen as N_2 is not directly available for uptake by most plants. Recall that legumes in symbiosis with particular species of the bacterial genus, *Rhizobium*, transform gaseous N_2 to plant available form, with capacities to fix N_2 ranging from about 40 to greater than 300 lbs N/acre/year. non-symbiotic fixation of N_2 by free living soil micro-organisms also occurs but quantities are usually less than 10 lbs N/acre/year. Lightning discharge in thunderstorms can oxidise atmospheric N_2 to plant available nitrate, but quantities are generally less than 20 lbs N/acre/year. N_2 can also be transformed to plant available forms through the fertilizer manufacturing process. The chemical triple bond that exists between the two N atoms in N_2 is very strong. All the above processes that convert N_2 to plant useable forms, therefore, are highly energy intensive.

Nitrogen is an extremely reactive element with many of its possible transformations being depicted in the nitrogen cycle. The nitrogen cycle is the dynamic system in which nitrogen is transformed, or cycled, from one form to another. Total nitrogen in the soil-plant-water-atmosphere continuum is conserved, but amounts existing in various "pools", or forms, of nitrogen change with time and environmental conditions. The vast majority of soil nitrogen (approximately 95 per cent) is found in soil organic matter. This organically-combined nitrogen is not immediately plant available, but can be converted to inorganic, available forms through the actions of soil micro-organisms. This process, as previously discussed, is termed nitrogen mineralisation. Prior to the production of inorganic nitrogen fertilizers, most nitrogen for plant growth was supplied through leguminous N_2 fixation and nitrogen mineralisation of added animal manures and indigenous soil organic nitrogen. Sole reliance on nitrogen mineralisation from soil organic matter normally is not sufficient for crop production and may result in soil deterioration associated with the loss of organic matter. Nitrogen supplementation, whether from organic or inorganic fertilizer sources, is normally necessary for crop production.

Nitrogen, primarily in the form of nitrate, may be lost from soils through leaching. Nitrate, being an anion, is repelled by the negatively-charged cation exchange sites in soils. Since the ion is not adsorbed and is highly water soluble, it will move downward in soils with percolating water. Nitrate leaching is most common in coarse, sandy soils receiving excess rainfall or irrigation. Nitrate losses may also be increased by applying nitrogen fertilizers, whether inorganic or organic, in excess of a crop's requirement. Nitrate leaching should be prevented not only from economic, but also from environmental and health standpoints. Ingestion of waters high in nitrate has been implicated in gastrointestinal problems in adults and methemoglobinemia ("blue baby syndrome") in infants, though confirmed cases are rare. Near-surface aquifers normally are the most susceptible. Nitrate contamination of waters is usually localised and can be decreased through proper management.

Denitrification is the bacterial reduction of nitrate under anaerobic conditions to N_2 or N_2O gases. Under anaerobic conditions, most nitrate in soils may be denitrified in a period of a few days. Some scientists theorise that atmospheric N_2 is the result of denitrification over geologic time. Evidence also indicates that N_2O may be partially responsible for depletion of the protective ozone layer and is also a potent "greenhouse gas". Thus, loss of nitrate through denitrification not only results in an economic loss of plant available nitrogen, but may also have other detrimental effects. Combustion of fossil fuels also produces nitrous oxides that may contribute to this effect.

Nitrogen is an essential ingredient for the production of sufficient food for an expanding world population. Proper nitrogen management can decrease the potential for negative environmental impacts.

Phosphorus

Phosphorus in soil organic matter accounts for about 20 to 65 per cent of the total phosphorus found in soils. Therefore, phosphorus mineralisation from soil organic matter is an important source of available phosphorus for plant growth. Phosphorous ranks second to nitrogen as a limiting nutrient for plant growth. Although plant available forms of this element are anionic, phosphorus is immobile in soils with appreciable colloid content because it tends to be tightly bound to these tiny particles. Phosphorus may also form water insoluble compounds such as insoluble calcium phosphates in alkaline soils and insoluble iron and aluminum phosphates in acid soils. The concentration of phosphorus in soil solution is normally much less than one part per million (ppm), even in fertilized soils, and often is only hundredths of a ppm in unfertilized soils.

Phosphorus fertilizers are normally produced through acidification of the mineral, apatite, found in high concentrations in some sedimentary deposits. Organic phosphorus sources, such as manure, may also be used. Manures, however, usually contain relatively large quantities of phosphorus relative to nitrogen. Care must be taken with manure additions so that excess phosphorus doesn't result in deficiencies of other nutrients, such as zinc, or contribute to soluble phosphorus in run-off waters.

Soluble phosphorus can be lost in surface run-off waters, but is usually found adsorbed to soil particles transported by erosion. Phosphorus in run-off has been implicated in eutrophication (excessive algal growth) of lakes and streams.

Potassium

Potassium is required by plants in amounts second only to nitrogen. Unlike nitrogen and phosphorus, potassium is not organically combined in soil organic matter. Different potassium-containing minerals, such as micas and feldspars, therefore, are the principal sources of potassium in soils. Clay-sised micas weather more rapidly to release potassium than feldspars because of their much greater surface area. Soils that contain considerable micaceous clays may be able to supply all of a crop's potassium requirement without fertilization. Acid, weathered soils are those most likely to be deficient in available potassium.

Calcium and Magnesium

Calcium is the predominant exchangeable cation in soils, even in the majority of acid soils, followed by magnesium. This occurs because of the large number of minerals in soils that contain calcium and/or magnesium. Actual plant deficiencies of these elements are infrequent because problems associated with soil acidity, such as aluminum toxicity, become limiting first.

Sulfur

Approximately 85 per cent of total soil sulfur is found in soil organic matter. Microbial mineralisation of the soil organic fraction is an important source of available sulfur for plant growth. Reactions of sulfur in soils are very similar to those of nitrogen. Sulfides, such as pyrite and other reduced forms of sulfur, are commonly unearthed in metal and coal mining. Upon exposure to oxygen, sulfuric acid which is produced through chemical and biological oxidation can result in soil acidification and acid mine drainage. Soils may also receive sulfur through atmospheric deposition. Soils near large metropolitan areas may receive greater than 150 lbs S/acre/year from the combustion of fossil fuels. Volcanic eruptions can also emit large quantities of sulfur gases. Soils most commonly deficient in available sulfur are sandy, leached soils that are low in organic matter.

Micro-nutrients

Iron, zinc, manganese, copper, chlorine, boron, and molybdenum are classified as micro-nutrients. Micro-nutrients are plant essential elements that are required by plants in much smaller amounts than the other essential nutrients. Generally, less than 1 lb/acre of each micro-nutrient will be present in the above-ground portion of crops. This small quantity contrasts with the 200 lb/acre or more of nitrogen.

The total quantity of many micro-nutrients in soils doesn't necessarily relate to plant availability. Most soils, for example, will contain from 20,000 to 200,000 lbs total iron/acre to a depth of six inches, but may not be able to supply a crop with sufficient available iron for uptake of 1 lb/acre. Iron deficiency is usually associated with highly alkaline soils because iron solubility roughly decreases 1000-fold for each one unit increase in soil pH.

Zinc, manganese, and copper availabilities are also decreased by alkalinity and by high organic matter concentrations (>10 per cent). These three elements form very stable bonds with soil organic matter which decrease their availability.

Plant micro-nutrient deficiencies are becoming more widespread because of greater quantities required by higher yields and decreasing micro-nutrient impurities in fertilizers.

SOIL MANAGEMENT

Most soils cannot provide one or more plant essential nutrients in sufficient available form for modern crop production. Soil samples are frequently tested to determine the quantities of nutrients and other amendments which should be applied. Soil testing normally involves extraction or reaction of a sample with a specific chemical solution(s) which removes essential elements in amounts related to those required for plant growth. Soil testing is recommended to prevent both under and over fertilization of crops, thereby providing economic crop production in an environmentally effective manner.

Lime

The main purpose of liming is to raise soil pH and supply calcium and sometimes magnesium for plant growth. Other benefits from liming acid soils include increased biotic activity, enhanced mineralisation of nutrients from soil organic matter, improved soil structure, decreased potential for aluminum toxicity, and increased availability of other nutrients, especially phosphorus.

Fertilization

A good fertilization programme is based on soil testing, especially over several years, so that changes in nutrient availability and other chemical properties can be determined over time. Nutrients can be added in either organic or inorganic forms. Remember that plants utilise only specific ionic forms of the essential nutrients. These specific ionic forms can be provided directly through inorganic fertilizers or indirectly through the decomposition of manures, composts, and other organic amendments. Thus, plant uptake of nutrients should be similar whether initially added through inorganic or organic sources. Additional benefits which may accrue from application of organic materials include improved soil structure and increased water holding and cation exchange capacities. The degree to which these additional benefits occur will be determined by the quantity and quality of organic material applied and its decomposition rate.

Municipal sewage sludge, food processing wastes, and other similar organic wastes are commonly added to soils. Besides the obvious need for disposal, these materials can provide nutrients for plant growth and improve soil properties. Sewage sludges can contain heavy metals, however, especially in industrialised areas. Land application of sewage sludge is usually regulated by state and federal guidelines. When properly utilised, the above materials can be suitable sources of plant nutrients.

Irrigation

Irrigation is practiced to at least a limited extent on each of the earth's continents. Irrigation is a method of at least partially overcoming problems in natural precipitation patterns. These problems may result from an overall lack of precipitation or poor seasonal distribution. Irrigation is normally practiced not only to increase yield, but to provide yield stability. Yields may be increased many fold by irrigation depending on climate and the crop produced. Large scale irrigation projects may also be one facet of an overall programme to provide hydroelectric power, flood control, municipal water supplies, and recreation.

Water sources for irrigation include lakes, streams, and groundwater. Numerous water delivery methods are used including furrow, flood, sprinkler, and drip irrigation. Each system is different in efficiency and cost. Both these factors are normally considered before a producer decides on the type of system to be used. Water districts and governmental agencies may also mandate the type of system used based on water quantity, quality, and environmental concerns.

Although development of irrigation capabilities almost immediately increases crop production, long-term effects must also be addressed. Irrigation water must be of reasonable quality (sufficiently low in dissolved salts and of proper ionic composition) to be used for an extended period of years. Many irrigation systems from ancient times through today have failed because of increasing soil salinity over time. Water applied through irrigation is partially transpired by plants or evaporated from the soil. Salts added in the water remain in the soil. Over time, salts can accumulate so as to decrease yields or prevent further crop production.

Providing adequate drainage is also imperative for irrigated lands. Proper drainage not only decreases salt accumulation, but prevents shallow water tables which may restrict plant rooting depth and contribute to upward salt migration during periods of high evaporative demand.

Irrigation has been extremely important throughout the world in providing a stable and abundant food supply, but proper planning and expertise are necessary to sustain economic and environmentally sound irrigated crop production.

Tillage Systems and Conservation Practices

Tillage systems vary widely from conventional, or "clean", tillage where all crop residues are incorporated into the soil with little residue remaining on the soil surface to no-tillage where crop residues are not disturbed and remain on the soil surface from one crop to the next. Conservation tillage denotes any tillage system in which at least 30 per cent of the soil surface is covered by residue after planting. Tillage is used to destroy weeds, incorporate residues from a previous crop, open the soil surface for increased infiltration and aeration, shatter compacted zones which might restrict root growth, and prepare seedbeds for optimal seed germination.

The principal advantage of conservation tillage is protection of surface soil from water and wind erosion and is frequently used to decrease soil loss on erodible lands. Secondary advantages include increased water storage and organic matter contents. Erosion not only decreases the productivity of soils, but additional economic loss is associated with siltation of reservoirs, dredging of waterways, and removal of particulates to meet drinking water standards. As plant residue cover of the soil surface nears 100 per cent , soil erosion decreases to near zero. Erosional loss can be reduced by more than 80 per cent with a residue cover of 50 per cent . Even a modest residue cover of 10 per cent can reduce erosion by about 30 per cent .

Other alternatives are available for erosion control. Construction of terraces, or berms, which slow and alter the movement of water across slopes has been used for centuries to decrease water erosion. Strip cropping, the planting of alternating strips of crops of different heights or seasonal maturities across a landscape, and contour farming, the production of crops across instead of with slopes, have also historically been used to reduce the erosional forces of wind and water.

Rotation, or planting different crop species, is normally beneficial compared to mono-culture, or continual production of the same crop species. Rotation often decreases problems from weeds, disease and insects and may improve soil chemical, physical, and microbiological properties, and crop yield.

Farmers must make many decisions each year. These decisions must result in both economic and environmentally prudent crop production if food production is to be sustainable over the long-term.

SOILS AND SOCIETAL ISSUES

An escalating world population will certainly result in increased demand for food, fibre, water, and other natural resources. Only one billion persons are estimated to have lived throughout all history up to the year 1850. Contrast this statistic with an estimated world population of seven billion by the year 2000 and the potential problem becomes more sharply focused. World population is currently doubling about every thirty-five years. Will we realistically be able to feed and clothe this burgeoning mass by the year 2025? By 2050?

As one of earth's most vital and fragile natural resources, soils come under ever increasing pressure. Quite simply, there is less soil per person with each passing day, making increased productivity per unit of land area a requirement. At the same time there are issues of appropriate land uses, sustainability, environmental protection and water rights.

The Soil Conservation Service (recently renamed the Natural Resources Conservation Service) was established in the United States in 1935 following catastrophic wind erosion of the plowed prairie lands of the Great Plains. Protection of arable land from the ravages of accelerated erosion has been its primary goal. Beginning in 1985, Farm Bills passed by the U.S. Congress have contained provisions that require use of erosion control measures by producers who participate in governmental price support programmes. For many years farmers and ranchers have been encouraged to make co-operative conservation agreements with the Soil Conservation Service. These agreements entitled them to financial assistance for implementing erosion control practices. In much of the developing world, where population growth has been greatest, organised conservation efforts have been weak and pressure on arable soil resources has been great. This growth has led to a tragic loss of valuable resources partially because resources not well suited for crop production are being used as food requirements increase.

Land use planning, not popular with persons who value individual rights in use of owned property, is becoming more widespread as more demands are placed on soil and water resources. Further, concern that land retain its productivity over time is increasing. "Wearing out the soils" on a farm and moving on to a virgin territory is no longer the option that it was well into the present century.

An additional land use issue which will take on increasing global importance is urban development. Many of the world's large cities developed where they did because of the stable agricultural production from productive soils in the area. As cities expand, formerly productive agricultural land is covered by highways, housing sub-divisions, shopping malls, and other commercial developments. Once appropriated for such uses, these lands cannot economically be returned to agricultural production. Responsible land use planning is expected to become increasingly important in the future.

Organic gardening and farming have attracted much attention in recent years. As emphasised, the organic component in soils has essential roles in determining the soil's physical, chemical and biological behaviour.

Increasingly, decisions that impact use and management of soil resources arise from discussion of problems such as sustainable use of finite resources, water quality, sedimentation in lakes, food quality, endangered species, preservation of wildlife habitats, and cultural practices. Every citisen, through the right to vote, can influence laws that determine how soil and water resources are used and managed. All citisens share the responsibility for ensuring that all natural resources are treated with respect and wisely used. Some citisens must devote their careers to developing still better technology for management and conservation that will ensure continued utility of these precious resources by our successors on planet earth.

Chapter 3

SOIL HORIZONS

SOIL HORIZON DEVELOPMENT OVER TIME

Because of the diversity of landscapes and controlling variables of soil formation — such as glaciations, flooding, erosion, and tectonics — soil horizons are often found in multiple combinations of vertical arrangement. To simplify the concept, consider a landscape where a river has deposited several meters of silty, gray, limestone-rich sediments (alluvium). At this stage, the profile would be C; since no soil forming processes have occurred yet, it is not yet a soil. The annual precipitation at this location is 75 to 100 cm (30 to 40 inches). The climate is temperate with warm summers and cool winters.

Over geologic time these sediments develop a soil profile :

- *0 – 50 years :* Plants (grasses and forbs) begin to establish on the surface. Plant growth is good, roots proliferate below ground and senesced plant material falls to the surface in the fall. Worms, ants, beetles, and other organisms break these leaves into smaller pieces and mix them in the upper layers of the soil. A thin, dark A horizon begins to form (addition) over the existing C horizon. Profile horizons are A-C.

- *50 – 500 years :* The plants continue to grow, and the A horizon becomes thicker (addition). The silty sediments contain calcium carbonate, $CaCO_3$, which is easily dissolved in the temperate, humid environment. The $CaCO_3$ begins to leach out of the upper part of the soil and move deeper as rain and snowmelt move down through the soil (translocation). Some of the $CaCO_3$ moves all the way through the soil profile and into the groundwater (loss). Some of the iron in of human beings : life, mind or reason, religion, property, and posterity or future generations.

SOIL PROFILES AND HORIZONS

Through the interactions of these four soil processes, the soil constituents are re-organised into visibly, chemically, and/or physically distinct layers, referred to as horizons. There are five soil horizons : O, A,E, B, and C. (R is used to denote bedrock.) There is no set order for these horizons within a soil. Some soil profiles have an A-C combination, some have an O-E-B, an O-A-B, or just an O. Some profiles may have all the horizons, O-A-E-B-C-R. And some profiles may have multiple varieties of one horizon, such as an AB-E-B. There are some generalised concepts of how soil layers develop with time; these are expressed below, but due to the variability of natural processes over geologic time, generalised concepts are sometimes overly general. Knowing something about the geomorphic history of the area being investigated helps unlock the landscape history the soils show.

- *A :* An A horizon is a mineral horizon. This horizon always forms at the surface and is what many people refer to as topsoil. Natural events, such as flooding, volcanic eruptions, landslides, and dust deposition can bury an A horizon so that it is no longer found at the surface. A buried A horizon is a clear indication that soil and landscape processes have changed some time in the past. Compared to other mineral horizons (E, B, or C) in the soil profile, they are rich in organic matter, giving them a darker colour. The A horizon, over time, is also a zone of loss – clays and easily dissolved compounds being leached out – and A horizons are typically more coarse (less clay) compared to underlying horizons. Additions and lossesare the dominant processes of A horizons.

- *B :* A B horizon is typically a mineral sub-surface horizon and is a zone of accumulation, called *illuviation*. Materials that commonly accumulate are clay, soluble salts, and/or iron. Minerals in the B horizon may be undergoing transformations such as chemical alteration of clay structure. In human modified landscapes, processes such as erosion can sometimes strip away overlying horizons and leave a B horizon at the surface. Such erosion is common in sloping, agricultural landscapes. A bulldozer preparing land for a new sub-division can also leave a B horizon at the surface. The dominant processes in a B horizon are *transformations and additions.*

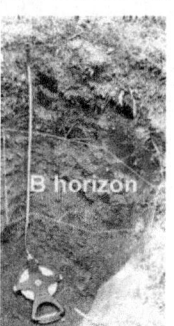

- *C:* A C horizon consists of parent material, such as glacial till or lake sediments that have little to no alteration due to the soil forming processes. Low intensity processes, such as movement of soluble salts or oxidasation and reduction of iron may occur. There are no dominant processes in the C horizon; minimal additions and losses of highly soluble material (*e.g.*, salts) may occur.

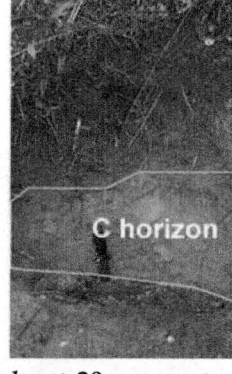

- *O :* An O horizon has at least 20 per cent organic matter by mass. Two main scenarios result in the formation of an O horizon : saturated, anaerobic conditions (wetlands) or high production of leaf litter in forested areas. Anaerobic conditions slow the decomposition process and allow organic material to accumulate. An O horizon can have various stages of decomposed organic matter : highly decomposed, sapric; moderately decomposed, hemic; and minimally decomposed, fibric. In a fibric O layer, plant matter is recognisable (*e.g.*, it is possible to identify a leaf). Sapric material is broken down into much finer matter and is unrecognisable as a plant part. Hemic is in between sapric and fibric, with some barely recognisable plant material present. It is possible to have multiple O horizons stacked upon one another exhibiting different decomposition stages. Because of their organic content, these horizons are typically black or dark brown in colour. The dominant processes of the O horizon are additions of organic matter, and transformations from fibric to sapric. translocation. The E horizon appears lighter in colour than an associated A horizon (above) or B horizon (below). An E horizon has a lower clay content than an underlying B horizon, and often has a lower clay content than an overlying A horizon, if an A is present. E horizons are more common in forested areas because forests are in regions with higher precipitation and forest litter is acidic. However, landscape hydrology, such as perched water tables, can result in the formation of an E horizon in the lower precipitation grasslands, as seen in the profile below. The dominant processes of an E horizon are losses.

- *R :* An R layer is bedrock. When a soil has direct contact with bedrock, especially close to the soil surface, the bedrock becomes a variable when developing land use management plans and its presence is noted in the soil profile description.

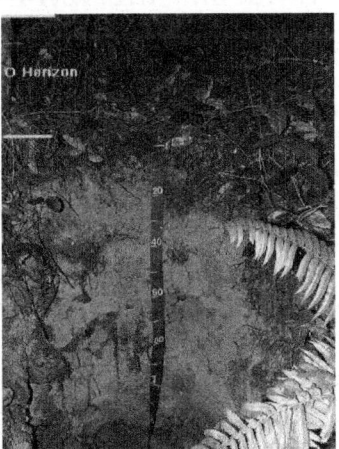

E : An E horizon is a zone of strong leaching, or eluviation. The chemistry and hydrology of this horizon are such that constituents such as clay, organic matter, and minerals like iron oxides are removed, leaving behind the un-pigmented soil particles. This represents a loss of soil constituents from the E horizon. If the materials enrich an underlying B horizon, that process is considered translocation. The E horizon appears lighter in colour than an associated A horizon (above) or B horizon (below). An E horizon has a lower clay content than an underlying B horizon, and often has a lower clay content than an overlying A horizon, if an A is present. E horizons are more common in forested areas because forests are in regions with higher precipitation and forest litter is acidic. However, landscape hydrology, such as perched water tables, can result in the formation of an E horizon in the lower precipitation grasslands, as seen in the profile below. The dominant processes of an E horizon are losses.

- *R :* An R layer is bedrock. When a soil has direct contact with bedrock, especially close to the soil surface, the bedrock becomes a variable when developing land use management plans and its presence is noted in the soil profile description.

SOIL HORIZONS AND NEW LAYERS

The definitions of classes in the Canadian system are based mainly on the kinds, degrees of development, and the sequence of soil horizons and other layers in pedons. Therefore, the clear definition and designation of soil horizons and other layers are basic to soil classification. A soil horizon is a layer of mineral or organic soil or soil material approximately parallel to the land surface that has characteristics altered by processes of soil formation. It differs from adjacent horizons in properties such as colour, structure, texture, and consistence, and in chemical, biological, and mineralogical composition. The other layers are either non-soil layers such as rock and water or layers of unconsolidated material considered to be unaffected by soil-forming processes.

The Major Mineral Horizons are A, B, and C. The major organic horizons are L, F, and H, which are mainly forest litter at various stages of decomposition, and O, which is derived mainly from bogs, marsh, or swamp vegetation. Sub-divisions of horizons are labelled by adding lower case suffixes to some of the major horizon symbols as with Ah or Re. Well-developed horizons are readily identified in the field. However, in cases of weak expression or of borderline properties, as between Ah and H, laboratory determinations are necessary before horizons can be designated positively. Many of the laboratory methods required are outlined in a manual prepared by a sub-committee of CSSC. Some other methods pertaining to organic horizons are outlined near the end of this chapter.

The layers defined are R, rock; W, water; and IIC or other non-conforming, unconsolidated mineral layers, IIIC, etc., below the control section that are unaffected by soil-forming processes. Theoretically a IIC affected by soil-forming processes is a horizon; for example a IICca is a horizon. In practice, it is usually difficult to determine the lower boundary of soil material affected by soil-forming processes. *Thus the following are considered as horizons :* C(IC), any unconforming layer within the control section, and any unconforming layer below the control section that has been affected by pedogenic processes (*e.g.* IIBc, IIIBtj). Unconforming layers below the control section that do not appear to have been affected by pedogenic processes are considered as layers. The tiers of Organic soils are also considered as layers.

Mineral Horizons and Layers

Mineral horizons contain 17 per cent or less organic C (about 30 per cent organic matter) by weight. This is a mineral horizon formed at or near the surface in the zone of leaching or eluviation of materials in solution or suspension, or of maximum *in situ* accumulation of organic matter or both. The accumulation of organic matter is usually expressed morphologically by a darkening of the surface soil (Ah), and conversely the removal of organic matter is usually expressed by a lightening of the soil colour usually in the upper part of the solum (Ae). The removal of clay from the upper part of the solum (Ae) is expressed by a coarser soil texture relative to the underlying subsoil layers. The removal of iron is indicated usually by a paler or less red soil colour in the upper part of the solum (Ae) relative to the lower part of the sub-soil.

This is a mineral horizon characterised by enrichment in organic matter, sesquioxides, or clay, or by the development of soil structure; or by a change of colour denoting hydrolysis, reduction, or oxidation. The accumulation in B horizons of organic matter (Bh) is evidenced usually by dark colours relative to the C horizon. Clay accumulation is indicated by finer soil textures and by clay cutans coating peds and lining pores (Bt). Soil structure developed in B horizons includes prismatic or columnar units with coatings or stainings and significant amounts of exchangeable sodium (Bn) and other changes of structure (Bm) from that of the parent material. Colour changes include relatively uniform browning due to oxidation of iron (Bm), and mottling and gleying of structurally altered material associated with periodic reduction (Bg). This is a mineral horizon comparatively unaffected by the pedogenic processes operative in A and B, (C), except the process of gleying (Cg), and the accumulation of calcium and magnesium carbonates (Cca) and more soluble salts (Cs, Csa). Marl, diatomaceous earth, and rock no harder than 3 on Mohs' scale are considered to be C horizons. This is a consolidated bedrock layer that is too hard to break with the hands (>3 on Mohs' scale) or to dig with a spade when moist and does not meet the requirements of a C horizon. The boundary between the R layer and any overlying unconsolidated material is called a lithic contact. This is a layer of water in Gleysolic, Organic, or Cryosolic soils. Hydric layers in Organic soils are a kind of W layer.

Lowercase Suffixes

b : A buried soil horizon.

c : A cemented (irreversible) pedogenic horizon. Ortstein, placic, and duric horizons of Podzolic soils, and a layer cemented by $CaCO_3$ are examples.

ca : A horizon of secondary carbonate enrichment in which the concentration of lime exceeds that in the unenriched parent material. It is more than 10 cm thick, and its CaCO, equivalent exceeds that of the parent material by at least 5 per cent if the CaCO, equivalent is less than 1 (13 per cent *vs* 8 per cent), or by at least 1/3 if the CaCO, equivalent of the horizon is 15 per cent or more (28 per cent *vs* 21 per cent). If no IC is present, this horizon is more than 10 cm thick and contains more than 5 per cent by volume of secondary carbonates in concretions or in soft, powdery forms, cc — Cemented (irreversible) pedogenic concretions.

e : A horizon characterised by the eluviation of clay, Fe, Al, or organic matter alone or in combination. When dry, it is usually higher in (colour value by one or more units than an underlying B horizon. It is used with A (Ae).

f : A horizon enriched with amorphous material, principally Al and Fe combined with organic matter. It must have a hue of 7.5YR or redder, or its hue must be 10YR near the upper boundary and become yellower with depth. When moist the chroma is higher than 3 or the value is 3 or less. It contains at least 0.6 per cent pyrophosphate-extractable Al + Fe in textures finer than sand and 0.4 per cent in sands (coarse sand, sand, fine sand, and very fine sand). The ratio of pyrophosphate-extractable Al + Fe to clay (< 0.002 mm) is more than 0.05 and organic C exceeds 0.5 o . Pyrophosphate- extractable Fe is at least 0.3 per cent, or the ratio of organic C to pyrophosphate-extractable -Fe is less than 20, or both are true. It is used with B alone (Bf), with B and h (Bhf), with B and g (Bfg), and with other suffixes. These criteria do not apply to Bgf horizons.

The following f horizons are differentiated on the basis of the organic C content :

Bf : 0.5-5 per cent organic C

Bhf : More than 5 per cent organic C

No minimum thickness is specified for a BF or a Bhf horizon. Thin Bf and Bhf horizons do not qualify as podzolic B horizons. Some Ah and Ap horizons contain sufficient pyrophosphate-extractable Al + Fe to satisfy this criterion of f but are designated Ah or Ap.

g : A horizon characterised by gray colours, or prominent mottling, or both, indicative of permanent or periodic intense reduction. Chromas of the matrix are generally :1 or less. T-t is used with A and e (Aeg); B alone (Bg); B and f (Bfg, Bgf); B, h, and f (Bhfg); B and t (Btg) C alone (Cg); C and k (Ckg); and several others. In some reddish parent materials matrix colours o reddish hues and high chromas may persist despite long periods of reduction. In these soils, horizons are designated as g if there is gray mottling or marked bleaching on ped faces or along cracks.

Aeg : This horizon must meet the definitions of A, e, and g. *Bg* : This horizon is analogous to a Em horizon but has colours indicative of poor drainage and periodic reduction. It includes horizons occurring between A and C horizons in which the main features are : (i) Colours of low chroma, that is : chromas of -1 or less, without mottles on ped surfaces or in the matrix if peds are lacking; or chromas of 2 or less in hues of 10YR or redder, on ped surfaces or in the matrix if peds are lacking, accompanied by more prominent mottles than those in the C horizon; or hues bluer than lOY, with or without mottles on ped surfaces or in the matrix if peds are lacking.

(ii) Colours indicated in (i) and a change in structure from that of the C horizon. (iii) Colours indicated in (i) and illuviation of clay too slight to meet the requirements of Bt, or an accumulation of iron oxide too slight to meet the limits of Bgf. (iv) Colours indicated in (i) and the removal of carbonates. Bg horizons occur in some Orthic Humic Gleysols and some Orthic Gleysols.

Bfg, Bhfg, Btg :When used in any of these combinations the limits set for f, hf, t, and others must be met.

Bgf : The dithionite-extractable Fe of this horizon exceeds that of the IC by 1 per cent or more. Pyre phosphate-extractable Al + Fe is less than the minimum limit specified for f horizons. This horizon occurs in Fera Gleysols and Fera Humic Gleysols and possibly below the Bfg of gleyed Podzols. It is distinguished from the Bfg of gleyed Podzols on the basis of the extractability of the Fe and Al. The Fe in the Bgf horizon is thought to have accumulated as a result of the oxidation of ferrous iron. The iron oxide formed is not associated intimately with organic matter or with Al and is sometimes crystalline. The Bgf horizons are usually prominently mottled, more than half of the soil material occurs as mottles of high chroma.

Cg, Ckg, Ccag, Csg, Csag : When g is used with C alone, or with C and one of the lowercase suffixes k, ca, s, or sa the horizon must meet the definition for C and for the particular suffix as well as for g.

h : A horizon enriched with organic matter. It is used with A alone (Ah), or with A and e (Ahe), or with B alone (Bh), or with -B and f (Bhf).

Ah : A horizon enriched with organic matter, it has a colour value at least one unit lower than the underlying horizon or 0.5.0 more organic C than the IC or both. It contains less than 17 per cent organic C by weight.

Ahe : An Ah horizon that has undergone eluviation as evidenced, under natural conditions, by streaks and splotches of different shades of gray and often by platy structure. It may be overlain by a dark-coloured Ah and underlain by a light-coloured Re.

Bh : This horizon contains more than 1 per cent organic C, less than 0.3 per cent pyrophosphate-extractable Fe, and has a ratio of organic C to pyrophosphate-extractable Fe of 20 or more. Generally the colour value and chroma are less than 3 when moist.

Bhf : Defined under f.

j : This is used as a modifier of suffixes e, f, g, n, and t to denote an expression of, but failure to meet, the specified limits of the suffix it modifies. It must be placed to the right and adjacent to the suffix it modifies. For example, Bfgj means a Bf horizon with a weak expression of gleying, Bfjgj means a B horizon with weak expression of both f and g features.

Aej : It denotes an eluvial horizon that is thin, discontinuous, or slightly discernible.

Btj : It is a horizon with some illuviation of clay but not enough to meet the limits of Bt.

Btgj, Bmgj : These are horizons that are mottled but do not meet the criteria of Bg.

Bfj It is a horizon with some accumulation of pyrophosphate-extractable Al + Fe but not enough to meet the limits of Bf. In addition, the colour of this horizon may not meet the colour criteria set for Bf.

Btnj or Bnj : These are horizons in which the development of solonetzic B properties is evident but insufficient to meet the limits for En or Bnt.

k : Denotes the presence of carbonate as indicated by visible effervescence when dilute HC1 is added. It is used mostly with B and m (Bmk) or C (Ck) and occasionally with Ah or Ap (Ahk, Apk), or organic horizons (Ofk, Omk).

m : A horizon slightly altered by hydrolysis, oxidation, or solution, or all three to give a change in colour or structure, or both. It has :

Evidence of alteration in one of the following forms :

- Higher chromas and redder hues than the underlying horizons.
- Removal of carbonates either partially (Bmk) or completely (Bm).
- A change in structure from that of the original material.
- Illuviation, if evident, too slight to meet the requirements of a Bt or a podzolic B 3.
- Some weatherable minerals.
- No cementation or induration and lacks a brittle consistence when moist This suffix can be used as Bm, Bmgj, Bmk, and Ems.

n : A horizon in which the ratio of exchangeable Ca to exchangeable Na is 10 or less. It must also have the following distinctive morphological characteristics : prismatic or columnar structure, dark coatings on ped surfaces, and hard to very hard consistence when dry. It is used with B as En or Bnt.

p : A horizon disturbed by man's activities such as cultivation, logging, and habitation. It is used with A and O.

s : A horizon with salts, including gypsum, which may be detected as crystals or veins, as surface crusts of salt crystals, by depressed crop growth, or by the presence of salt-tolerant plants. It is commonly used with C and k (Csk), but can be used with any horizon or combination of horizon and lowercase suffix.

sa : A horizon with secondary enrichment of salts more soluble than Ca and Mg carbonates; the concentration of salts exceeds that in the unenriched parent material. The horizon is at least 10 cm thick. The conductivity of the saturation extract must be at least 4 mS/cm and exceed that of the C horizon by at least one-third. (The unit mho has been replaced by siemens [S].) t -An illuvial horizon enriched with silicate clay. It is used with B alone (Bt), with B and g (Btg), with B and n (Bnt), etc.

Bt : A Bt horizon is one that contains illuvial layer-lattice clays. It forms below an eluvial horizon but may occur at the surface of a soil that has been partially truncated. It usually has a higher ratio of fine clay to total clay than the IC. It has the following properties :

- If any part of an eluvial horizon remains and there is no lithologic discontinuity between it and the Bt horizon, the Bt horizon contains more total clay than the eluvial horizon as follows :

- If any part of the eluvial horizon has less than 15 per cent total clay in the fine earth fraction (<2 mm), the Bt horizon must contain at least 3 per cent more clay, *e.g.* Re 10 per cent clay; Bt minimum 13 per cent clay.

- If the eluvial horizon has more than 15 per cent and less than 40 per cent total clay in the fine earth fraction, the ratio of the clay in the Bt horizon to that in the eluvial horizon must be 1.2 or more, *e.g.* Re 25 per cent clay; Bt at least 30 per cent clay.

- If the eluvial horizon has more than 40 per cent total clay in the fine earth fraction, the Bt horizon must contain at least 8 per cent more clay,

- *e.g.* Re 50.0 clay; Bt at least 58 per cent clay.

- A Bt horizon must be at least 5 cm thick. In some sandy soils where clay accumulation occurs in the lamellae, the total thickness of the lamellae should be more than 10 cm in the upper 150 cm of the profile.

- In massive soils the Bt horizon should have oriented clay in some pores and also as bridges between the sand grains.

- If peds are present, a Bt horizon has clay skins on some of the vertical and horizontal ped surfaces and in the fine pores or has illuvial oriented clays in 1 per cent or more of the cross-section as viewed in thin section.

- If a soil shows a lithologic discontinuity between the eluvial horizon and the Bt horizon, or if only a plow layer overlies the Bt horizon, the Bt horizon need show only clay skins in some part, either in some fine pores or on some vertical and horizontal ped surfaces. Thin sections should show that the horizon has about 1 per cent or more of oriented clay bodies Btj and Btg are defined under j and g.

u : A horizon that is markedly disrupted by physical or faunal processes other than cryoturbation. Evidence of marked disruption such as the inclusion of material from other horizons or the absence of the horizon must be evident in at least half of the cross section of the pedon. Such turbation can result from a blowdown of trees, mass movement of soil on slopes, and burrowing animals. The u can be used with any horizon or subhorizon with the exception of A or B alone; *e.g.* Aeu, Bfu, BCu.

x : A horizon of fragipan character. A fragipan is a loamy sub-surface horizon of high bulk density and very low organic matter content. When dry, it has a hard consistence and seems to be cemented. When moist, it has moderate to weak brittleness. It frequently has bleached fracture planes and is overlain by a friable B horizon. Air-dry clods of fragic horizons slake in water.

y : A horizon affected by cryoturbation as manifested by disrupted and broken horizons, incorporation of materials from other horizons, and mechanical sorting in at least half of the cross section of the pedon. It is used with A, B, and C alone or in combination with other subscripts, *e.g.* Ahy, Ahgy, Bmy, Cy, Cgy, Cygj.

z : A frozen layer. It may be used with any horizon or layer, *e.g.* Ohz, Bmz, Cz, Wz.

SOIL HORIZON LAYERS

Along with wind and water, soil is the major natural resource that supports life on earth. It is so important that a whole branch of study called soil science has developed around it. It is made of three components minerals, organic matter and the living organisms that live in its upper layers. Soil is formed by the weathering action of natural elements like wind, water, glaciers and change in temperature. Soil can be formed from the rocks lying below or from rocks present somewhere far away. These agents of weathering progressively break rocks into finer grains that are laid in layers to form the soil.

Soil is made of distinct layers that lie one above the other, parallel to the soil surface. Each distinct layer is called a soil horizon. A vertical cross-section of a soil known as the soil profile reveals the various horizons of the soil.

Each horizon is the result of a number of geological, chemical and biological processes that have been taking place for over thousands of years. Hence, the soil horizons are best formed and delineated from each other in older soils. The various soil horizons are identified on the basis of physical features, mainly their colour, texture and particle size. Though the soil composition varies from place to place, most soils conform to a general pattern consisting of six horizons.

The soil contains the following physically distinct horizons from top to bottom :

O Horizon

The letter 'O' stands for organic. As the name suggests, this horizon is rich in organic material of plant and animal origin. These materials are generally in various stages of decomposition. This decomposed organic material is called the humus that gives this horizon its characteristic dark colour.

A Horizon

This is also known as the 'top-soil', and it is the topmost layer of the mineral soil. However, as it lies just below the O horizon, this layer also has some amount of humus in it. Hence, it is darker in colour than the layers lying below it. This layer is also known as the 'biomantle' as it is the A horizon in which most of the biological activities take place. Soil organisms like earthworms, fungi and bacteria are mainly concentrated in this layer. The soil particles in this region are smallest and finest as compared to the lower horizons of the soil.

E Horizon

This layer lies below the A horizon and above the B horizon. It is light in colour and contains mainly sand and silt. It is poor in mineral and clay content as these are lost to the lower layer by the process of leaching. Hence, this horizon is also called the layer of eluviation (leaching).The soil particles of this layer are larger in size than those in the A horizon but smaller than those in the underlying B horizon.

B Horizon

This is referred to as the 'sub-soil'. This lies just below the E horizon and is rich in clay and minerals like iron or aluminum. Though this layer has a higher mineral content than the topsoil, some organic material may reach this layer from the layers above by the process of leaching. Plant roots may reach this layer. However, the B horizon is reddish or brownish due to the oxides of iron and clay.

C Horizon

This layer is also known as regolith. The C horizon is mainly made of large rocks or lumps of partially broken bedrock. This layer is least affected by weathering as it lies deep within the soil and is inaccessible to the soil-forming agents. Hence, the rocks in this layer have changed very little since their origin. Plant roots do not reach so deep down to this layer. The C horizon is typically devoid of organic matter.

R Horizon

This is the bedrock. It is the deepest soil horizon in the soil profile. Unlike the above layers, this horizon does not consist of rocks or even boulders. It is made of continuous mass of bedrock. Digging through this layer is very difficult.

Study of the soil horizon is the first step towards soil taxonomy. These layers help us to understand the geological events of the past, and the properties of the soil in general.

DESIGNATIONS FOR HORIZONS AND OTHER LAYERS

Soils vary widely in the degree to which horizons are expressed. Relatively fresh geologic formations, such as fresh alluvium, sand dunes, or blankets of volcanic ash, may have no recognisable genetic horizons, although they may have distinct layers that reflect different modes of deposition. As soil formation proceeds, horizons may be detected in their early stages only by very careful examination. As age increases, horizons generally are more easily identified in the field. Only one or two different horizons may be readily apparent in some very old, deeply weathered soils in tropical areas where annual precipitation is high.

Layers of different kinds are identified by symbols. Designations are provided for layers that have been changed by soil formation and for those that have not. Each horizon designation indicates either that the original material has been changed in certain ways or that there has been little or no change. The designation is assigned after comparison of the observed properties of the layer with properties inferred for the material before it was affected by soil formation. The processes that have caused the change need not be known; properties of soils relative to those of an estimated parent material are the criteria for judgement. The parent material inferred for the horizon in question, not the material below the solum, is used as the basis of comparison. The inferred parent material commonly is very similar to, or the same as, the soil material below the solum.

Designations show the investigator's interpretations of genetic relationships among the layers within a soil. Layers need not be identified by symbols for a good description; yet, the usefulness of soil descriptions is greatly enhanced by the proper use of designations. Designations are not substitutes for descriptions. If both designations and adequate descriptions of a soil are provided, the reader has the interpretation made by the person who described the soil and also the evidence on which the interpretation was based.

Genetic horizons are not equivalent to the diagnostic horizons of Soil Taxonomy. Designations of genetic horizons express a qualitative judgement about the kind of changes that are believed to have taken place. Diagnostic horizons are quantitatively defined features used to differentiate among taxa. Changes implied by genetic horizon designations may not be large enough to justify recognition of diagnostic criteria. For example, a designation of Bt does not always indicate an argillic horizon. Furthermore, the diagnostic horizons may not be co-extensive with genetic horizons. Three kinds of symbols are used in various combinations to designate horizons and layers.

These are capital letters, lower case letters, and Arabic numerals. Capital letters are used to designate the master horizons and layers; lower case letters are used as suffixes to indicate specific characteristics of master horizons and layers; and Arabic numerals are used both as suffixes to indicate vertical sub-divisions within a horizon or layer and as prefixes to indicate discontinuities.

Master Horizons and Layers

The capital letters O, A, E, B, C, and R represent the master horizons and layers of soils. The capital letters are the base symbols to which other characters are added to complete the designations. Most horizons and layers are given a single capital letter symbol; some require two.

O horizons or layers : Layers dominated by organic material. Some are saturated with water for long periods or were once saturated but are now artificially drained;

others have never been saturated. Some O layers consist of undecomposed or partially decomposed litter, such as leaves, needles, twigs, moss, and lichens, that has been deposited on the surface; they may be on top of either mineral or organic soils. Other O layers, are organic materials that were deposited under saturated conditions and have decomposed to varying stages. The mineral fraction of such material is only a small percentage of the volume of the material and generally is much less than half of the weight. Some soils consist entirely of material designated as O horizons or layers. An O layer may be on the surface of a mineral soil or at any depth beneath the surface, if it is buried. A horizon formed by illuviation of organic material into a mineral subsoil is not an O horizon, although some horizons that formed in this manner contain much organic matter.

A horizons : Mineral horizons that formed at the surface or below an O horizon, that exhibit obliteration of all or much of the original rock structure, and that show one or more of the following :

- An accumulation of humified organic matter intimately mixed with the mineral fraction and not dominated by properties characteristic of E or B horizons or

- Properties resulting from cultivation, pasturing, or similar kinds of disturbance.

If a surface horizon has properties of both A and E horizons but the feature emphasised is an accumulation of humified organic matter, it is designated an A horizon. In some places, as in warm arid climates, the undisturbed surface horizon is less dark than the adjacent underlying horizon and contains only small amounts of organic matter. It has a morphology distinct from the C layer, although the mineral fraction is unaltered or only slightly altered by weathering. Such a horizon is designated A because it is at the surface; however, recent alluvial or eolian deposits that retain rock structure7 are not considered to be an A horizon unless cultivated.

E horizons : Mineral horizons in which the main feature is loss of silicate clay, iron, aluminum, or some combination of these, leaving a concentration of sand and silt particles. These horizons exhibit obliteration of all or much of the original rock structure.

An E horizon is usually, but not necessarily, lighter in colour than an underlying B horizon. In some soils the colour is that of the sand and silt particles, but in many soils coatings of iron oxides or other compounds mask the colour of the primary particles. An E horizon is most commonly differentiated from an overlying A horizon by its lighter colour.

It generally has less organic matter than the A horizon. An E horizon is most commonly differentiated from an underlying B horizon in the same sequum by colour of higher value, by lower chroma or both, by coarser texture, or by a combination of these properties. An E horizon is commonly near the surface below an O or A horizon and above a B horizon, but the symbol E can be used for eluvial horizons within or between parts of the B horizon or for those that extend to depths greater than normal observation if the horizon has resulted from soil genesis.

B horizons : Horizons that formed below an A, E,, or O horizon and are dominated by obliteration of all or much of the original rock structure and show one or more of the following :

- Illuvial concentration of silicate clay, iron, aluminum, humus, carbonates, gypsum, or silica, alone or in combination;
- Evidence of removal of carbonates;
- Residual concentration of sesquioxides;
- Coatings of sesquioxides that make the horizon conspicuously lower in value, higher in chroma, or redder in hue than overlying and underlying horizons without apparent illuviation of iron;
- Alteration that forms silicate clay or liberates oxides or both and that forms granular, blocky, or prismatic structure if volume changes accompany changes in moisture content; or
- Brittleness.

All kinds of B horizons are sub-surface horizons or were originally. Included as B horizons where contiguous to another genetic horizon are layers of illuvial concentration of carbonates, gypsum, or silica that are the result of pedogenic processes (these layers may or may not be cemented) and brittle layers that have other evidence of alteration, such as prismatic structure or illuvial accumulation of clay.

Examples that are not B horizons are layers in which clay films coat rock fragments or are on finely stratified unconsolidated sediments, whether the films were formed in place or by illuviation, layers into which carbonates have been illuviated but are not contiguous to an overlying genetic horizon, and layers with gleying but no other pedogenic changes.

C horizons or layers : Horizons or layers, excluding hard bedrock, that are little affected by pedogenic processes and lack properties of O, A, E, or B horizons. The material of C layers may be either like or unlike that from which the solum presumably formed. The C horizon may have been modified even if there is no evidence of pedogenesis.

Included as C layers are sediment, saprolite, unconsolidated bedrock, and other geologic materials that commonly are uncemented and exhibit low or moderate excavation difficulty. Some soils form in material that is already highly weathered. If such material does not meet the requirements of A, E, or B horizons, it is designated C. Changes not considered pedogenic are those not related to overlying horizons. Layers that have accumulations of silica, carbonates, or gypsum or more soluble salts are included in C horizons, even if indurated. If the indurated layers are obviously affected by pedogenic processes, they are a B horizon.

R layers : Hard Bedrock, Granite, basalt, quartzite and indurated limestone or sandstone are examples of bedrock that are designated R. These layers are cemented and excavation difficulty exceeds moderate. The R layer is sufficiently coherent when moist to make hand digging with a spade impractical, although it may be chipped or scraped. Some R layers can be ripped with heavy power equipment. The bedrock may contain cracks that generally are too few and too small to allow roots to penetrate at intervals of less than 10 cm. The cracks may be coated or filled with clay or other material.

Transitional and Combination Horizons

Horizons dominated by properties of one master horizon but having subordinate properties of another. Two capital letter symbols are used, as AB, EB, BE, or BC. The master horizon symbol that is given first designates the kind of horizon whose properties dominate the transitional horizon. An AB horizon, for example, has characteristics of both an overlying A horizon and an underlying B horizon, but it is more like the A than like the B.

In some cases, a horizon can be designated as transitional even if one of the master horizons to which it is apparently transitional is not present. A BE horizon may be recognised in a truncated soil if its properties are similar to those of a BE horizon in a soil in which the overlying E horizon has not been removed by erosion. A BC horizon may be recognised even if no underlying C horizon is present; it is transitional to assumed parent material.

Horizons in which distinct parts have recognisable properties of the two kinds of master horizons indicated by the capital letters. The two capital letters are separated by a virgule (/), as E/B, B/E, or B/C. Most of the individual parts of one of the components are surrounded by the other.

The designation may be used even though horizons similar to one or both of the components are not present, if the separate components can be recognised. The first symbol is that of the horizon that makes up the greater volume.

Single sets of designators do not cover all situations; therefore, some improvising may be necessary. For example, Alfic Udipsamments have lamellae that are separated from each other by eluvial layers. Because it is generally not practical to describe each lamellae and eluvial layer as a separate horizon, the horizons are combined but the components are described separately. One horizon would then contain several lamellae and eluvial layers and might be designated as an E and Bt horizon. The complete horizon sequence for this soil could be : Ap-Bw-E and Bt1-E and Bt2-C. r material.

Subordinate Distinctions Within Master Horizons and Layers

Lower case letters are used as suffixes to designate specific kinds of master horizons and layers. The word "accumulation" is used in many of the definitions in the sense that the horizon must have more of the material in question than is presumed to have been present in the parent material.

The symbols and their meanings are as follows :

- *Highly decomposed organic material :* This symbol is used with "O" to indicate the most highly decomposed of the organic materials. The rubbed fibre content is less than about 17 per cent of the volume.

- *Buried genetic horizon :* This symbol is used in mineral soils to indicate identifiable buried horizons with major genetic features that were formed before burial. Genetic horizons may or may not have formed in the overlying material, which may be either like or unlike the assumed parent material of the buried soil. The symbol is not used in organic soils or to separate an organic layer from a mineral layer.

- *Concretions or nodules :* This symbol is used to indicate a significant accumulation of concretions or of nodules. Cementation is required. The cementing agent is not specified except it cannot be silica. This symbol is not used if concretions or nodules are dolomite or calcite or more soluble salts, but it is used if the nodules or concretions are enriched in minerals that contain iron, aluminum, manganese, or titanium.

- *Physical root restriction :* This symbol is used to indicate root restricting layers in naturally occurring or manmade unconsolidated sediments or materials such as dense basal till, plow pans, and other mechanically compacted zones.

- *Organic material of intermediate decomposition :* This symbol is used with "O" to indicate organic materials of intermediate decomposition. Rubbed fibre content is 17 to 40 per cent of the volume.

- *Frozen soil :* This symbol is used to indicate that the horizon or layer contains permanent ice. Symbol is not used for seasonally frozen layers or for "dry permafrost" (material that is colder than 0°C but does not contain ice).

- *Strong gleying :* This symbol is used to indicate either that iron has been reduced and removed during soil formation or that saturation with stagnant water has preserved a reduced state. Most of the affected layers have chroma of 2 or less and many have redox concentrations. The low chroma can be the colour of reduced iron or the colour of uncoated sand and silt particles from which iron has been removed. Symbol "g" is not used for soil materials of low chroma, such as some shales or E horizons, unless they have a history of wetness. If "g" is used with "B," pedogenic change in addition to gleying is implied. If no other pedogenic change in addition to gleying has taken place, the horizon is designated Cg.

- *Illuvial accumulation of organic matter :* This symbol used with "B" to indicate the accumulation of illuvial, amorphous, dispersible organic matter-sesquioxides complexes. The sesquioxide component coats sand and silt particles. In some horizons, coatings have coalesced, filled pores, and cemented the horizon. The symbol "h" is also used in combination with "s" as "Bhs" if the amount of sesquioxide component is significant but value and chroma of the horizon are 3 or less.

- *Slightly decomposed organic material* : This symbol is used with "O" to indicate the least decomposed of the organic materials. Rubbed fibre content is more than about 40 per cent of the volume.

- *Accumulation of carbonates* : This symbol is used to indicate the accumulation of alkaline earth carbonates, commonly calcium carbonate.

- *Cementation or induration* : This symbol is used to indicate continuous or nearly continuous cementation. The symbol is used only for horizons that are more than 90 per cent cemented, although they may be fractured. The layer is physically root restrictive. The single predominant or codominant cementing agent may be indicated by using defined letter suffixes, singly or in pairs. If the horizon is cemented by carbonates, "km" is used; by silica, "qm"; by iron, "sm"; by gypsum, "ym"; by both lime and silica, "kqm"; by salts more soluble than gypsum, "zm."

- *Accumulation of sodium* : This symbol is used to indicate an accumulation of exchangeable sodium.

- *Residual accumulation of sesquioxides* : This symbol is used to indicate residual accumulation of sesquioxides.

- *Tillage or other disturbance* : This symbol is used to indicate a disturbance of the surface layer by mechanical means, pasturing, or similar uses. A disturbed organic horizon is designated Op. A disturbed mineral horizon is designated Ap even though clearly once an E, B, or C horizon.

- *Accumulation of silica* : This symbol is used to indicate an accumulation of secondary silica.

- *Weathered or soft bedrock* : This symbol is used with "C" to indicate root restrictive layers of soft bedrock or saprolite, such as weathered igneous rock; partly consolidated soft sandstone; siltstone; and shale. Excavation difficulty is low or moderate.

- *Illuvial accumulation of sesquioxides and organic matter* : This symbol is used with "B" to indicate the accumulation of illuvial, amorphous, dispersible organic matter-sesquioxide complexes if both the organic matter and sesquioxide components are significant and the value and chroma of the horizon is more than 3. The symbol is also used in combination with "h" as "Bhs" if both the organic matter and sesquioxide components are significant and the value and chroma are 3 or less.

- *Presence of slickensides* : This symbol is used to indicate the presence of slickensides. Slickensides result directly from the swelling of clay minerals and shear failure, commonly at angles of 20 to 60 degrees above horizontal. They are indicators that other vertic characteristics, such as wedge-shaped peds and surface cracks, may be present.

- *Accumulation of silicate clay :* This symbol is used to indicate an accumulation of silicate clay that has formed and subsequently translocated within the horizon or has been moved into the horizon by illuviation, or both. At least some part should show evidence of clay accumulation in the form of coatings on surfaces of peds or in pores, or as lamellae, or bridges between mineral grains.

- *Plinthite :* This symbol is used to indicate the presence of iron-rich, humus-poor, reddish material that is firm or very firm when moist and that hardens irreversibly when exposed to the atmosphere and to repeated wetting and drying.

- *Development of colour or structure :* This symbol is used with "B" to indicate the development of colour or structure, or both, with little or no apparent illuvial accumulation of material. It should not be used to indicate a transitional horizon.

- *Fragipan character :* This symbol is used to indicate genetically developed layers that have a combination of firmness, brittleness, very coarse prisms with few to many bleached vertical faces, and commonly higher bulk density than adjacent layers. Some part is physically root restrictive.

- *Accumulation of gypsum :* This symbol is used to indicate the accumulation of gypsum.

- *Accumulation of salts more soluble than gypsum :* This symbol is used to indicate an accumulation of salts more soluble than gypsum.

Conventions for using letter suffixes : Many master horizons and layers that are symbolised by a single capital letter will have one or more lower case letter suffixes.

The following rules apply :

- Letter suffixes should immediately follow the capital letter.

- More than three suffixes are rarely used.

- When more than one suffix is needed, the following letters, if used, are written first : a, e, h, i, r, s, t, and w. Except for the Bhs or Crt8 horizons, none of these letters are used in combination in a single horizon.

- If more than one suffix is needed and the horizon is not buried, these symbols, if used, are written last : c, d, f, g, m, v, and x. Some examples : Btg, Bkm, and Bsm.

- If a horizon is buried, the suffix "b" is written last. Suffix "b" is used only for buried mineral soils.

- A B horizon that has significant accumulation of clay and also shows evidence of development of colour or structure, or both, is designated Bt ("t" has precedence over "w," "s," and "h"). A B horizon that is gleyed or that has accumulations of carbonates, sodium, silica, gypsum, salts more soluble than gypsum, or residual accumulation or sesquioxides carries the appropriate symbol – g, k, n, q, y, z, or o. If illuvial clay is also present, "t" precedes the other symbol : Btg.

• Suffixes "h," "s," and "w" are not normally used with g, k, n, q, y, z, or o.

Vertical sub-division : Commonly a horizon or layer designated by a single letter or a combination of letters needs to be sub-divided. The Arabic numerals used for this purpose always follow all letters. Within a C, for example, successive layers could be C1, C2, C3, and so on; or, if the lower part is gleyed and the upper part is not, the designations could be C1-C2-Cg1-Cg2 or C-Cg1-Cg2-R.

These conventions apply whatever the purpose of sub-division. In many soils, horizons that would be identified by one unique set of letters are sub-divided on the basis of evident morphological features, such as structure, colour, or texture. These divisions are numbered consecutively. The numbering starts with 1 at whatever level in the profile any element of the letter symbol changes. Thus Bt1-Bt2-Btk1-Btk2 is used, not Bt1-Bt2-Btk3Btk4. The numbering of vertical sub-divisions within a horizon is not interrupted at a discontinuity (indicated by a numerical prefix) if the same letter combination is used in both materials : Bs1-Bs2-2Bs3-2Bs4 is used, not Bs1-Bs2-2Bs1-2Bs2.

Sometimes, thick layers are sub-divided during sampling for laboratory analyses even though differences in morphology are not evident in the field. These layers need to be identified. This is done by following the convention of using Arabic numerals to identify the sub-division. The Arabic numerals would follow the letter designations and be a part of the horizon designation. For example, four layers of a Bt2 horizon sampled by 10-cm increments would be designated Bt21, Bt22, Bt23, and Bt24. The Bt2 horizon is sub-divided for sampling purposes only.

Discontinuities : In mineral soils Arabic numerals are used as prefixes to indicate discontinuities. Wherever needed, they are used preceding A, E, B, C, and R. These prefixes are distinct from Arabic numerals used as suffixes to denote vertical sub-divisions. A discontinuity is a significant change in particle-size distribution or mineralogy that indicates a difference in the material from which the horizons formed and/or a significant difference in age, unless that difference in age is indicated by the suffix "b." Symbols to identify discontinuities are used only when they will contribute substantially to the reader's understanding of relationships among horizons. Stratification common to soils formed in alluvium is not designated as discontinuity, unless particle size distribution differs markedly (strongly contrasting particle-size class, as defined by Soil Taxonomy) from layer to layer even though genetic horizons have formed in the contrasting layers.

Where a soil has formed entirely in one kind of material, a prefix is omitted from the symbol; the whole profile is material 1. Similarly, the uppermost material in a profile having two or more contrasting materials is understood to be material 1, but the number is omitted. Numbering starts with the second layer of contrasting material, which is designated "2." Underlying contrasting layers are numbered consecutively. Even though a layer below material 2 is similar to material 1, it is designated "3" in the sequence. The numbers indicate a change in the material, not the type of material. Where two or more consecutive horizons formed in one kind of material, the same prefix number is applied to all of the horizon designations in that material : Ap-E-Bt1-2Bt2-2Bt3-2BC. The number of suffixes designating sub-divisions of the Bt horizon continue in consecutive order across the discontinuity.

If an R layer is below a soil that formed in residuum and the material of the R layer is judged to be like that from which the material of the soil weathered, the Arabic number prefix is not used. If it is thought that the R layer would not produce material like that in the solum, the number prefix is used, as in A-Bt-C-2R or A-Bt-2R. If part of the solum formed in residuum, "R" is given the appropriate prefix : Ap-Bt1-2Bt2-2Bt3-2C1-2C2-2R. Buried horizons (designated "b") are special problems. A buried horizon is obviously not in the same deposit as horizons in the overlying deposit. Some buried horizons, however, formed in material lithologically like that of the overlying deposit. A prefix is not used to distinguish material of such buried horizons. If the material in which a horizon of a buried soil formed is lithologically unlike that of the overlying material, the discontinuity is designated by number prefixes and the symbol for a buried horizon is used as well : ApBt1-Bt2-BC-C-2ABb-2Btb1-2Btb2-2C.

In organic soils, discontinuities between different kinds of layers are not identified. In most cases, the differences are shown by the letter suffix designations if the different layers are organic or by the master symbol if the different layers are mineral.

Use of the prime. — Identical letter and numerical designations may be appropriate for two or more horizons separated by at least one horizon or layer of a different kind in the same pedon. The sequence A-E-Bt-E-Btx-C is an example : the soil has two E horizons. To make communication easier, the prime is used with the master horizon symbol of the lower of two horizons having identical designations : A-E-Bt-E'-Btx-C. The prime is applied to the capital letter designation and any lower-case symbols follow it : B't. The prime is not used unless all letters of the designations of two different layers are identical. Rarely, three layers have identical letter symbols; a double prime can be used : E''.

The same principle applies in designating layers of organic soils. The prime is used only to distinguish two or more horizons that have identical symbols : Oi-C-O'i-C' or Oi-C-Oe-C'. The prime is added to the lower C layer to differentiate it from the upper.

Sample Horizons and Sequences

The following examples illustrate some common horizon and layer sequences of important soils and the use of Arabic numerals to identify their sub-divisions. The examples were selected from soil descriptions on file and modified to reflect present conventions.

- *Mineral soils :*
- Typic Hapludoll : A1-A2-Bw-BC-C
- Typic Haploboroll : Ap-A-Bw-Bk-Bky1-Bky2-C
- Cumulic Haploboroll : Ap-A-Bw1-Bw2-BC-Ab-Bwb1-Bwb2-2C

- Typic Argialboll : Ap-A-E-Bt1-Bt2-BC-C
- Typic Argiaquoll : A-AB-BA-Btg-BCg-Cg
- Entic Haplorthod : Oi-Oa-E-Bs1-Bs2-BC-C
- Typic Haplorthod : Ap-E-Bhs-Bs-BC-C1-C2
- Typic Fragiudalf : Oi-A-E-BE-Bt1-Bt2-B/E-Btx1-Btx2-C
- Typic Haploxeralf : A1-A2-A3-2Bt1-2Bt2-2Bt3-2BC-2C
- Glossoboric Hapludalf : Ap-E-B/E-Bt1-Bt2-C
- Typic Paleudult : A-E-Bt1-Bt2-B/E-B't1-B't2-B't3
- Typic Hapludult : 0i-A1-A2-BA-Bt1-Bt2-BC-C
- Arenic Plinthic Paleudult : Ap-E-Bt-Btc-Btv1-Btv2-BC-C
- Typic Haplargid : A-Bt-Bk1-Bk2-C
- Entic Durorthid : A-Bw-Bq-Bqm-2Ab-2Btkb-3Byb-3Bqmb-3Bqkb
- Typic Dystrochrept : Ap-Bw1-Bw2-C-R
- Typic Fragiochrept : Ap-Bw-E-Bx1-Bx2-C
- Typic Haplaquept : Ap-AB-Bg1-Bg2-BCg-Cg
- Typic Udifluvent : Ap-C-Ab-C'
- Typic Haplustert : Ap-A-AC-C1-C2
- • *Organic soils :*
- Typic Medisaprist : Op-Oa1-Oa2-Oa3-C
- Typic Sphagnofibrist : Oi1-Oi2-Oi3-Oe
- Limnic Borofibrist : Oi-C-O'i1-O'i2-C'-Oe-C'
- Lithic Cryofolist : Oi-Oa-R

Cyclic and Intermittent Horizons and Layers

A profile of a soil having cyclic horizons exposes layers whose boundaries are near the surface at one point and extend deep into the soil at another. At one place the aggregate horizon thickness may be only 50 cm; two meters away, the same horizons may be more than 125 cm thick. The cycle is repeated, commonly with considerable variation in both depth and horizontal interval, but still with some degree of regularity. If the soil is visualised in three dimensions instead of two, some cyclic horizons extend downward in inverted cones. The cone of the lower horizon fits around the cone of the horizon above. Other cyclic horizons would appear wedge-shaped.

A profile of a soil having an intermittent horizon shows that the horizon extends horizontally for some distance, ends, and reappears again some distance away. A B horizon interrupted at intervals by upward extensions of bedrock into the A horizon is an example. The distance between places where the horizon is absent is commonly variable, yet it has some degree of regularity. The distances range from less than one meter to several meters. Obviously, a soil profile at one place could be unlike a profile only a few meters away for soils with cyclic or intermittent horizons or layers. The order of the variations of these soils are given in soil descriptions. Descriptions of the order of horizontal variation within a pedon include the kind of variation, the spacing of cycles or interruptions, and the amplitude of depth variation of cyclic horizons.

Near Surface Sub-zones

The morphology of the uppermost few centimeters is subject in many soils to strong control by antecedent weather and by soil use. A soil may be freshly tilled today and have a loose surface. Tomorrow it may have a strong crust because of a heavy rain. Or, in one place soil may be highly compacted by livestock and have a firm near surface even though over most of its extent the same uppermost few centimeters are little disturbed and very friable. There is a need for a set of terms to describe sub-zones of the near surface and, in particular, the near surface of tilled soils. Five sub-zones of the near surface are recognised.

The mechanically bulked sub-zone has undergone through mechanical manipulation a reduction in bulk density and an increase in discreteness of structural units, if present. Usually the mechanical manipulation is the consequence of tillage operations. Rupture resistance of the mass overall, inclusive of a number of structural units, is loose or very friable and occasionally friable. Individual structural units may be friable or even firm. Mechanical continuity among structural units is low. Structure grade, if the soil material exhibits structural units < 20 mm across, is moderate or strong. Strain that results from contraction on drying of individual structural units may not extend among structural units. Hence, internally initiated desiccation cracks may be weak or absent even though the soil material in a consolidated condition has considerable potential extensibility. Cracks may be present, however, if they are initiated deeper in the soil. The mechanically compacted sub-zone has been subject to compaction, usually in tillage operations but possibly by animals. Commonly, mechanical continuity of the fabric and bulk density are increased. Rupture resistance depends on texture and degree of compaction. Generally, friable is the minimum class. Mechanical continuity of the fabric permits propagation of strain that results on drying only over several centimeters. Internally initiated cracks appear if the soil material has appreciable extensibility and drying has been sufficient. In some soils this sub-zone restricts root growth. The suffix "d" may be used if compaction results in a strong plow pan.

The water-compacted sub-zone has been compacted by repetitive large changes in water state without mechanical load except for the weight of the soil. Repetitive occurrence of free water is particularly conducive to compaction. Depending on texture, moist rupture resistance ranges from very friable through firm. Structural units, if present, are less discrete than for the same soil material if mechanically bulked. Structure generally would be weak or the condition would be massive. Mechanical continuity of the fabric is sufficient that strain which originates on drying propagates appreciable distances. As a consequence, if extensibility is sufficient, cracks develop on drying. In many soils, over time the water-compacted sub-zone replaces the mechanically bulked sub-zone. The replacement can occur in a single year if the sub-zone is subject to periodic occurrence of free water with intervening periods when slightly moist or dry. The presence of a water-compacted sub-zone and the absence of the mechanically bulked sub-zone is an important consequence of no-till farming systems.

The surficial bulked sub-zone occurs in the very near surface. Continuity of the fabric is low. Cracks are not initiated in this sub-zone, although they may be present if initiated in underlying more compacted soil. The sub-zone is formed by various processes. Frost action under conditions where the soil is drier than wet is a mechanism. Wetting and drying of soil material with high extensibility is another origin; certain Vertisols are illustrative. Crust is a surficial sub-zone, usually less than 50 mm thick, that exhibits markedly more mechanical continuity of the soil fabric than the zone immediately beneath. Commonly, the original soil fabric has been reconstituted by water action and the original structure has been replaced by a massive condition. While the material is wet, raindrop impact and freeze-thaw cycles are mechanisms leading to reconstitution. Crusting related to raindrop-impact and freeze-thaw are recognised.

A fluventic zone may be formed by local transport and deposition of soil material in tilled fields. Such a feature has weaker mechanical continuity than a crust. The rupture resistance is lower, and the reduction in infiltration may be less than for crusts of similar texture. A raindrop-impact crust may occur on a fluventic zone. Crusts and a fluventic zone may be described in terms of thickness in millimeters, structure and other aspects of the fabric, and by consistence, including rupture resistance while dry and micropenetration resistance while wet. Thickness pertains to the zone where reconstitution of the fabric has been pronounced. Also, the distance between surface-initiated cracks may be a useful observation for seedling emergence considerations. If the distance is short, the weight of the crust slabs is low.

Soil material with little apparent reconstitution commonly adheres beneath the crust and is removed with the crust. This soil material that shows little or no reconstitution is not part of the crust and does not contribute to the thickness. Identification of sub-zones is not clear cut. Morphological expression of bulking and compaction may be quite different among soils dependent on particle size distribution, organic matter content, clay mineralogy, water regime, and possibly other factors.

The distinction between a bulked and compacted state for soil material with appreciable extensibility is made in part on the potential for the transmission of strain on drying over distances greater than the horizontal dimensions of the larger structural units. In a bulked sub-zone little or no strain is propagated; in a compacted sub-zone the strain would be propagated over distances greater than the horizontal dimensions of the larger structural units. Many soils have low extensibility because of texture, clay mineralogy, or both. For these soils, the expression of cracks cannot be used to distinguish between a bulked and compacted state.

The distinction between compaction and bulking is subjective. It is useful to establish a concept of a normal degree of compaction of the near surface to which the actual degree of compaction is compared. The concept for tilled soils should be the compaction of soil material on level or convex parts of the tillage determined relief. The soil should have been subject to the bulking action of conventional tillage without the subsequent mechanical compaction. The sub-zone in question should have been brought to a wet or very moist water state from an appreciably drier condition followed by drying to slightly moist or drier at least once. It should not have been subject, however, to a large number of wetting and drying cycles where the maximum wetness involves the presence of free water. If the soil material has a degree of compaction similar to what would be expected, then the term normal compaction is employed.

Boundaries of Horizons and Layers

A boundary is a surface or transitional layer between two adjoining horizons or layers. Most boundaries are zones of transition rather than sharp lines of division. Boundaries vary in distinctness and in topography.

Distinctness : Distinctness refers to the thickness of the zone within which the boundary can be located. The distinctness of a boundary depends partly on the degree of contrast between the adjacent layers and partly on the thickness of the transitional zone between them.

Distinctness is defined in terms of thickness of the transitional zone :

- Abrupt : Less than 2 cm thick
- Clear : 2 to 5 cm thick
- Gradual : 5 to 15 cm thick
- Diffuse : More than 15 cm thick.

Abrupt soil boundaries, such as those between the E and Bt horizons in many soils, are easily determined. Some boundaries are not readily seen but can be located by testing the soil above and below the boundary. Diffuse boundaries, such as those in many old soils in tropical areas, are most difficult to locate and require time-consuming comparisons of small specimens of soil from various parts of the profile until the midpoint of the transitional zone is determined. For soils that have nearly uniform properties or that change very gradually as depth increases, horizon boundaries are imposed more or less arbitrarily without clear evidence of differences.

Topography : Topography refers to the irregularities of the surface that divides the horizons. Even though soil layers are commonly seen in vertical section, they are three-dimensional.

Topography of boundaries is described with the following terms :

• Smooth : The boundary is a plane with few or no irregularities.

• Wavy : The boundary has undulations in which depressions are wider then they are deep.

• Irregular : The boundary has pockets that are deeper than they are wide.

• Broken : One or both of the horizons or layers separated by the boundary are discontinuous and the boundary is interrupted.

HORIZONS AND CHARACTERISTICS DIAGNOSTIC FOR MINERAL SOILS

The criteria for some of the following horizons and characteristics, such as histic and folistic epipedons, can be met in organic soils. They are diagnostic, however, only for the mineral soils.

Diagnostic Surface Horizons : The Epipedon

The epipedon (Gr. epi, over, upon, and pedon, soil) is a horizon that forms at or near the surface and in which most of the rock structure has been destroyed. It is darkened by organic matter or shows evidence of eluviation, or both. Rock structure as used here and in other places in this taxonomy includes fine stratification (less than 5 mm) in unconsolidated sediments (eolian, alluvial, lacustrine, or marine) and saprolite derived from consolidated rocks in which the unweathered minerals and pseudomorphs of weathered minerals retain their relative positions to each other.

Any horizon may be at the surface of a truncated soil. The following section, however, is concerned with eight diagnostic horizons that have formed at or near the soil surface. These horizons can be covered by a surface mantle of new soil material. If the surface mantle has rock structure, the top of the epipedon is considered the soil surface unless the mantle meets. If the soil includes a buried soil, the epipedon, if any, is at the soil surface and the epipedon of the buried soil is considered a buried epipedon and is not considered in selecting taxa unless the keys specifically indicate buried horizons, such as those in Thapto-Histic subgroups. A soil with a mantle thick enough to have a buried soil has no epipedon if the soil has rock structure to the surface or has an Ap horizon less than 25 cm thick that is underlain by soil material with rock structure. The melanic epipedon (defined below) is unique among epipedons. It forms commonly in volcanic deposits and can receive fresh deposits of ash. Therefore, this horizon is permitted to have layers within and above the epipedon that are not part of the melanic epipedon. A recent alluvial or eolian deposit that retains stratifications (5 mm or less thick) or an Ap horizon directly underlain by such stratified material is not included in the concept of the epipedon because time has not been sufficient for soil-forming processes to erase these transient marks of deposition and for diagnostic and accessory properties to develop. An epipedon is not the same as an A horizon. It may include part or all of an illuvial B horizon if the darkening by organic matter extends from the soil surface into or through the B horizon.

Anthropic Epipedon

The anthropic epipedon shows some evidence of disturbance by human activity and meets all of the requirements for a mollic epipedon, except for one or both of the following :

- 1,500 milligrams per kilogram or more P_2O_5 soluble in 1 per cent citric acid and a regular decrease in P_2O_5 to a depth of 125 cm; or
- If the soil is not irrigated, all parts of the epipedon are dry for 9 months or more in normal years.

Melanic Epipedon

The melanic epipedon has both of the following :

- An upper boundary at, or within 30 cm of, either the mineral soil surface or the upper boundary of an organic layer with andic soil properties (defined below), whichever is shallower; and
- In layers with a cumulative thickness of 30 cm or more within a total thickness of 40 cm, *all of the following :*
- Andic soil properties throughout; and
- A colour value, moist, and chroma (Munsell designations) of 2 or less throughout and a melanic index of 1.70 or less throughout; and
- 6 per cent or more organic carbon as a weighted average and 4 per cent or more organic carbon in all layers.

Mollic Epipedon

The mollic epipedon consists of mineral soil materials and has the following properties :

- When dry, either or both :
- Structural units with a diameter of 30 cm or less or secondary structure with a diameter of 30 cm or less; or
- A moderately hard or softer rupture-resistance class; and
- Rock structure, including fine (less than 5 mm) stratifications, in less than one-half of the volume of all parts; and
- One of the following
- *– All of the following :*
- a. Colours with a value of 3 or less, moist, and of 5 or less, dry; and
- b. Colours with chroma of 3 or less, moist; and
- c. If the soil has a C horizon, the mollic epipedon has a colour value at least 1 Munsell unit lower or chroma at least 2 units lower (both moist

and dry) than that of the C horizon or the epipedon has at least 0.6 per cent more organic carbon than the C horizon; or

- – A fine-earth fraction that has a calcium carbonate equivalent of 15 to 40 per cent and colours with a value and chroma of 3 or less, moist; or

- – A fine-earth fraction that has a calcium carbonate equivalent of 40 per cent or more and a colour value, moist, of 5 or less; and

- A base saturation (by NH4OAc) of 50 per cent or more; and

- *An organic-carbon content of :*

- – 2.5 per cent or more if the epipedon has a colour value, moist, of 4 or 5; or

- – 0.6 per cent more than that of the C horizon (if one occurs) if the mollic epipedon has a colour value less than 1 Munsell unit lower or chroma less than 2 units lower (both moist and dry) than the C horizon; or

- – 0.6 per cent or more; and

- After mixing of the upper 18 cm of the mineral soil or of the whole mineral soil if its depth to a densic, lithic, or paralithic contact, petrocalcic horizon, or duripan (all defined below) is less than 18 cm, *the minimum thickness of the epipedon is as follows :*

- – 10 cm or the depth of the non-cemented soil if the epipedon is loamy very fine sand or finer and is directly above a densic, lithic, or paralithic contact, a petrocalcic horizon, or a duripan that is within 18 cm of the mineral soil surface; or

- – 25 cm or more if the epipedon is loamy fine sand or coarser throughout or if there are no underlying diagnostic horizons (defined below) and the organic-carbon content of the underlying materials decreases irregularly with increasing depth; or

- – 25 cm or more if all of the following are 75 cm or more below the mineral soil surface.

- a. The upper boundary of any pedogenic lime that is present as filaments, soft coatings, or soft nodules; and

- b. The lower boundary of any argillic, cambic, natric, oxic, or spodic horizon (defined below); and

- c. The upper boundary of any petrocalcic horizon, duripan, or fragipan;

- – 18 cm if the epipedon is loamy very fine sand or finer in some part and one-third or more of the total thickness between the top of the epipedon and the shallowest of any features listed in item 6-c is less than 75 cm below the mineral soil surface; or

- – 18 cm or more if none of the above conditions apply; and

- *Phosphate :*

- Content less than 1,500 milligrams per kilogram soluble in 1 per cent citric acid; or

- Content decreasing irregularly with increasing depth below the epipedon; or

- Nodules are within the epipedon; and

- Some part of the epipedon is moist for 90 days or more (cumulative) in normal years during times when the soil temperature at a depth of 50 cm is 5 °C or higher, if the soil is not irrigated; and

- The n value (defined below) is less than 0.7.

Ochric Epipedon

The ochric epipedon fails to meet the definitions for any of the other seven epipedons because it is too thin or too dry, has too high a colour value or chroma, contains too little organic carbon, has too high an n value or melanic index, or is both massive and hard or harder when dry . Many ochric epipedons have either a Munsell colour value of 4 or more, moist, and 6 or more, dry, or chroma of 4 or more, or they include an A or Ap horizon that has both low colour values and low chroma but is too thin to be recognised as a mollic or umbric epipedon (and has less than 15 per cent calcium carbonate equivalent in the fine-earth fraction). Ochric epipedons also include horizons of organic materials that are too thin to meet the requirements for a histic or folistic epipedon.

The ochric epipedon includes eluvial horizons that are at or near the soil surface, and it extends to the first underlying diagnostic illuvial horizon (defined below as an argillic, kandic, natric, or spodic horizon). If the underlying horizon is a B horizon of alteration (defined below as a cambic or oxic horizon) and there is no surface horizon that is appreciably darkened by humus, the lower limit of the ochric epipedon is the lower boundary of the plow layer or an equivalent depth (18 cm) in a soil that has not been plowed. Actually, the same horizon in an unplowed soil may be both part of the epipedon and part of the cambic horizon; the ochric epipedon and the sub-surface diagnostic horizons are not all mutually exclusive. The ochric epipedon does not have rock structure and does not include finely stratified fresh sediments, nor can it be an Ap horizon directly overlying such deposits.

Plaggen Epipedon

The plaggen epipedon is a human-made surface layer 50 cm or more thick that has been produced by long-continued manuring.

A plaggen epipedon can be identified by several means. Commonly, it contains artifacts, such as bits of brick and pottery, throughout its depth. There may be chunks of diverse materials, such as black sand and light gray sand, as large as the size held by a spade. The plaggen epipedon normally shows spade marks throughout its depth and also remnants of thin stratified beds of sand that were probably produced on the soil surface by beating rains and were later buried by spading. A map unit delineation of soils with plaggen epipedons would tend to have straight-sided rectangular bodies that are higher than the adjacent soils by as much as or more than the thickness of the plaggen epipedon.

Umbric Epipedon

The umbric epipedon consists of mineral soil materials and has the following properties :

- *When dry, either or both :*
- Structural units with a diameter of 30 cm or less or secondary structure with a diameter of 30 cm or less; or
- A moderately hard or softer rupture-resistance class; and
- *All of the following :*
- Colours with a value of 3 or less, moist, and of 5 or less, dry; and
- Colours with chroma of 3 or less, moist; and
- If the soil has a C horizon, the umbric epipedon has a colour value at least 1 Munsell unit lower or chroma at least 2 units lower (both moist and dry) than that of the C horizon or the epipedon has at least 0.6 per cent more organic carbon than that of the C horizon; and
- A base saturation (by NH_4OAc) of less than 50 per cent in some or all parts; and
- *An organic-carbon content of :*
- 0.6 per cent more than that of the C horizon (if one occurs) if the umbric epipedon has a colour value less than 1 Munsell unit lower or chroma less than 2 units lower (both moist and dry) than the C horizon; or
- 0.6 per cent or more; and
- After mixing of the upper 18 cm of the mineral soil or of the whole mineral soil if its depth to a densic, lithic, or paralithic contact or a duripan (all defined below) is less than 18 cm, *the minimum thickness of the epipedon is as follows :*
- 10 cm or the depth of the non-cemented soil if the epipedon is loamy very fine sand or finer and is directly above a densic, lithic, or paralithic contact or a duripan that is within 18 cm of the mineral soil surface; or
- 25 cm or more if the epipedon is loamy fine sand or coarser throughout or if there are no underlying diagnostic horizons (defined below) and the organic-carbon content of the underlying materials decreases irregularly with increasing depth; or
- 25 cm or more if the lower boundary of any argillic, cambic, natric, oxic, or spodic horizon (defined below) is 75 cm or more below the mineral soil surface; or
- 18 cm if the epipedon is loamy very fine sand or finer in some part and one-third or more of the total thickness between the top of the epipedon and the shallowest of any features listed in item 5-c is less than 75 cm below the mineral soil surface; or
- 18 cm or more if none of the above conditions apply; and

- *Phosphate :*
- Content less than 1,500 milligrams per kilogram soluble in 1 per cent citric acid; or
- Content decreasing irregularly with increasing depth below the epipedon; or
- Nodules are within the epipedon; and
- Some part of the epipedon is moist for 90 days or more (cumulative) in normal years during times when the soil temperature at a depth of 50 cm is 5 °C or higher, if the soil is not irrigated; and
- The n value (defined below) is less than 0.7; and
- The umbric epipedon does not have the artifacts, spade marks, and raised surfaces that are characteristic of the plaggen epipedon.

Diagnostic Sub-surface Horizons

The horizons described in this section form below the surface of the soil, although in some areas they form directly below a layer of leaf litter. They may be exposed at the surface by truncation of the soil. Some of these horizons are generally regarded as B horizons, some are considered B horizons by many but not all pedologists, and others are generally regarded as parts of the A horizon.

Agric Horizon

The agric horizon is an illuvial horizon that has formed under cultivation and contains significant amounts of illuvial silt, clay, and humus.

The agric horizon is directly below an Ap horizon and has the following properties :

- A thickness of 10 cm or more and either :
- 5 per cent or more (by volume) wormholes, including coatings that are 2 mm or more thick and have a value, moist, of 4 or less and chroma of 2 or less; or
- 5 per cent or more (by volume) lamellae that have a thickness of 5 mm or more and have a value, moist, of 4 or less and chroma of 2 or less.

Albic Horizon

The albic horizon is an eluvial horizon, 1.0 cm or more thick, that has 85 per cent or more (by volume) albic materials (defined below). It generally occurs below an A horizon but may be at the mineral soil surface. Under the albic horizon there generally is an argillic, cambic, kandic, natric, or spodic horizon or a fragipan (defined below). The albic horizon may lie between a spodic horizon and either a fragipan or an argillic horizon, or it may be between an argillic or kandic horizon and a fragipan. It may lie between a mollic epipedon and an argillic or natric horizon or between a cambic horizon and an argillic, kandic, or natric horizon or a fragipan. The albic horizon may separate horizons that, if they were together, would meet the requirements for a mollic epipedon. It may separate lamellae that together meet the requirements for an argillic horizon. These lamellae are not considered to be part of the albic horizon.

Argillic Horizon

An argillic horizon is normally a sub-surface horizon with a significantly higher percentage of phyllosilicate clay than the overlying soil material. It shows evidence of clay illuviation. The argillic horizon forms below the soil surface, but it may be exposed at the surface later by erosion.

- *All argillic horizons must meet both of the following requirements :*
  - *One of the following :*
    a. If the argillic horizon is coarse-loamy, fine-loamy, coarse-silty, fine-silty, fine, or very-fine or is loamy or clayey, including skeletal counterparts, it must be at least 7.5 cm thick or at least one-tenth as thick as the sum of the thickness of all overlying horizons, whichever is greater; or
    b. If the argillic horizon is sandy or sandy-skeletal, it must be at least 15 cm thick; or
    c. If the argillic horizon is composed entirely of lamellae, the combined thickness of the lamellae that are 0.5 cm or more thick must be 15 cm or more; and
  - *Evidence of clay illuviation in at least one of the following forms :*
    a. Oriented clay bridging the sand grains; or
    b. Clay films lining pores; or
    c. Clay films on both vertical and horizontal surfaces of peds; or
    d. Thin sections with oriented clay bodies that are more than 1 per cent of the section; or
    e. If the coefficient of linear extensibility is 0.04 or higher and the soil has distinct wet and dry seasons, then the ratio of fine clay to total clay in the illuvial horizon is greater by 1.2 times or more than the ratio in the eluvial horizon; and
- If an eluvial horizon remains and there is no lithologic discontinuity between it and the illuvial horizon and no plow layer directly above the illuvial layer, then the illuvial horizon must contain more total clay than the eluvial horizon within a vertical distance of 30 cm or less, *as follows :*
- If any part of the eluvial horizon has less than 15 per cent total clay in the fine-earth fraction, the argillic horizon must contain at least 3 per cent (absolute) more clay (10 per cent versus 13 per cent, for example); or

- If the eluvial horizon has 15 to 40 per cent total clay in the fine-earth fraction, the argillic horizon must have at least 1.2 times more clay than the eluvial horizon; or

- If the eluvial horizon has 40 per cent or more total clay in the fine-earth fraction, the argillic horizon must contain at least 8 per cent (absolute) more clay (42 per cent *versus* 50 per cent, for example).

Calcic Horizon

The calcic horizon is an illuvial horizon in which secondary calcium carbonate or other carbonates have accumulated to a significant extent.

The calcic horizon has all of the following properties :

- Is 15 cm or more thick; and

- Is not indurated or cemented to such a degree that it meets the requirements for a petrocalcic horizon; and

- *Has one or more of the following :*

- – 15 per cent or more $CaCO_3$ equivalent, and its $CaCO_3$ equivalent is 5 per cent or more (absolute) higher than that of an underlying horizon; or

- – 15 per cent or more $CaCO_3$ equivalent and 5 per cent or more (by volume) identifiable secondary carbonates; or

- *– 5 per cent or more calcium carbonate equivalent and has :*

- a. Less than 18 per cent clay in the fine-earth fraction; and

- b. A sandy, sandy-skeletal, coarse-loamy, or loamy-skeletal particle-size class; and

- c. 5 per cent or more (by volume) identifiable secondary carbonates or a calcium carbonate equivalent (by weight) that is 5 per cent or more (absolute) higher than that of an underlying horizon.

Cambic Horizon

A cambic horizon is the result of physical alterations, chemical transformations, or removals or of a combination of two or more of these processes.

The cambic horizon is an altered horizon 15 cm or more thick. If it is composed of lamellae, the combined thickness of the lamellae must be 15 cm or more. *In addition, the cambic horizon must meet all of the following :*

- Has a texture of very fine sand, loamy very fine sand, or finer; and

- *Shows evidence of alteration in one of the following forms :*

- *– Aquic conditions within 50 cm of the soil surface or artificial drainage and all of the following :*

- a. Soil structure or the absence of rock structure in more than one-half of the volume; and

- b. Colours that do not change on exposure to air; and

- c. Dominant colour, moist, on faces of peds or in the matrix as follows :
- i. Value of 3 or less and chroma of 0; or
- ii. Value of 4 or more and chroma of 1 or less; or
- iii. Any value, chroma of 2 or less, and redox concentrations; or
- – Does not have the combination of aquic conditions within 50 cm of the soil surface or artificial drainage and colours, moist, as defined in item 2-a-(3) above, and has soil structure or the absence of rock structure in more than one-half of the volume and, *one or more of the following properties :*
    - a. Higher chroma, higher value, redder hue, or higher clay content than the underlying horizon or an overlying horizon; or
    - b. Evidence of the removal of carbonates or gypsum; and
- Has properties that do not meet the requirements for an anthropic, histic, folistic, melanic, mollic, plaggen, or umbric epipedon, a duripan or fragipan, or an argillic, calcic, gypsic, natric, oxic, petrocalcic, petrogypsic, placic, or spodic horizon; and
- Is not part of an Ap horizon and does not have a brittle manner of failure in more than 60 per cent of the matrix.

Duripan

A duripan is a silica-cemented sub-surface horizon with or without auxiliary cementing agents. It can occur in conjunction with a petrocalcic horizon.

A duripan must meet all of the following requirements :

- The pan is cemented or indurated in more than 50 per cent of the volume of some horizon; and
- The pan shows evidence of the accumulation of opal or other forms of silica, such as laminar caps, coatings, lenses, partly filled interstices, bridges between sand-sised grains, or coatings on rock and pararock fragments; and
- Less than 50 per cent of the volume of air-dry fragments slakes in 1N HCl even during prolonged soaking, but more than 50 per cent slakes in concentrated KOH or NaOH or in alternating acid and alkali; and
- Because of lateral continuity, roots can penetrate the pan only along vertical fractures with a horizontal spacing of 10 cm or more.

Fragipan

To be identified as a fragipan, a layer must have all of the following characteristics :

- The layer is 15 cm or more thick; and
- The layer shows evidence of pedogenesis within the horizon or, at a minimum, on the faces of structural units; and

- The layer has very coarse prismatic, columnar, or blocky structure of any grade, has weak structure of any size, or is massive. Separations between structural units that allow roots to enter have an average spacing of 10 cm or more on the horizontal dimensions; and

- Air-dry fragments of the natural soil fabric, 5 to 10 cm in diameter, from more than 50 per cent of the horizon slake when they are sub-merged in water; and

- The layer has, in 60 per cent or more of the volume, a firm or firmer rupture-resistance class, a brittle manner of failure at or near field capacity, and virtually no roots.

Glossic Horizon

The glossic horizon (Gr. glossa, tongue) develops as a result of the degradation of an argillic, kandic, or natric horizon from which clay and free iron oxides are removed.

The glossic horizon is 5 cm or more thick and consists of :

- An eluvial part, *i.e.*, albic materials, which constitute 15 to 85 per cent (by volume) of the glossic horizon; and

- An illuvial part, *i.e.*, remnants (pieces) of an argillic, kandic, or natric horizon.

Gypsic Horizon

The gypsic horizon is an illuvial horizon in which secondary gypsum has accumulated to a significant extent.

A gypsic horizon has all of the following properties :

- Is 15 cm or more thick; and

- Is not cemented or indurated to such a degree that it meets the requirements for a petrogypsic horizon; and

- Is 5 per cent or more gypsum and 1 per cent or more (by volume) secondary visible gypsum; and

- Has a product of thickness, in cm, multiplied by the gypsum content percentage of 150 or more.

Thus, a horizon 30 cm thick that is 5 per cent gypsum qualifies as a gypsic horizon if it is 1 per cent or more (by volume) visible gypsum and is not cemented or indurated to such a degree that it meets the requirements for a petrogypsic horizon.

The gypsum percentage can be calculated by multiplying the milliequivalents of gypsum per 100 g soil by the milli equivalent weight of $CaSO_4.2H_2O$, which is 0.086.

Kandic Horizon

- Is a vertically continuous sub-surface horizon that underlies a coarser textured surface horizon. The minimum thickness of the surface horizon is 18 cm after mixing or 5 cm if the textural transition to the kandic horizon is abrupt and there is no densic, lithic, paralithic, or petroferric contact (defined below) within 50 cm of the mineral soil surface; and

- *Has its upper boundary :*
- – At the point where the clay percentage in the fine-earth fraction, increasing with depth within a vertical distance of 15 cm or less, is either :
- a. 4 per cent or more (absolute) higher than that in the surface horizon if that horizon has less than 20 per cent total clay in the fine-earth fraction; or
- b. 20 per cent or more (relative) higher than that in the surface horizon if that horizon has 20 to 40 per cent total clay in the fine-earth fraction; or
- c. 8 per cent or more (absolute) higher than that in the surface horizon if that horizon has more than 40 per cent total clay in the fine-earth fraction; and

- – *At a depth :*
- a. Between 100 cm and 200 cm from the mineral soil surface if the particle-size class is sandy or sandy-skeletal throughout the upper 100 cm; or
- b. Within 100 cm from the mineral soil surface if the clay content in the fine-earth fraction of the surface horizon is 20 per cent or more; or

a. Within 125 cm from the mineral soil surface for all other soils; and

• *Has a thickness of either :*
- 30 cm or more; or
- 15 cm or more if there is a densic, lithic, paralithic, or petroferric contact within 50 cm of the mineral soil surface and the kandic horizon constitutes 60 per cent or more of the vertical distance between a depth of 18 cm and the contact; and
- Has a texture of loamy very fine sand or finer; and
- Has an apparent CEC of 16 cmol(+) or less per kg clay (by 1N NH$_4$OAc pH 7) and an apparent ECEC of 12 cmol(+) or less per kg clay (sum of bases extracted with 1N NH$_4$OAc pH 7 plus 1N KCl-extractable Al) in 50 per cent or more of its thickness between the point where the clay increase requirements are met and either a depth of 100 cm below that point or a densic, lithic, paralithic, or petroferric contact if shallower. (The percentage of clay is either measured by the pipette method or estimated to be 2.5 times [per cent water retained at 1500 kPa tension minus per cent organic carbon], whichever is higher, but no more than 100); and
- Has a regular decrease in organic-carbon content with increasing depth, no fine stratification, and no overlying layers more than 30 cm thick that

have fine stratification and/or an organic-carbon content that decreases irregularly with increasing depth.

Natric Horizon

The natric horizon has, in addition to the properties of the argillic horizon :

- *Either :*
- Columns or prisms in some part (generally the upper part), which may break to blocks; or
- Both blocky structure and eluvial materials, which contain uncoated silt or sand grains and extend more than 2.5 cm into the horizon; and
- *Either :*
- An exchangeable sodium percentage (ESP) of 15 per cent or more (or a sodium adsorption ratio [SAR] of 13 or more) in one or more horizons within 40 cm of its upper boundary; or
- More exchangeable magnesium plus sodium than calcium plus exchange acidity (at pH 8.2) in one or more horizons within 40 cm of its upper boundary if the ESP is 15 or more (or the SAR is 13 or more) in one or more horizons within 200 cm of the mineral soil surface.

Ortstein

Ortstein has all of the following :

- Consists of spodic materials; and
- Is in a layer that is 50 per cent or more cemented; and
- Is 25 mm or more thick.

Oxic Horizon

The oxic horizon is a sub-surface horizon that does not have andic soil properties (defined below) and has all of the following characteristics :

- A thickness of 30 cm or more; and
- A texture of sandy loam or finer in the fine-earth fraction; and
- Less than 10 per cent weatherable minerals in the 50- to 200-micron fraction; and
- Rock structure in less than 5 per cent of its volume, unless the lithorelicts with weatherable minerals are coated with sesquioxides; and
- A diffuse upper boundary, *i.e.,* within a vertical distance of 15 cm, *a clay increase with increasing depth of :*
- a. Less than 4 per cent (absolute) in its fine-earth fraction if the fine-earth fraction of the surface horizon contains less than 20 per cent clay; or
- b. Less than 20 per cent (relative) in its fine-earth fraction if the fine-earth fraction of the surface horizon contains 20 to 40 per cent clay; or

- c. Less than 8 per cent (absolute) in its fine-earth fraction if the fine-earth fraction of the surface horizon contains 40 per cent or more clay); and

- An apparent CEC of 16 cmol(+) or less per kg clay (by 1N NH_4OAc pH 7) and an apparent ECEC of 12 cmol(+) or less per kg clay (sum of bases extracted with 1N NH_4OAc pH 7 plus 1N KCl-extractable Al). (The percentage of clay is either measured by the pipette method or estimated to be 3 times [per cent water retained at 1500 kPa tension minus per cent organic carbon], whichever value is higher, but no more than 100).

Petrocalcic Horizon

The petrocalcic horizon is an illuvial horizon in which secondary calcium carbonate or other carbonates have accumulated to the extent that the horizon is cemented or indurated.

A petrocalcic horizon must meet the following requirements :

- The horizon is cemented or indurated by carbonates, with or without silica or other cementing agents; and

- Because of lateral continuity, roots can penetrate only along vertical fractures with a horizontal spacing of 10 cm or more; and

- *The horizon has a thickness of :*

- 10 cm or more; or

- 1 cm or more if it consists of a laminar cap directly underlain by bedrock.

Petrogypsic Horizon

The petrogypsic horizon is an illuvial horizon, 10 cm or more thick, in which secondary gypsum has accumulated to the extent that the horizon is cemented or indurated.

A petrogypsic horizon must meet the following requirements :

- The horizon is cemented or indurated by gypsum, with or without other cementing agents; and

- Because of lateral continuity, roots can penetrate only along vertical fractures with a horizontal spacing of 10 cm or more; and

- The horizon is 10 cm or more thick; and

- The horizon is 5 per cent or more gypsum, and the product of its thickness, in cm, multiplied by the gypsum content percentage is 150 or more.

Placic Horizon

The placic horizon (Gr. base of plax, flat stone; meaning a thin cemented pan) is a thin, black to dark reddish pan that is cemented by iron (or iron and manganese) and organic matter.

A placic horizon must meet the following requirements :

- The horizon is cemented or indurated with iron or iron and manganese and organic matter, with or without other cementing agents; and
- Because of lateral continuity, roots can penetrate only along vertical fractures with a horizontal spacing of 10 cm or more; and
- The horizon has a minimum thickness of 1 mm and, where associated with spodic materials, is less than 25 mm thick.

Salic Horizon

A salic horizon is a horizon of accumulation of salts that are more soluble than gypsum in cold water.

Required Characteristics

A salic horizon is 15 cm or more thick and has, for 90 consecutive days or more in normal years :

- An electrical conductivity (EC) equal to or greater than 30 dS/m in the water extracted from a saturated paste; and
- A product of the EC, in dS/m, and thickness, in cm, equal to 900 or more.

Sombric Horizon

A sombric horizon is a sub-surface horizon in mineral soils that has formed under free drainage. It contains illuvial humus that is neither associated with aluminum, as is the humus in the spodic horizon, nor dispersed by sodium, as is common in the natric horizon. Consequently, the sombric horizon does not have the high cation-exchange capacity in its clay that characterises a spodic horizon and does not have the high base saturation of a natric horizon. It does not underlie an albic horizon.

Sombric horizons are thought to be restricted to the cool, moist soils of high plateaus and mountains in tropical or sub-tropical regions. Because of strong leaching, their base saturation is low (less than 50 per cent by NH_4OAc).

The sombric horizon has a lower colour value or chroma, or both, than the overlying horizon and commonly contains more organic matter. It may have formed in an argillic, cambic, or oxic horizon. If peds are present, the dark colours are most pronounced on surfaces of peds. In the field a sombric horizon is easily mistaken for a buried A horizon. It can be distinguished from some buried epipedons by lateral tracing. In thin sections the organic matter of a sombric horizon appears more concentrated on peds and in pores than uniformly dispersed throughout the matrix.

Spodic Horizon

- A spodic horizon is an illuvial layer with 85 per cent or more spodic materials.

- A spodic horizon is normally a sub-surface horizon underlying an O, A, Ap, or E horizon. It may, however, meet the definition of an umbric epipedon.

- A spodic horizon must have 85 per cent or more spodic materials in a layer 2.5 cm or more thick that is not part of any Ap horizon.

Other Diagnostic Soil Characteristics (Mineral Soils)

Diagnostic soil characteristics are features of the soil that are used in various places in the keys or in definitions of diagnostic horizons.

Abrupt Textural Change

An abrupt textural change is a specific kind of change that may occur between an ochric epipedon or an albic horizon and an argillic horizon. It is characterised by a considerable increase in clay content within a very short vertical distance in the zone of contact. If the clay content in the fine-earth fraction of the ochric epipedon or albic horizon is less than 20 per cent, it doubles within a vertical distance of 7.5 cm or less. If the clay content in the fine-earth fraction of the ochric epipedon or the albic horizon is 20 per cent or more, there is an increase of 20 per cent or more (absolute) within a vertical distance of 7.5 cm or less (*e.g.*, an increase from 22 to 42 per cent) and the clay content in some part of the argillic horizon is 2 times or more the amount contained in the overlying horizon.

Normally, there is no transitional horizon between an ochric epipedon or an albic horizon and an argillic horizon, or the transitional horizon is too thin to be sampled. Some soils, however, have a glossic horizon or interfingering of albic materials (defined below) in parts of the argillic horizon. The upper boundary of such a horizon is irregular or even discontinuous. Sampling this mixture as a single horizon might create the impression of a relatively thick transitional horizon, whereas the thickness of the actual transition at the contact may be no more than 1 mm.

Albic Materials

Albic (L. albus, white) materials are soil materials with a colour that is largely determined by the colour of primary sand and silt particles rather than by the colour of their coatings. This definition implies that clay and/or free iron oxides have been removed from the materials or that the oxides have been segregated to such an extent that the colour of the materials is largely determined by the colour of the primary particles.

Albic materials have one of the following colours :

- Chroma of 2 or less; and either
- A colour value, moist, of 3 and a colour value, dry, of 6 or more; or
- A colour value, moist, of 4 or more and a colour value, dry, of 5 or more; or
- Chroma of 3 or less; and either
- A colour value, moist, of 6 or more; or
- A colour value, dry, of 7 or more; or
- Chroma that is controlled by the colour of uncoated grains of silt or sand, hue of 5YR or redder, and the colour values listed in item 1-a or 1-b above.

Relatively unaltered layers of light coloured sand, volcanic ash, or other materials deposited by wind or water are not considered albic materials, although they may have the same colour and apparent morphology. These deposits are parent materials that are not characterised by the removal of clay and/or free iron and do not overlie an illuvial horizon or other soil horizon, except for a buried soil. Light coloured krotovinas or filled root channels should be considered albic materials only if they have no fine stratifications or lamellae, if any sealing along the krotovina walls has been destroyed, and if these intrusions have been leached of free iron oxides and/or clay after deposition.

Andic Soil Properties

Andic soil properties result mainly from the presence of significant amounts of allophane, imogolite, ferrihydrite, or aluminum-humus complexes in soils. These materials, originally termed "amorphous" (but understood to contain allophane) in the 1975 edition of Soil Taxonomy (USDA, SCS, 1975), are commonly formed during the weathering of tephra and other parent materials with a significant content of volcanic glass. Although volcanic glass is or was a common component in many Andisols, it is not a requirement of the Andisol order.

To be recognised as having andic soil properties, soil materials must contain less than 25 per cent (by weight) organic carbon and meet one or both of the following requirements :

- *In the fine-earth fraction, all of the following :*
- Aluminum plus 1/2 iron percentages (by ammonium oxalate) totaling 2.0 per cent or more; and
- A bulk density, measured at 33 kPa water retention, of 0.90 g/cm^3 or less; and
- A phosphate retention of 85 per cent or more; or
- In the fine-earth fraction, a phosphate retention of 25 per cent or more, 30 per cent or more particles 0.02 to 2.0 mm in size, *and one of the following :*
- Aluminum plus ½ iron percentages (by ammonium oxalate) totaling 0.40 or more and, in the 0.02 to 2.0 mm fraction, 30 per cent or more volcanic glass; or

- Aluminum plus ½ iron percentages (by ammonium oxalate) totaling 2.0 or more and, in the 0.02 to 2.0 mm fraction, 5 per cent or more volcanic glass; or

- Aluminum plus ½ iron percentages (by ammonium oxalate) totaling between 0.40 and 2.0 and, in the 0.02 to 2.0 mm fraction, enough volcanic glass so that the glass percentage, when plotted against the value obtained by adding aluminum plus ½ iron percentages in the fine-earth fraction, falls within the shaded area.

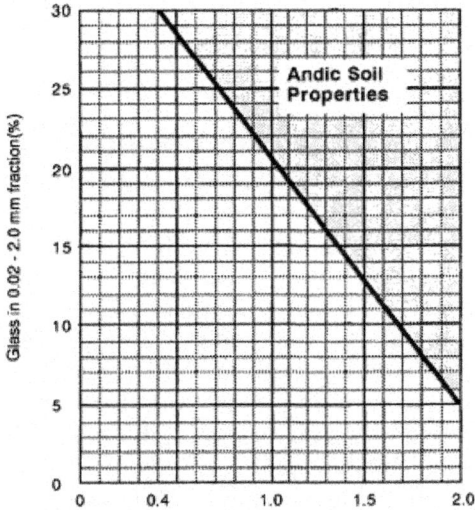

Fig. Soils that are Plotted in the Shaded Area have Andic Soil Properties. A Soil has these Properties if the Fraction Less than 2.0 mm in Size has Phosphate Retention of more than 25 Per cent and the 0.02 to 2.0 mm Fraction is at Least 30 Per cent of the Fraction less than 2.0 mm in Size.

Anhydrous Conditions

Anhydrous conditions refer to the active layer in soils of cold deserts and other areas with permafrost (often dry permafrost) and low precipitation (usually less than 50 mm water equivalent). Anhydrous soil conditions are similar to the aridic (torric) soil moisture regimes, except that the soil temperature is less than 0 °C.

Coefficient of Linear Extensibility (COLE)

The coefficient of linear extensibility (COLE) is the ratio of the difference between the moist length and dry length of a clod to its dry length. It is $(Lm-Ld)/Ld$, where Lm is the length at 33 kPa tension and Ld is the length when dry. COLE can be calculated from the differences in bulk density of the clod when moist and when dry. An estimate of COLE can be calculated in the field by measuring the distance between two pins in a clod of undisturbed soil at field capacity and again after the clod has dried. COLE does not apply if the shrinkage is irreversible.

Durinodes

Durinodes (L. *durus, hard,* and nodus, *knot*) are weakly cemented to indurated nodules with a diameter of 1 cm or more. The cement is SiO_2, presumably opal and micro-crystalline forms of silica. Durinodes break down in hot concentrated KOH after treatment with HCl to remove carbonates but do not break down with concentrated HCl alone. Dry durinodes do not slake appreciably in water, but prolonged soaking can result in spalling of very thin platelets. Durinodes are firm or firmer and brittle when wet, both before and after treatment with acid. Most durinodes are roughly concentric when viewed in cross section, and concentric stringers of opal are visible under a hand lens.

Fragic Soil Properties

Fragic soil properties are the essential properties of a fragipan. They have neither the layer thickness nor volume requirements for the fragipan. Fragic soil properties are in sub-surface horizons, although they can be at or near the surface in truncated soils. Aggregates with fragic soil properties have a firm or firmer rupture-resistance class and a brittle manner of failure when soil water is at or near field capacity. Air-dry fragments of the natural fabric, 5 to 10 cm in diameter, slake when they are sub-merged in water. Aggregates with fragic soil properties show evidence of pedogenesis, including one or more of the following : oriented clay within the matrix or on faces of peds, redoximorphic features within the matrix or on faces of peds, strong or moderate soil structure, and coatings of albic materials or uncoated silt and sand grains on faces of peds or in seams. Peds with these properties are considered to have fragic soil properties regardless of whether or not the density and brittleness are pedogenic.

Soil aggregates with fragic soil properties must :

- Show evidence of pedogenesis within the aggregates or, at a minimum, on the faces of the aggregates; and
- Slake when air-dry fragments of the natural fabric, 5 to 10 cm in diameter, are sub-merged in water; and
- Have a firm or firmer rupture-resistance class and a brittle manner of failure when soil water is at or near field capacity; and
- Restrict the entry of roots into the matrix when soil water is at or near field capacity.

Identifiable Secondary Carbonates

The term "identifiable secondary carbonates" is used in the definitions of a number of taxa. It refers to translocated authigenic calcium carbonate that has been precipitated in place from the soil solution rather than inherited from a soil parent material, such as a calcareous loess or till. Identifiable secondary carbonates either may disrupt the soil structure or fabric, forming masses, nodules, concretions, or spheroidal aggregates (white eyes) that are soft and powdery when dry, or may be present as coatings in pores, on structural faces, or on the undersides of rock or pararock fragments. If present as coatings, the secondary carbonates cover a significant part of the surfaces. Commonly, they coat all of the surfaces to a thickness of 1 mm or more. If little calcium carbonate is present in the soil, however, the surfaces may be only partially coated. The coatings must be thick enough to be visible when moist. Some horizons are entirely engulfed by carbonates. The colour of these horizons is largely determined by the carbonates. The carbonates in these horizons are within the concept of identifiable secondary carbonates.

The filaments commonly seen in a dry calcareous horizon are within the meaning of identifiable secondary carbonates if the filaments are thick enough to be visible when the soil is moist. Filaments commonly branch on structural faces.

Interfingering of Albic Materials

The term "interfingering of albic materials" refers to albic materials that penetrate 5 cm or more into an underlying argillic, kandic, or natric horizon along vertical and, to a lesser degree, horizontal faces of peds. There need not be a continuous overlying albic horizon. The albic materials constitute less than 15 per cent of the layer that they penetrate, but they form continuous skeletans (ped coatings of clean silt or sand defined by Brewer, 1976) 1 mm or more thick on the vertical faces of peds, which means a total width of 2 mm or more between abutting peds. Because quartz is such a common constituent of silt and sand, these skeletans are usually light gray when moist and nearly white when dry, but their colour is determined in large part by the colour of the sand or silt fraction.

Interfingering of albic materials is recognised if albic materials :

* Penetrate 5 cm or more into an underlying argillic or natric horizon; and
* Are 2 mm or more thick between vertical faces of abutting peds; and
* Constitute less than 15 per cent (by volume) of the layer that they penetrate.

Lamellae

A lamella is an illuvial horizon less than 7.5 cm thick. Each lamella contains an accumulation of oriented silicate clay on or bridging sand and silt grains (and rock fragments if any are present). A lamella has more silicate clay than the overlying eluvial horizon. A lamella is an illuvial horizon less than 7.5 cm thick formed in unconsolidated regolith more than 50 cm thick. Each lamella contains an accumulation of oriented silicate clay on or bridging the sand and silt grains (and coarse fragments if any are present). Each lamella is required to have more silicate clay than the overlying eluvial horizon.

Lamellae occur in a vertical series of two or more, and each lamella must have an overlying eluvial horizon. (An eluvial horizon is not required above the uppermost lamella if the soil is truncated). Lamellae may meet the requirements for either a cambic or an argillic horizon. A combination of two or more lamellae

15 cm or more thick is a cambic horizon if the texture is very fine sand, loamy very fine sand, or finer. A combination of two or more lamellae meets the requirements for an argillic horizon if there is 15 cm or more cumulative thickness of lamellae that are 0.5 cm or more thick and, *that have a clay content of either* :

- 3 per cent or more (absolute) higher than in the overlying eluvial horizon (*e.g.*, 13 per cent *versus* 10 per cent) if any part of the eluvial horizon has less than 15 per cent clay in the fine-earth fraction; or

- 20 per cent or more (relative) higher than in the overlying eluvial horizon (*e.g.*, 24 per cent *versus* 20 per cent) if all parts of the eluvial horizon have more than 15 per cent clay in the fine-earth fraction.

Linear Extensibility (LE)

Linear extensibility (LE) helps to predict the potential of a soil to shrink and swell. The LE of a soil layer is the product of the thickness, in cm, multiplied by the COLE of the layer in question. The LE of a soil is the sum of these products for all soil horizons. Lithologic discontinuities are significant changes in particle-size distribution or mineralogy that represent differences in lithology within a soil. A lithologic discontinuity can also denote an age difference. For information on using horizon designations for lithologic discontinuities.

Not everyone agrees on the degree of change required for a lithologic discontinuity. No attempt is made to quantify lithologic discontinuities. The discussion below is meant to serve as a guideline. Several lines of field evidence can be used to evaluate lithologic discontinuities. In addition to mineralogical and textural differences that may require laboratory studies, certain observations can be made in the field.

These include but are not limited to the following :

- *Abrupt textural contacts* : An abrupt change in particle-size distribution, which is not solely a change in clay content resulting from pedogenesis, can often be observed.

- *Contrasting sand sizes* : Significant changes in sand size can be detected. For example, if material containing mostly medium sand or finer sand abruptly overlies material containing mostly coarse sand and very coarse sand, one can assume that there are two different materials. Although the materials may be of the same mineralogy, the contrasting sand sizes result from differences in energy at the time of deposition by water and/ or wind.

- *Bedrock lithology vs. rock fragment lithology in the soil* : If a soil with rock fragments overlies a lithic contact, one would expect the rock fragments to have a lithology similar to that of the material below the lithic contact. If many of the rock fragments do not have the same lithology as the underlying bedrock, the soil is not derived completely from the underlying bedrock.

- *Stone lines* : The occurrence of a horizontal line of rock fragments in the vertical sequence of a soil indicates that the soil may have developed in more than one kind of parent material. The material above the stone line is most likely transported, and the material below may be of different origin.

- *Inverse distribution of rock fragments* : A lithologic discontinuity is often indicated by an erratic distribution of rock fragments. The percentage of rock fragments decreases with increasing depth. This line of evidence is useful in areas of soils that have relatively unweathered rock fragments.

- *Rock fragment weathering rinds* : Horizons containing rock fragments with no rinds that overlie horizons containing rocks with rinds suggest that the upper material is in part depositional and not related to the lower part in time and perhaps in lithology.

- *Shape of rock fragments* : A soil with horizons containing angular rock fragments overlying horizons containing well rounded rock fragments may indicate a discontinuity. This line of evidence represents different mechanisms of transport (colluvial vs. alluvial) or even different transport distances.

- *Soil colour* : Abrupt changes in colour that are not the result of pedogenic processes can be used as indicators of discontinuity.

- *Micromorphological features* : Marked differences in the size and shape of resistant minerals in one horizon and not in another are indicators of differences in materials.

Use of Laboratory Data

Discontinuities are not always readily apparent in the field. In these cases laboratory data are necessary. Even with laboratory data, detecting discontinuities may be difficult. The decision is a qualitative or perhaps a partly quantitative judgement.

General concepts of lithology as a function of depth might include :

- *Laboratory data – visual scan* : The array of laboratory data is assessed in an attempt to determine if a field-designated discontinuity is corroborated and if any data show evidence of a discontinuity not observed in the field. One must sort changes in lithology from changes caused by pedogenic processes. In most cases the quantities of sand and coarser fractions are not altered significantly by soil-forming processes. Therefore, an abrupt change in sand size or sand mineralogy is a clue to lithologic change. Gross soil mineralogy and the resistant mineral suite are other clues.

- *Data on a clay-free basis* : A common manipulation in assessing lithologic change is computation of sand and silt separates on a carbonate-free, clay-free basis (per cent fraction, *e.g.*, fine sand and very fine sand, divided by per cent sand plus silt, times 100). Clay distribution is subject to pedogenic change and may either mask inherited lithologic differences or produce differences that are not inherited from lithology. The numerical array computed on a clay-free basis can be inspected visually or plotted as a function of depth.

Another aid used to assess lithologic changes is computation of the ratios of one sand separate to another. The ratios can be computed and examined as a numerical array, or they can be plotted. The ratios work well if sufficient quantities of the two fractions are available. Low quantities magnify changes in ratios, especially if the denominator is low.

n Value

The n value (Pons and Zonneveld, 1965) characterises the relation between the percentage of water in a soil under field conditions and its percentages of inorganic clay and humus. The n value is helpful in predicting whether a soil can be grazed by livestock or can support other loads and in predicting what degree of subsidence would occur after drainage.

For mineral soil materials that are not thixotropic, the n value can be calculated by the following formula :

$$n = (A - 0.2R)/(L + 3H)$$

In this formula, A is the percentage of water in the soil in field condition, calculated on a dry-soil basis; R is the percentage of silt plus sand; L is the percentage of clay; and H is the percentage of organic matter (per cent organic carbon multiplied by 1.724).

Few data for calculations of the n value are available in the United States, but the critical n value of 0.7 can be approximated closely in the field by a simple test of squeesing a soil sample in the hand. If the soil flows between the fingers with difficulty, the n value is between 0.7 and 1.0 (slightly fluid manner of failure class); if the soil flows easily between the fingers, the n value is 1 or more (moderately fluid or very fluid manner of failure class).

Petroferric Contact

A petroferric (Gr. petra, rock, and L. ferrum, iron; implying ironstone) contact is a boundary between soil and a continuous layer of indurated material in which iron is an important cement and organic matter is either absent or present only in traces. The indurated layer must be continuous within the limits of each pedon, but it may be fractured if the average lateral distance between fractures is 10 cm or more. The fact that this ironstone layer contains little or no organic matter distinguishes it from a placic horizon and an indurated spodic horizon (ortstein), both of which contain organic matter.

Several features can aid in making the distinction between a lithic contact and a petroferric contact. First, a petroferric contact is roughly horizontal. Second, the material directly below a petroferric contact contains a high amount of iron (normally 30 per cent or more Fe_2O_3). Third, the ironstone sheets below a petroferric contact are thin; their thickness ranges from a few centimeters to very few meters. Sandstone, on the other hand, may be thin or very thick, may be level-bedded or tilted, and may contain only a small percentage of Fe_2O_3. In the Tropics, the ironstone is generally more or less vesicular.

Plinthite

Plinthite (Gr. plinthos, brick) is an iron-rich, humus-poor mixture of clay with quartz and other minerals. It commonly occurs as dark red redox concentrations that usually form platy, polygonal, or reticulate patterns. Plinthite changes irreversibly to an ironstone hardpan or to irregular aggregates on exposure to repeated wetting and drying, especially if it is also exposed to heat from the sun. The lower boundary of a zone in which plinthite occurs generally is diffuse or gradual, but it may be abrupt at a lithologic discontinuity.

Generally, plinthite forms in a horizon that is saturated with water for some time during the year. Initially, iron is normally segregated in the form of soft, more or less clayey, red or dark red redox concentrations. These concentrations are not considered plinthite unless there has been enough segregation of iron to permit their irreversible hardening on exposure to repeated wetting and drying. Plinthite is firm or very firm when the soil moisture content is near field capacity and hard when the moisture content is below the wilting point. Plinthite does not harden irreversibly as a result of a single cycle of drying and rewetting. After a single drying, it will remoisten and then can be dispersed in large part if one shakes it in water with a dispersing agent. In a moist soil, plinthite is soft enough to be cut with a spade. After irreversible hardening, it is no longer considered plinthite but is called ironstone. Indurated ironstone materials can be broken or shattered with a spade but cannot be dispersed if one shakes tham in water with a dispersing agent.

Resistant Minerals

Several references are made to resistant minerals in this taxonomy. Obviously, the stability of a mineral in the soil is a partial function of the soil moisture regime. Where resistant minerals are referred to in the definitions of diagnostic horizons and of various taxa, a humid climate, past or present, is always assumed.

Resistant minerals are durable minerals in the 0.02 to 2.0 mm fraction. Quartz is the most common resistant mineral in soils. The less common ones include sphene, rutile, zircon, tourmaline, and beryl.

Slickensides

Slickensides are polished and grooved surfaces and generally have dimensions exceeding 5 cm. They are produced when one soil mass slides past another. Some slickensides occur at the lower boundary of a slip surface where a mass of soil moves downward on a relatively steep slope. Slickensides result directly from the swelling of clay minerals and shear failure. They are very common in swelling clays that undergo marked changes in moisture content.

Spodic Materials

Spodic materials form in an illuvial horizon that normally underlies a histic, ochric, or umbric epipedon or an albic horizon. In most undisturbed areas, spodic materials underlie an albic horizon. They may occur within an umbric epipedon or an Ap horizon.

A horizon consisting of spodic materials normally has an optical-density-of-oxalate-extract (ODOE) value of 0.25 or more, and that value is commonly at least 2 times as high as the ODOE value in an overlying eluvial horizon. This increase in ODOE value indicates an accumulation of translocated organic materials in an illuvial horizon. Soils with spodic materials show evidence that organic materials and aluminum, with or without iron, have been moved from an eluvial horizon to an illuvial horizon.

Definition of Spodic Materials

Spodic materials are mineral soil materials that do not have all of the properties of an argillic or kandic horizon; are dominated by active amorphous materials that are illuvial and are composed of organic matter and aluminum, with or without iron; *and have both of the following :*

- A pH value in water (1 : 1) of 5.9 or less and an organic-carbon content of 0.6 per cent or more; and

- • *One or both of the following :*

- – An overlying albic horizon that extends horizontally through 50 per cent or more of each pedon and, directly under the albic horizon, colours, moist (crushed and smoothed sample), *as follows :*

- a. Hue of 5YR or redder; or

- b. Hue of 7.5YR, colour value of 5 or less, and chroma of 4 or less; or

- c. Hue of 10YR or neutral and a colour value and chroma of 2 or less; or

- d. A colour of 10YR 3/1; or

- – With or without an albic horizon and one of the colours listed above or hue of 7.5YR, colour value, moist, of 5 or less, chroma of 5 or 6 (crushed and smoothed sample), and *one or more of the following morphological or chemical properties :*

- a. Cementation by organic matter and aluminum, with or without iron, in 50 per cent or more of each pedon and a very firm or firmer rupture-resistance class in the cemented part; or

- b. 10 per cent or more cracked coatings on sand grains; or

- c. Aluminum plus ½ iron percentages (by ammonium oxalate) totaling 0.50 or more, and half that amount or less in an overlying umbric (or subhorizon of an umbric) epipedon, ochric epipedon, or albic horizon; or

- d. An optical-density-of-oxalate-extract (ODOE) value of 0.25 or more, and a value half as high or lower in an overlying umbric (or subhorizon of an umbric) epipedon, ochric epipedon, or albic horizon.

Weatherable Minerals

Several references are made to weatherable minerals in this taxonomy. Obviously, the stability of a mineral in a soil is a partial function of the soil moisture regime. Where weatherable minerals are referred to in the definitions of diagnostic horizons and of various taxa in this taxonomy, a humid climate, either present or past, is always assumed.

The minerals that are included in the meaning of weatherable minerals are as follows :

- *Clay minerals :* All 2 : 1 lattice clays, except for one that is currently considered to be an aluminum-interlayered chlorite. Sepiolite, talc, and glauconite are also included in this group of weatherable clay minerals, although they are not everywhere of clay size.
- *Silt- and sand-sised minerals (0.02 to 0.2 mm in diameter) :* Feldspars, feldspathoids, ferromagnesian minerals, glass, micas, zeolites, and apatite.

Obviously, this definition of the term "weatherable minerals" is restrictive. The intent is to include, in the definitions of diagnostic horizons and various taxa, only those weatherable minerals that are unstable in a humid climate compared to other minerals, such as quartz and 1 : 1 lattice clays, but that are more resistant to weathering than calcite.

Characteristics Diagnostic for Organic Soils

Following is a description of the characteristics that are used only with organic soils.

KINDS OF ORGANIC SOIL MATERIALS

Three different kinds of organic soil materials are distinguished in this taxonomy, based on the degree of decomposition of the plant materials from which the organic materials are derived.

The three kinds are :

- Fibric,
- Hemic, and
- Sapric.

Because of the importance of fibre content in the definitions of these materials, fibres are defined before the kinds of organic soil materials.

Fibres

Fibres are pieces of plant tissue in organic soil materials (excluding live roots) that :

- Are large enough to be retained on a 100-mesh sieve (openings 0.15 mm across) when the materials are screened; and

- Show evidence of the cellular structure of the plants from which they are derived; and
- Either are 2 cm or less in their smallest dimension or are decomposed enough to be crushed and shredded with the fingers.

Pieces of wood that are larger than 2 cm in cross-section and are so undecomposed that they cannot be crushed and shredded with the fingers, such as large branches, logs, and stumps, are not considered fibres but are considered coarse fragments (comparable to gravel, stones, and boulders in mineral soils).

Fibric Soil Materials

Fibric soil materials are organic soil materials that either :

- Contain three-fourths or more (by volume) fibres after rubbing, excluding coarse fragments; or
- Contain two-fifths or more (by volume) fibres after rubbing, excluding coarse fragments, and yield colour values and chromas of 7/1, 7/2, 8/1, 8/2, or 8/3 on white chromatographic or filter paper that is inserted into a paste made of the soil materials in a saturated sodium-pyrophosphate solution.

Hemic Soil Materials

Hemic soil materials are intermediate in their degree of decomposition between the less decomposed fibric and more decomposed sapric materials. Their morphological features give intermediate values for fibre content, bulk density, and water content. Hemic soil materials are partly altered both physically and biochemically.

Sapric Soil Materials

Sapric soil materials (Gr. sapros, rotten) are the most highly decomposed of the three kinds of organic soil materials. They have the smallest amount of plant fibre, the highest bulk density, and the lowest water content on a dry-weight basis at saturation. Sapric soil materials are commonly very dark gray to black. They are relatively stable; *i.e.*, they change very little physically and chemically with time in comparison to other organic soil materials.

Sapric materials have the following characteristics :

- The fibre content, after rubbing, is less than one-sixth (by volume), excluding coarse fragments; and
- The colour of the sodium-pyrophosphate extract on white chromatographic or filter paper is below or to the right of a line drawn to exclude blocks 5/1, 6/2, and 7/3. If few or no fibres can be detected and the colour of the pyrophosphate extract is to the left of or above this line, the possibility that the material is limnic must be considered.

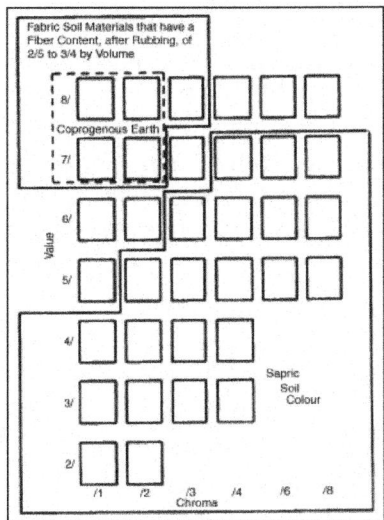

Fig. Value and Chroma of Pyrophosphate Solution of Fibric and Sapric Materials.

Humilluvic Material

Humilluvic material, *i.e.*, illuvial humus, accumulates in the lower parts of some organic soils that are acid and have been drained and cultivated. The humilluvic material has a C^{14} age that is not older than the overlying organic materials. It has very high solubility in sodium pyrophosphate and rewets very slowly after drying. Most commonly, it accumulates near a contact with a sandy mineral horizon. To be recognised as a differentia in classification, the humilluvic material must constitute one-half or more (by volume) of a layer 2 cm or more thick.

Limnic Materials

The presence or absence of limnic deposits is taken into account in the higher categories of Histosols but not Histels. The nature of such deposits is considered in the lower categories of Histosols. Limnic materials include both organic and inorganic materials that were either.

- Deposited in water by precipitation or through the action of aquatic organisms, such as algae or diatoms, or
- Derived from underwater and floating aquatic plants and subsequently modified by aquatic animals. They include coprogenous earth (sedimentary peat), diatomaceous earth, and marl.

Coprogenous Earth

A layer of coprogenous earth (sedimentary peat) is a limnic layer that :

- Contains many fecal pellets with diameters between a few hundredths and a few tenths of a millimeter; and

- Has a colour value, moist, of 4 or less; and
- Either forms a slightly viscous water suspension and is non-plastic or slightly plastic but not sticky, or shrinks upon drying, forming clods that are difficult to rewet and often tend to crack along horizontal planes; and
- Either yields a saturated sodium-pyrophosphate extract on white chromatographic or filter paper that has a colour value of 7 or more and chroma of 2 or less or has a cation-exchange capacity of less than 240 cmol(+) per kg organic matter, or both.

Diatomaceous Earth

A layer of diatomaceous earth is a limnic layer that :
- If not previously dried, has a matrix colour value of 3, 4, or 5, which changes irreversibly on drying as a result of the irreversible shrinkage of organic-matter coatings on diatoms; and
- Either yields a saturated sodium-pyrophosphate extract on white chromatographic or filter paper that has a colour value of 8 or more and chroma of 2 or less or has a cation-exchange capacity of less than 240 cmol(+) per kg organic matter, or both.

Marl

A layer of marl is a limnic layer that :
- Has a colour value, moist, of 5 or more; and
- Reacts with dilute HCl to evolve CO_2.

The colour of marl usually does not change irreversibly on drying because a layer of marl contains too little organic matter, even before it has been shrunk by drying, to coat the carbonate particles.

Thickness of Organic Soil Materials

The thickness of organic materials over limnic materials, mineral materials, water, or permafrost is used to define the Histosols and Histels. For practical reasons, an arbitrary control section has been established for the classification of Histosols and Histels. Depending on the kinds of soil material in the surface layer, the control section has a thickness of either 130 cm or 160 cm from the soil surface if there is no densic, lithic, or paralithic contact, thick layer of water, or permafrost within the respective limit. The thicker control section is used if the surface layer to a depth of 60 cm either contains three-fourths or more fibres derived from Sphagnum, Hypnum, or other mosses or has a bulk density of less than 0.1. Layers of water, which may be between a few centimeters and many meters thick in these soils, are considered to be the lower boundary of the control section only if the water extends below a depth of 130 or 160 cm, respectively. A densic, lithic, or paralithic contact, if shallower than 130 or 160 cm, constitutes the lower boundary of the control section. In some soils the lower boundary is 25 cm below the upper limit of permafrost. An unconsolidated mineral substratum shallower than those limits does not change the lower boundary of the control section. The control section of Histosols and Histels is divided somewhat arbitrarily into three tiers—surface, sub-surface, and bottom tiers.

Sur-face Tier

The surface tier of a Histosol or Histel extends from the soil surface to a depth of 60 cm if either.

- The materials within that depth are fibric and three-fourths or more of the fibre volume is derived from Sphagnum or other mosses or
- The materials have a bulk density of less than 0.1. Otherwise, the surface tier extends from the soil surface to a depth of 30 cm.

Some organic soils have a mineral surface layer less than 40 cm thick as a result of flooding, volcanic eruptions, additions of mineral materials to increase soil strength or reduce the hazard of frost, or other causes. If such a mineral layer is less than 30 cm thick, it constitutes the upper part of the surface tier; if it is 30 to 40 cm thick, it constitutes the whole surface tier and part of the sub-surface tier.

Sub-surface Tier

The sub-surface tier is normally 60 cm thick. If the control section ends at a shallower depth (at a densic, lithic, or paralithic contact or a water layer or in permafrost), however, the sub-surface tier extends from the lower boundary of the surface tier to the lower boundary of the control section. It includes any unconsolidated mineral layers that may be present within those depths.

Bottom Tier

The bottom tier is 40 cm thick unless the control section has its lower boundary at a shallower depth (at a densic, lithic, or paralithic contact or a water layer or in permafrost). Thus, if the organic materials are thick, there are two possible thicknesses of the control section, depending on the presence or absence and the thickness of a surface mantle of fibric moss or other organic material that has a low bulk density (less than 0.1). If the fibric moss extends to a depth of 60 cm and is the dominant material within this depth (three-fourths or more of the volume), the control section is 160 cm thick. If the fibric moss is thin or absent, the control section extends to a depth of 130 cm.

HORIZONS AND CHARACTERISTICS DIAGNOSTIC FOR BOTH MINERAL AND ORGANIC SOILS

Following are descriptions of the horizons and characteristics that are diagnostic for both mineral and organic soils.

Aquic Conditions

Soils with aquic (L. aqua, water) conditions are those that currently undergo continuous or periodic saturation and reduction. The presence of these conditions is indicated by redoximorphic features, except in Histosols and Histels, and can be verified by measuring saturation and reduction, except in artificially drained soils. Artificial drainage is defined here as the removal of free water from soils having aquic conditions by surface mounding, ditches, or sub-surface tiles to the extent that water table levels are changed significantly in connection with specific types of land use. In the keys, artificially drained soils are included with soils that have aquic conditions.

Elements of aquic conditions are as follows :

- Saturation is characterised by zero or positive pressure in the soil water and can generally be determined by observing free water in an unlined auger hole. Problems may arise, however, in clayey soils with peds, where an unlined auger hole may fill with water flowing along faces of peds while the soil matrix is and remains unsaturated (bypass flow). Such free water may incorrectly suggest the presence of a water table, while the actual water table occurs at greater depth. Use of well sealed piezometers or tensiometers is therefore recommended for measuring saturation. Problems may still occur, however, if water runs into piezometer slits near the bottom of the piezometer hole or if tensiometers with slowly reacting manometers are used. The first problem can be overcome by using piezometers with smaller slits and the second by using transducer tensiometry, which reacts faster than manometers. Soils are considered wet if they have pressure heads greater than -1 kPa. Only macropores, such as cracks between peds or channels, are then filled with air, while the soil matrix is usually still saturated. Obviously, exact measurements of the wet state can be obtained only with tensiometers. For operational purposes, the use of piezometers is recommended as a standard method. The duration of saturation required for creating aquic conditions varies, depending on the soil environment, and is not specified.

Three types of saturation are defined :

- *Endosaturation :* The soil is saturated with water in all layers from the upper boundary of saturation to a depth of 200 cm or more from the mineral soil surface.

- *Episaturation :* The soil is saturated with water in one or more layers within 200 cm of the mineral soil surface and also has one or more unsaturated layers, with an upper boundary above a depth of 200 cm, below the saturated layer. The zone of saturation, *i.e.*, the water table, is perched on top of a relatively impermeable layer.

- *Anthric saturation :* This term refers to a special kind of aquic conditions that occur in soils that are cultivated and irrigated (flood irrigation). Soils with anthraquic conditions must meet the requirements for aquic conditions and *in addition have both of the following :*

- A tilled surface layer and a directly underlying slowly permeable layer that has, for 3 months or more in normal years, *both :*

- Saturation and reduction; and

- Chroma of 2 or less in the matrix.

A sub-surface horizon with one or more of the following :

○ Redox depletions with a colour value, moist, of 4 or more and chroma of 2 or less in macropores; or

○ Redox concentrations of iron; or

○ 2 times or more the amount of iron contained in the tilled surface layer.

- The degree of reduction in a soil can be characterised by the direct measurement of redox potentials. Direct measurements should take into account chemical equilibria as expressed by stability diagrams in standard soil textbooks. Reduction and oxidation processes are also a function of soil pH. Obtaining accurate measurements of the degree of reduction in a soil is difficult. In the context of this taxonomy, however, only a degree of reduction that results in reduced iron is considered, because it produces the visible redoximorphic features that are identified in the keys. A simple field test is available to determine if reduced iron ions are present. A freshly broken surface of a field-wet soil sample is treated with alpha, alpha-dipyridyl in neutral, 1-normal ammonium-acetate solution. The appearance of a strong red colour on the freshly broken surface indicates the presence of reduced iron ions. A positive reaction to the alpha, alpha-dipyridyl field test for ferrous iron may be used to confirm the existence of reducing conditions and is especially useful in situations where, despite saturation, normal morphological indicators of such conditions are either absent or obscured (as by the dark colours characteristic of melanic great groups). A negative reaction, however, does not imply that reducing conditions are always absent. It may only mean that the level of free iron in the soil is below the sensitivity limit of the test or that the soil is in an oxidised phase at the time of testing. Use of alpha, alpha-dipyridyl in a 10 per cent acetic-acid solution is not recommended because the acid is likely to change soil conditions, for example, by dissolving $CaCO_3$. The duration of reduction required for creating aquic conditions is not specified.

- Redoximorphic features associated with wetness result from alternating periods of reduction and oxidation of iron and manganese compounds in the soil. Reduction occurs during saturation with water, and oxidation occurs when the soil is not saturated. The reduced iron and manganese ions are mobile and may be transported by water as it moves through the soil. Certain redox patterns occur as a function of the patterns in which the ion-carrying water moves through the soil and as a function of the location of aerated zones in the soil. Redox patterns are also affected by the fact that manganese is reduced more rapidly than iron, while iron oxidises more rapidly upon aeration. Characteristic colour patterns are created by these processes. The reduced iron and manganese ions may be removed from a soil if vertical or lateral fluxes of water occur, in which case there is no iron or manganese precipitation in that soil. Wherever the iron and manganese are oxidised and precipitated, they form either soft masses or hard concretions or nodules. Movement of iron and manganese as a result of redox processes in a soil may result in redoximorphic features.

That are defined as follows :

- *Redox concentrations :* These are zones of apparent accumulation of Fe-Mn oxides, *including :*
  - o Nodules and concretions, which are cemented bodies that can be removed from the soil intact. Concretions are distinguished from nodules on the basis of internal organisation. A concretion typically has concentric layers that are visible to the naked eye. Nodules do not have visible organised internal structure. Boundaries commonly are diffuse if formed *in situ* and sharp after pedoturbation. Sharp boundaries may be relict features in some soils; and
  - o Masses, which are non-cemented concentrations of substances within the soil matrix; and
  - o Pore linings, *i.e.,* zones of accumulation along pores that may be either coatings on pore surfaces or impregnations from the matrix adjacent to the pores.
- *Redox depletions :* These are zones of low chroma (chromas less than those in the matrix) where either Fe-Mn oxides alone or both Fe-Mn oxides and clay have been stripped out, *including :*
  - o Iron depletions, *i.e.,* zones that contain low amounts of Fe and Mn oxides but have a clay content similar to that of the adjacent matrix (often referred to as albans or neo-albans); and
  - o Clay depletions, *i.e.,* zones that contain low amounts of Fe, Mn, and clay (often referred to as silt coatings or skeletans).
- *Reduced matrix :* This is a soil matrix that has low chroma in situ but undergoes a change in hue or chroma within 30 minutes after the soil material has been exposed to air.
- In soils that have no visible redoximorphic features, a reaction to an alpha, alpha-dipyridyl solution satisfies the requirement for redoximorphic features.

Field experience indicates that it is not possible to define a specific set of redoximorphic features that is uniquely characteristic of all of the taxa in one particular category. Therefore, colour patterns that are unique to specific taxa are referenced in the keys. Anthraquic conditions are a variant of episaturation and are associated with controlled flooding (for such crops as wetland rice and cranberries), which causes reduction processes in the saturated, puddled surface soil and oxidation of reduced and mobilised iron and manganese in the unsaturated subsoil.

Cryoturbation

Cryoturbation (frost churning) is the mixing of the soil matrix within the pedon that results in irregular or broken horizons, involutions, accumulation of organic matter on the permafrost table, oriented rock fragments, and silt caps on rock fragments.

Densic Contact

A densic contact (L. densus, thick) is a contact between soil and densic materials. It has no cracks, or the spacing of cracks that roots can enter is 10 cm or more.

Densic Materials

Densic materials are relatively unaltered materials (do not meet the requirements for any other named diagnostic horizons or any other diagnostic soil characteristic) that have a non-cemented rupture-resistance class. The bulk density or the organisation is such that roots cannot enter, except in cracks. These are mostly earthy materials, such as till, volcanic mudflows, and some mechanically compacted materials, for example, mine spoils. Some non-cemented rocks can be densic materials if they are dense or resistant enough to keep roots from entering, except in cracks.

Densic materials are non-cemented and thus differ from paralithic materials and the material below a lithic contact, both of which are cemented. Densic materials have, at their upper boundary, a densic contact if they have no cracks or if the spacing of cracks that roots can enter is 10 cm or more. These materials can be used to differentiate soil series if the materials are within the series control section.

Gelic Materials

Gelic materials are mineral or organic soil materials that show evidence of cryoturbation (frost churning) and/or ice segregation in the active layer (seasonal thaw layer) and/or the upper part of the permafrost. Cryoturbation is manifested by irregular and broken horizons, involutions, accumulation of organic matter on top of and within the permafrost, oriented rock fragments, and silt-enriched layers. The characteristic structures associated with gelic materials include platy, blocky, or granular macrostructures; the structural results of sorting; and orbiculic, conglomeric, banded, or vesicular microfabrics. Ice segregation is manifested by ice lenses, vein ice, segregated ice crystals, and ice wedges. Cryopedogenic processes that lead to gelic materials are driven by the physical volume change of water to ice, moisture migration along a thermal gradient in the frozen system, or thermal contraction of the frozen material by continued rapid cooling.

Glacic Layer

A glacic layer is massive ice or ground ice in the form of ice lenses or wedges. The layer is 30 cm or more thick and contains 75 per cent or more visible ice.

Lithic Contact

A lithic contact is the boundary between soil and a coherent underlying material. Except in Ruptic-Lithic sub-groups, the underlying material must be virtually continuous within the limits of a pedon. Cracks that can be penetrated by roots are few, and their horizontal spacing is 10 cm or more. The underlying material must be sufficiently coherent when moist to make hand-digging with a spade impractical, although the material may be chipped or scraped with a spade. The material below a lithic contact must be in a strongly cemented or more cemented rupture-resistance class. Commonly, the material is indurated. The underlying material considered here does not include diagnostic soil horizons, such as a duripan or a petrocalcic horizon. A lithic contact is diagnostic at the sub-group level if it is within 125 cm of the mineral soil surface in Oxisols and within 50 cm of the mineral soil surface in all other mineral soils. In organic soils the lithic contact must be within the control section to be recognised at the sub-group level.

Paralithic Contact

A paralithic (lithiclike) contact is a contact between soil and paralithic materials where the paralithic materials have no cracks or the spacing of cracks that roots can enter is 10 cm or more.

Paralithic Materials

Paralithic materials are relatively unaltered materials (do not meet the requirements for any other named diagnostic horizons or any other diagnostic soil characteristic) that have an extremely weakly cemented to moderately cemented rupture-resistance class. Cementation, bulk density, and the organisation are such that roots cannot enter, except in cracks. Paralithic materials have, at their upper boundary, a paralithic contact if they have no cracks or if the spacing of cracks that roots can enter is 10 cm or more. Commonly, these materials are partially weathered bedrock or weakly consolidated bedrock, such as sandstone, siltstone, or shale. Paralithic materials can be used to differentiate soil series if the materials are within the series control section. Fragments of paralithic materials 2.0 mm or more in diameter are referred to as pararock fragments.

Permafrost

Permafrost is defined as a thermal condition in which a material (including soil material) remains below 0 °C for 2 or more years in succession. Those gelic materials having permafrost contain the unfrozen soil solution that drives cryopedogenic processes. Permafrost may be cemented by ice or, in the case of insufficient interstitial water, may be dry. The frozen layer has a variety of ice lenses, vein ice, segregated ice crystals, and ice wedges. The permafrost table is in dynamic equilibrium with the environment.

Soil Moisture Regimes

The term "soil moisture regime" refers to the presence or absence either of ground water or of water held at a tension of less than 1500 kPa in the soil or in specific horizons during periods of the year. Water held at a tension of 1500 kPa or more is not available to keep most mesophytic plants alive. The availability of water is also affected by dissolved salts. If a soil is saturated with water that is too salty to be available to most plants, it is considered salty rather than dry. Consequently, a horizon is considered dry when the moisture tension is 1500 kPa or more and is considered moist if water is held at a tension of less than 1500 kPa but more than zero. A soil may be continuously moist in some or all horizons either throughout the year or for some part of the year. It may be either moist in winter and dry in summer or the reverse. In the Northern Hemisphere, summer refers to June, July, and August and winter refers to December, January, and February.

Normal Years

In the discussions that follow and throughout the keys, the term "normal years" is used. A normal year is defined as a year that has plus or minus one standard deviation of the long-term mean annual precipitation. (Long-term refers to 30 years or more.) Also, the mean monthly precipitation during a normal year must be plus or minus one standard deviation of the long-term monthly precipitation for 8 of the 12 months. For the most part, normal years can be calculated from the mean annual precipitation. When catastrophic events occur during a year, however, the standard deviations of the monthly means should also be calculated. The term "normal years" replaces the terms "most years" and "6 out of 10 years," which were used in the 1975 edition of Soil Taxonomy.

Soil Moisture Control Section

The intent in defining the soil moisture control section is to facilitate estimation of soil moisture regimes from climatic data. The upper boundary of this control section is the depth to which a dry (tension of more than 1500 kPa, but not air-dry) soil will be moistened by 2.5 cm of water within 24 hours. The lower boundary is the depth to which a dry soil will be moistened by 7.5 cm of water within 48 hours. These depths do not include the depth of moistening along any cracks or animal burrows that are open to the surface.

If 7.5 cm of water moistens the soil to a densic, lithic, paralithic, or petroferric contact or to a petrocalcic or petrogypsic horizon or a duripan, the contact or the upper boundary of the cemented horizon constitutes the lower boundary of the soil moisture control section. If a soil is moistened to one of these contacts or horizons by 2.5 cm of water, the soil moisture control section is the boundary or the contact itself. The control section of such a soil is considered moist if the contact or upper boundary of the cemented horizon has a thin film of water. If that upper boundary is dry, the control section is considered dry.

The moisture control section of a soil extends approximately.

- From 10 to 30 cm below the soil surface if the particle-size class of the soil is fine-loamy, coarse-silty, fine-silty, or clayey;
- From 20 to 60 cm if the particle-size class is coarse-loamy; and
- From 30 to 90 cm if the particle-size class is sandy.

If the soil contains rock and pararock fragments that do not absorb and release water, the limits of the moisture control section are deeper. The limits of the soil moisture control section are affected not only by the particle-size class but also by differences in soil structure or pore-size distribution or by other factors that influence the movement and retention of water in the soil.

Classes of Soil Moisture Regimes

The soil moisture regimes are defined in terms of the level of ground water and in terms of the seasonal presence or absence of water held at a tension of less than 1500 kPa in the moisture control section. It is assumed in the definitions that the soil supports whatever vegetation it is capable of supporting, *i.e.*, crops, grass, or native vegetation, and that the amount of stored moisture is not being increased by irrigation or fallowing. These cultural practices affect the soil moisture conditions as long as they are continued.

Aquic moisture regime : The aquic moisture regime is a reducing regime in a soil that is virtually free of dissolved oxygen because it is saturated by water. Some soils are saturated with water at times while dissolved oxygen is present, either because the water is moving or because the environment is unfavourable for micro-organisms (*e.g.*, if the temperature is less than 1 °C); such a regime is not considered aquic.

It is not known how long a soil must be saturated before it is said to have an aquic moisture regime, but the duration must be at least a few days, because it is implicit in the concept that dissolved oxygen is virtually absent. Because dissolved oxygen is removed from ground water by respiration of micro-organisms, roots, and soil fauna, it is also implicit in the concept that the soil temperature is above biologic zero for some time while the soil is saturated. Biologic zero is defined as 5 °C in this taxonomy. In some of the very cold regions of the world, however, biological activity occurs at temperatures below 5 °C.

Very commonly, the level of ground water fluctuates with the seasons; it is highest in the rainy season or in fall, winter, or spring if cold weather virtually stops evapotranspiration. There are soils, however, in which the ground water is always at or very close to the surface. Examples are soils in tidal marshes or in closed, landlocked depressions fed by perennial streams. Such soils are considered to have a peraquic moisture regime.

Aridic and torric (L. aridus, dry, and L. torridus, hot and dry) moisture regimes. These terms are used for the same moisture regime but in different categories of the taxonomy.

In the aridic (torric) moisture regime, the moisture control section is, in normal years :

- Dry in all parts for more than half of the cumulative days per year when the soil temperature at a depth of 50 cm from the soil surface is above 5 °C; and

- Moist in some or all parts for less than 90 consecutive days when the soil temperature at a depth of 50 cm is above 8 °C.

Soils that have an aridic (torric) moisture regime normally occur in areas of arid climates. A few are in areas of semi-arid climates and either have physical properties that keep them dry, such as a crusty surface that virtually precludes the infiltration of water, or are on steep slopes where run-off is high. There is little or no leaching in this moisture regime, and soluble salts accumulate in the soils if there is a source. The limits set for soil temperature exclude from these moisture regimes soils in the very cold and dry polar regions and in areas at high elevations. Such soils are considered to have anhydrous conditions (defined earlier).

Udic moisture regime : The udic moisture regime is one in which the soil moisture control section is not dry in any part for as long as 90 cumulative days in normal years. If the mean annual soil temperature is lower than 22 °C and if the mean winter and mean summer soil temperatures at a depth of 50 cm from the soil surface differ by 6 °C or more, the soil moisture control section, in normal years, is dry in all parts for less than 45 consecutive days in the 4 months following the summer solstice. In addition, the udic moisture regime requires, except for short periods, a three-phase system, solid-liquid-gas, in part or all of the soil moisture control section when the soil temperature is above 5 °C.

The udic moisture regime is common to the soils of humid climates that have well distributed rainfall; have enough rain in summer so that the amount of stored moisture plus rainfall is approximately equal to, or exceeds, the amount of evapotranspiration; or have adequate winter rains to recharge the soils and cool, foggy summers, as in coastal areas. Water moves downward through the soils at some time in normal years.

In climates where precipitation exceeds evapotranspiration in all months of normal years, the moisture tension rarely reaches 100 kPa in the soil moisture control section, although there are occasional brief periods when some stored moisture is used. The water moves through the soil in all months when it is not frozen. Such an extremely wet moisture regime is called perudic. In the names of most taxa, the formative element "ud" is used to indicate either a udic or a perudic regime; the formative element "per" is used in selected taxa.

Ustic moisture regime : The ustic moisture regime is intermediate between the aridic regime and the udic regime. Its concept is one of moisture that is limited but is present at a time when conditions are suitable for plant growth. The concept of the ustic moisture regime is not applied to soils that have permafrost or a cryic soil temperature regime. If the mean annual soil temperature is 22 °C or higher or if the mean summer and winter soil temperatures differ by less than 6 °C at a depth of 50 cm below the soil surface, the soil moisture control section in areas of the ustic moisture regime is dry in some or all parts for 90 or more cumulative days in normal years. It is moist, however, in some part either for more than 180 cumulative days per year or for 90 or more consecutive days.

If the mean annual soil temperature is lower than 22 °C and if the mean summer and winter soil temperatures differ by 6 °C or more at a depth of 50 cm from the soil surface, the soil moisture control section in areas of the ustic moisture regime is dry in some or all parts for 90 or more cumulative days in normal years, but it is not dry in all parts for more than half of the cumulative days when the soil temperature at a depth of 50 cm is higher than 5 °C. If in normal years the moisture control section is moist in all parts for 45 or more consecutive days in the 4 months following the winter solstice, the moisture control section is dry in all parts for less than 45 consecutive days in the 4 months following the summer solstice. In tropical and sub-tropical regions that have a monsoon climate with either one or two dry seasons, summer and winter seasons have little meaning. In those regions the moisture regime is ustic if there is at least one rainy season of 3 months or more. In temperate regions of sub-humid or semi-arid climates, the rainy seasons are usually spring and summer or spring and fall, but never winter. Native plants are mostly annuals or plants that have a dormant period while the soil is dry.

Xeric moisture regime : The xeric moisture regime is the typical moisture regime in areas of Mediterranean climates, where winters are moist and cool and summers are warm and dry. The moisture, which falls during the winter, when potential evapotranspiration is at a minimum, is particularly effective for leaching. In areas of a xeric moisture regime, the soil moisture control section, in normal years, is dry in all parts for 45 or more consecutive days in the 4 months following the summer solstice and moist in all parts for 45 or more consecutive days in the 4 months following the winter solstice. Also, in normal years, the moisture control section is moist in some part for more than half of the cumulative days per year when the soil temperature at a depth of 50 cm from the soil surface is higher than 6 °C or for 90 or more consecutive days when the soil temperature at a depth of 50 cm is higher than 8 °C. The mean annual soil temperature is lower than 22 °C, and the mean summer and mean winter soil temperatures differ by 6 °C or more either at a depth of 50 cm from the soil surface or at a densic, lithic, or paralithic contact if shallower.

Soil Temperature Regimes

Classes of Soil Temperature Regimes

Following is a description of the soil temperature regimes used in defining classes at various categoric levels in this taxonomy.

Cryic (Gr. kryos, coldness; meaning very cold soils) : Soils in this temperature regime have a mean annual temperature lower than 8 °C but do not have permafrost.

- In mineral soils the mean summer soil temperature (June, July, and August in the Northern Hemisphere and December, January, and February in the Southern Hemisphere) either at a depth of 50 cm from the soil surface or at a densic, lithic, or paralithic contact, whichever is shallower, *is as follows* :
  - o If the soil is not saturated with water during some part of the summer and
  - o *If there is no O horizon* : lower than 15 °C; or
  - o *If there is an O horizon* : lower than 8 °C; or
  - o If the soil is saturated with water during some part of the summer and
  - o *If there is no O horizon* : lower than 13 °C; or
  - o *If there is an O horizon or a histic epipedon* : lower than 6 °C.
- In organic soils the mean annual soil temperature is lower than 6 °C. Cryic soils that have an aquic moisture regime commonly are churned by frost.

Isofrigid soils could also have a cryic temperature regime. A few with organic materials in the upper part are exceptions. The concepts of the soil temperature regimes described below are used in defining classes of soils in the low categories.

- *Frigid* : A soil with a frigid temperature regime is warmer in summer than a soil with a cryic regime, but its mean annual temperature is lower than 8 °C and the difference between mean summer (June, July, and August) and mean winter (December, January, and February) soil temperatures is more than 6 °C either at a depth of 50 cm from the soil surface or at a densic, lithic, or paralithic contact, whichever is shallower.

- *Mesic* : The mean annual soil temperature is 8 °C or higher but lower than 15 °C, and the difference between mean summer and mean winter soil temperatures is more than 6 °C either at a depth of 50 cm from the soil surface or at a densic, lithic, or paralithic contact, whichever is shallower.

- *Thermic* : The mean annual soil temperature is 15 °C or higher but lower than 22 °C, and the difference between mean summer and mean winter soil temperatures is more than 6 °C either at a depth of 50 cm from the soil surface or at a densic, lithic, or paralithic contact, whichever is shallower.

- *Hyperthermic* : The mean annual soil temperature is 22 °C or higher, and the difference between mean summer and mean winter soil temperatures is more than 6 °C either at a depth of 50 cm from the soil surface or at a densic, lithic, or paralithic contact, whichever is shallower.

If the name of a soil temperature regime has the prefix iso, the mean summer and mean winter soil temperatures differ by less than 6 °C at a depth of 50 cm or at a densic, lithic, or paralithic contact, whichever is shallower.

- *Isofrigid* : The mean annual soil temperature is lower than 8 °C.
- *Isomesic* : The mean annual soil temperature is 8 °C or higher but lower than 15 °C.

- *Isothermic* : The mean annual soil temperature is 15 °C or higher but lower than 22 °C.

- *Isohyperthermic* : The mean annual soil temperature is 22 °C or higher.

Sulfidic Materials

Sulfidic materials contain oxidisable sulfur compounds. They are mineral or organic soil materials that have a pH value of more than 3.5 and that, if incubated as a layer 1 cm thick under moist aerobic conditions (field capacity) at room temperature, show a drop in pH of 0.5 or more units to a pH value of 4.0 or less (1 : 1 by weight in water or in a minimum of water to permit measurement) within 8 weeks.

Sulfidic materials accumulate as a soil or sediment that is permanently saturated, generally with brackish water. The sulfates in the water are biologically reduced to sulfides as the materials accumulate. Sulfidic materials most commonly accumulate in coastal marshes near the mouth of rivers that carry non-calcareous sediments, but they may occur in freshwater marshes if there is sulfur in the water. Upland sulfidic materials may have accumulated in a similar manner in the geologic past. If a soil containing sulfidic materials is drained or if sulfidic materials are otherwise exposed to aerobic conditions, the sulfides oxidise and form sulfuric acid. The pH value, which normally is near neutrality before drainage or exposure, may drop below 3. The acid may induce the formation of iron and aluminum sulfates. The iron sulfate, jarosite, may segregate, forming the yellow redoximorphic concentrations that commonly characterise a sulfuric horizon.

The transition from sulfidic materials to a sulfuric horizon normally requires very few years and may occur within a few weeks. A sample of sulfidic materials, if air-dried slowly in shade for about 2 months with occasional remoistening, becomes extremely acid.

Chapter 4

ROCK, MINERAL AND WEATHERING OF SOILS

WEATHERING OF ROCKS AND MINERALS

Rocks and minerals are formed under a very high temperature and pressure, exposed to atmospheric conditions of low pressure and low temperature and they become unstable and weather. Soils are formed from rocks through the intermediate stage of formation of Regolith which is the resultant of weathering.

The sequence of processes in the formation of soils is :

- Weathering of rocks and minerals
- Formation of regolith or parent material
- Formation of true soil from regolith Rock
- Weathering
- Regolith
- Soil forming factors and processes
- True soil (otherwise)

Two processes involved in the formation of soil are :

1. Formation of regolith by breaking down (weathering) of the bed rock.
2. The addition of organic matter through the decomposition of plant and animal tissues, and re-organisation of these components by soil forming processes to form soil.

Weathering : A process of disintegration and decomposition of rocks and minerals which are brought about by physical agents and chemical processes, leading to the formation of Regolith (unconsolidated residues of the weathering rock on the earth's surface or above the solid rocks).

(OR)

The process by which the earth's crust or lithosphere is broken down by the activities of the atmosphere, with the aid of the hydrosphere and biosphere

(OR)

The process of transformation of solid rocks into parent material or Regolith

Parent material : It is the regolith or at least its upper portion. May be defined as the unconsolidated and more or less chemically weathered mineral materials from which soil are developed

Two basic processes of Weathering :

Physical (or) mechanical - disintegration

Chemical – decomposition

In addition, another process : Biological and all these processes are work hand in hand

Depending up on the agents taking part in weathering processes, it is classified into three types.

Diffrerent agents of weathering

Different agents of weathering

Physical/ Mechanical (disintegration)	Chemical (decomposition)	Biological (disint + decomp)
1. Physical condition of rock	1. Hydration	1. Man & animals
2. Change in temperature	2. Hydrolysis	2. higher plants & their roots
3. Action of H20	3. Solution	3 .Micro organisms
-fragment & transport	4. Carbonation	
- action of freezing	5. Oxidation	
- alter. Wet & drying	6. Reduction	
- action of glaciers		
4. Action ofwind		
5. Atmosp. electric pheno		

PHYSICAL WEATHERING OF ROCKS

The rocks are disintegrated and are broken down to comparatively smaller pieces, with out producing any new substances :

• *Physical condition of rocks :* The permeability of rocks is the most important single factor.

    – Coarse textured (porous) sand stone weather more readily than a fine textured (almost solid) basalt.

    – Unconsolidated volcanic ash weather quickly as compared to unconsolidated coarse deposits such as gravels.

- *Action of Temperature* : The variations in temperature exert great influence on the disintegration of rocks.
    - During day time, the rocks get heated up by the sun and expand. At night, the temperature falls and the rocks get cooled and contract.
    - This alternate expansion and contraction weakens the surface of the rock and crumbles it because the rocks do not conduct heat easily.
    - The minerals with in the rock also vary in their rate of expansion and contraction.
    - The cubical expansion of quartz is twice as feldspar.
    - Dark coloured rocks are subjected to fast changes in temperature as compared to light coloured rocks.
    - The differential expansion of minerals in a rock surface generates stress between the heated surface and cooled unexpanded parts resulting in fragmentation of rocks.
    - This process causes the surface layer to peel off from the parent mass and the rock ultimately disintegrates. This process is called Exfoliation.
- *Action of Water* : Water acts as a disintegrating, transporting and depositing agent.
    - *Fragmentation and transport :* Water beats over the surface of the rock when the rain occurs and starts flowing towards the ocean :
        a. Moving water has the great cutting and carrying force.
        b. It forms gullies and ravines and carries with the suspended soil material of variable sizes.
        c. Transporting power of water varies. It is estimated that the transporting power of stream varies as the sixth power of its velocity *i.e* the greater the speed of water, more is the transporting power and carrying capacity.

Speed/Sec	Carrying Capacity
15 cm	Fine sand
30 cm	Gravel
1.2 m	Stones (1kg)
9.0 m	Boulders (several tons)

The disintegration is greater near the source of river than its mouth :
- *Action of freesing :* Frost is much more effective than heat in producing physical weathering :
    a. In cold regions, the water in the cracks and crevices freezes into ice and the volume increases to one tenth.
    b. As the freesing starts from the top there is no possibility of its upward expansion. Hence, the increase in volume creates enormous out ward pressure which breaks apart the rocks.

- *Alternate wetting and Drying* : Some natural substances increase considerably in volume on wetting and shrink on drying.(*e.g.*) smectite, montmorilonite :

    a. During dry summer/dry weather – these clays shrink considerably forming deep cracks or wide cracks.

    b. On subsequent wetting, it swells.

    c. This alternate swelling and shrinking/wetting or drying of clay enriched rocks make them loose and eventually breaks.

- *Action of glaciers* :

    a. In cold regions, when snow falls, it accumulates and changes into ice sheet.

    b. These big glaciers start moving owing to the change in temperature and/or gradient.

    c. On moving, these exert tremendous pressure over the rock on which they pass and carry the loose materials.

    d. These materials get deposited on reaching the warmer regions, where its movement stops with the melting of ice.

- *Action of wind* :

    - Wind has an erosive and transporting effect. Often when the wind is laden with fine material *viz.*, fine sand, silt or clay particles, it has a serious abrasive effect and the sand laden winds itch the rocks and ultimately breaks down under its force

    - The dust storm may transport tons of material from one place to another. The shifting of soil causes serious wind erosion problem and may render cultivated land as degraded (*e.g.*) Rajasthan deserts

- Atmospheric electrical phenomenon : It is an important factor causing break down during rainy season and lightning breaks up rocks and or widens cracks.

CHEMICAL WEATHERING OF SILICATES

The most important silicates are quartz, feldspar and certain Ferro magnesium minerals.

Weathering Products of Common Silicate Minerals			
Minerals	**Composition**	**Decomposition Products**	
		Minerals	Others
Olivine	(Fe, Mg)2SiO2	serpentine, limonitehaematite, quartz	Some Si in solution, carbonates of Fe and Mg
Pyroxenes	Fe, Mg	Clay, calcite, limonite	Some Si in solution, carbonates of Ca and Mg
Amphibole	Ca- silicates	Haematite, quartz	-do -
Biotite	Al		-do -
Plagioclase	Calcic	Clay, quartz, calcite	Some Si in solution, Na and Ca carbonates
	Sodic		-do-
Orthoclase	Potassic	Clay, quartz	Some Si in solution, potassium carbonate
Quartz		Quartz grains	Some Si in solution

BIOLOGICAL WEATHERING OF ROCKS

Unlike physical and chemical weathering, the biological or living agents are responsible for both decomposition and disintegration of rocks and minerals. The biological life is mainly controlled largely by the prevailing environment.

- *Man and Animals :*
  - The action of man in disintegration of rocks is well known as he cuts rocks to build dams, channels and construct roads and buildings. All these activities result in increasing the surface area of the rocks for attack of chemical agents and accelerate the process of rock decomposition.
  - A large number of animals, birds, insects and worms, by their activities they make holes in them and thus aids for weathering.
  - In tropical and sub-tropical regions, ants and termites build galleries and passages and carry materials from lower to upper surface and excrete acids. The oxygen and water with many dissolved substances, reach every part of the rock through the cracks, holes and galleries, and thus brings about speedy disintegration.
  - Rabbits, by burrowing in to the ground, destroy soft rocks. Moles, ants and bodies of the dead animals, provides substances which react with minerals and aid in decaying process.
  - The earthworms pass the soil through the alimentary canal and thus bring about physical and chemical changes in soil material.
- *Higher Plants and Roots :* The roots of trees and other plants penetrates into the joints and crevices of the rocks. As they grew, they exert a great disruptive force and the hard rock may break apart. (*e.g.*) pipal tree growing on walls/ rocks

The grass root form a sponge like mass prevents erosion and conserve moisture and thus allowing moisture and air to enter in to the rock for further action.

Some roots penetrate deep into the soil and may open some sort of drainage channel. The roots running in crevices in lime stone and marble produces acids. These acids have a solvent action on carbonates.

The dead roots and plant residues decompose and produce carbon dioxide which is of great importance in weathering.

- *Micro- organisms :* In early stages of mineral decomposition and soil formation, the lower forms of plants and animals like, mosses, bacteria and fungi and actinomycetes play an important role. They extract nutrients from the rock and N from air and live with a small quantity of water. In due course of time, the soil develops under the cluster of these micro-organisms.

This organism closely associated with the decay of plant and animal remains and thus liberates nutrients for the use of next generation plants and also produces CO_2 and organic compounds which aid in mineral decomposition.

WEATHERING AND SURFACE PROCESSES

Types of surface processes :

- *Weathering :* Is the *in situ* disintegration of rocks at the surface of the earth.
- *Erosion :* Is the transportation of the products of weathering by some transporting agent (*e.g.* wind, water, ... etc.).
- *Mass movement :* Is the movement of rock material down-slope by gravity.

Weathering Area

Weathering in an area and the amount and nature of weathering products will depend on :

- The type of rocks exposed on the surface
- The nature and number of planes of weakness in these rocks
- The relief of the area
- The climate.

Weathering is of two kinds, mechanical and chemical. Mechanical weathering is the physical breakup of rocks into smaller pieces that have the same chemical composition as the original rock (*i.e.* disintegration without any changes in chemical composition). Chemical weathering is the breakdown of rocks into smaller particles through one or more chemical reactions that take place between minerals and weathering agents as water or solutions. Each of these two types involves a number of processes.

Mechanical weathering is greatly facilitated if the rock originally had one or more planes of weakness. These planes of weakness could be joints (particularly when they occur in sets or systems), bedding planes (in sedimentary rocks) or foliation planes (planes of preferred orientation in metamorphic rocks). By breaking up the rock into smaller particles, mechanical weathering will lead to an increase in the surface area of the total rocks exposed to surface conditions, and will therefore facilitate the process of chemical weathering. Joints are particularly important for mechanical weathering because (a) they effectively cut large blocks of rock into smaller ones thus increasing the surface area where chemical reaction can take place, and (b) they provide channel ways through which water can pass, thus increasing the chance of frost wedging or chemical weathering. Therefore, while mechanical weathering will always lead to an increase in the surface area of the rocks affected, weathering in general will slow down if the rocks have the smallest surface area. For a given volume, the smallest surface area is achieved if the rock is spherical in shape. Therefore, weathering will always try to change rocks with many edges into rounded or spherically shaped masses.

Types of Mechanical Weathering

If water percolates into a joint or fracture in a rock where it gets trapped, and the surface temperature then drops below freesing, this water will change to ice.

Because ice occupies a larger volume compared to water (freesing water involves an increase in volume of 9 per cent), freesing is accompanied by expansion which exerts a significant force on the rock, widening the fracture. When the temperature rises, the ice melts, and more water is allowed to occupy the same fracture. When the temperature drops again below freesing, the water is converted to ice with yet a larger volume, which again exerts more force on the rock. Repeated freesing and thawing will therefore eventually cause the rock to break along these fractures. Frost wedging is very important in mountains where at night the temperature falls below freesing, but during the day ice may melt, thus giving rise to a daily freesing thawing cycle.

The daily cycle of temperature change, particularly in dry hot climates (*e.g.* in deserts) where such change may be as much as 30°C, will cause the different minerals in a rock to expand (during the day) and contract (at night) at different rates. This repeated expansion and contraction is thought to weaken the rock substantially, and was believed by some scientists to cause jointing to develop. However, recent lab experiments have shown that this process may not be as important as once thought.

Sheet jointing or exfoliation is the process by which rocks break into thin sheets that are parallel to their external surfaces, which are usually rounded or domal in shape. Exfoliation probably results from the expansion of rocks as the confining pressure on them is released. This release takes place when the overlying rocks under which such rocks were buried are removed (possibly by weathering and erosion).

Living organisms have the ability of disintegrating rocks. Examples include the roots of trees, which during growth push rocks apart, or burrowing organisms which produce small "tunnel" like features in the rocks, or human beings who carry out excavations in search of precious metals, or to construct roads, dams, bridges or houses.

Chemical Weathering

Chemical weathering is the breakdown of rocks through the reaction of one or more of their constituent minerals with water, other chemicals dissolved in water, or atmospheric gases. Accordingly, the products of chemical weathering will have a chemical composition that is different from the original rock. Note that the process by which one mineral changes to another in the weathering environment is often known as alteration.

Chemical weathering takes place by one or more of the following processes :

Dissolution is the process whereby a mineral dissolves in a solvent as a result of the freeing up of its ions (*i.e.* transformation of the compound into free ions). The most important solvent in nature is water, and the minerals which dissolve most readily or easily are the halides, nitrates, carbonates and sulfates. Dissolution is therefore an important process in the weathering of limestones (made predominantly of calcite) and evaporites (which consist of halides and sulfates).

For example :

$$NaCl + H_2O = Na^+ + Cl^- \text{ (both in aqueous solution)}$$

Hydrolysis is the process in which a water molecule dissociates into the very reactive H^+ and $(OH)^-$ ions while attempting to "dissolve" a mineral or compound. The hydrogen ions may then replace other positive ions in the structure of the mineral, causing it to breakdown into other components, some of which may be soluble. An example of this type of chemical weathering is the hydrolysis of K-feldspars to give clay minerals :

$$2\ KAlSi_3O_8 + H_2O + 2\ H^+ = Al_2Si_2O_5(OH)_4 + 2\ K^+ + 4\ SiO_2$$
K-feldspar   kaolinite   aqueous   silica

Note that both the K^+ ions and the SiO_2 can thus be carried in the aqueous solution, leaving behind the clay mineral kaolinite ($Al_2Si_2O_5(OH)_4$). In nature, water usually contains other substances that contribute additional hydrogen ions or which cause the dissociation of H_2O to release H^+. The most common of these substances is atmospheric CO_2, which dissolves in H_2O to form H_2CO_3 which then dissociates into H^+ and HCO_3^- *according to the reaction* :

$$CO_2 + H_2O = H^+ + HCO_3^-$$

Weak organic acids in soils can also supply H^+ to meteoric water, making such water slightly acidic and facilitating hydrolysis. Hydrolysis is the main process by which silicates are chemically weathered.

In terms of weathering, oxidation is the process of combination of atmospheric oxygen (or oxygen dissolved in meteoric water) with a mineral to form one or more minerals that are more "oxidised" and hence more stable in the weathering environment. Examples of oxidation include the oxidation of the Fe-end-member of olivine (fayalite) to produce Fe-oxides *according to the reaction* :

$$Fe_2SiO_4 + 1/2\ O_2 + H_2O = Fe_2O_3 + H_4SiO_4$$
Fayalite   Hematite   (in solution)

Similarly, a clinopyroxene upon oxidation gives Fe-oxides, and Ca is released in solution according to the reaction :

$$CaFeSi_2O_6 + 1/2\ O_2 + 10\ H_2O + 4\ CO_2 = Fe_2O_3 + 4\ H_4SiO_4 + 2\ Ca^{+2} + 4\ HCO_3^-$$

Clinopyroxene   Hematite

The Fe_2O_3 produced by these reactions is relatively insoluble, and precipitates to form hematite, which is red in colour. This gives rise to the characteristic reddish or rust-like colours on the weathered surfaces of Fe-rich rocks (*e.g.* basalts). Note that oxidation of Fe in a silicate will take place only after this Fe is released from the silicate structure by hydrolysis. From the above discussion, it can be seen that chemical weathering requires a solvent. Because almost all chemical reactions become faster with increasing temperature, chemical weathering will take place more easily in warm wet climates, as those which occur in the tropics. The rate of chemical weathering will also increase if the rain water becomes more acidic (which will facilitate hydrolysis). Therefore, in areas close to large industrial complexes that release large amounts of CO_2 or SO_2 in the atmosphere, rain water may become more acidic (acid rain), which increases the intensity of chemical weathering.

A quick examination of the 3 different processes involved in chemical weathering shows that 2 of these processes (hydrolysis and oxidation) usually produce minerals that have a higher volume : weight ratio than the original minerals of the unweathered rock. This will result in an overall increase in the volume of the rock due to weathering, causing the rock to "expand". Such an expansion will be accompanied by the fracturing of the rock providing a larger surface area for more chemical weathering. Eventually, the sharp edges will be rounded, and the rock acquires the shape of spherical masses that are characterised by the smallest surface area for a given volume.

This type of weathering is known as spheroidal weathering. Once a spherical shape has been attained, agents of chemical weathering continue to act uniformly on the entire surface of the spherical body, peeling off layers in a process termed "chemical exfoliation". Although the final appearance of the rock may be to some extent similar to one that is "exfoliated", the difference between exfoliation and spheroidal weathering lies in the process : the former is a product of mechanical weathering, whereas the latter results from chemical weathering.

The final products of chemical weathering are also a function of the original composition of the weathered rock. A rock consisting predominantly of the minerals calcite or halite will weather more easily than one consisting of quartz and other silicates. Therefore, if we had different rock types with different resistances to weathering interlayered in an area, these rocks would weather differently : the hardest units which are most resistant to weathering will appear least affected by this process, whereas the softer units will be more strongly weathered. This process is known as differential weathering, and results in a variety of characteristic geomorphological features, some of which will be described in detail in forthcoming chapters. In order to fully understand differential weathering and erosion, we need to examine the stability of the different minerals in the weathering environment, and in particular silicates, as these are the most abundant rock-forming minerals.

The most stable minerals during weathering are the oxides of Fe and Al. Next to these is quartz, which is the most resistant among silicates to weathering. This is followed by micas and alkali feldspars, then amphiboles, followed by pyroxenes and calcic feldspars, and finally olivines. This sequence seems to be the exact opposite of the order of crystallisation of silicates from a melt according to the Bowen reaction series. This relationship is not surprising; the minerals which form at the highest pressures and temperatures (*e.g.* olivines) are the ones least stable on the surface (where P and T are much lower), and will be more likely to alter to other more stable minerals at the surface than a mineral which formed at lower P and T.

Talus : Blocks and rock fragments which accumulate at the base of a cliff or a steep slope and which result from the weathering (mainly mechanical) of rocks on top of that cliff. Talus therefore either forms in place or represents fragments transported over very short distances by gravity.

Regolith : A general name given to a blanket of loose rock fragments forming a discontinuous cover to the solid bedrock. Regolith is therefore a term used for material that has not been transported.

Debris or detritus : Are general names given to rock fragments produced by weathering (regardless of their size or distance of transport).

Soil : A blanket of loose material that includes mineral grains and organic material (known as humus), and which can, at least to some extent, sustain plant life. Soils are highly porous (with as much as 50 per cent pore space about half of which could be filled with water). Mineral grains in many "good quality soils" are predominated by clay minerals, but this really depends on the climate, nature of weathered bed-rock, topography, ... etc. Organic matter (~ 5 per cent of most soils) in a soil is supplied by the decomposition of dead animals and plants.

Since soil - forming processes operate from the surface downward, there is a gradual variation in composition, texture, structure, and colour of soil with depth. These vertical differences, which normally become more pronounced with time, allow for the identification of several "layers" or "horizons".

A horizon : darkest in colour, containing the largest amount of organic matter or humus. As water percolates downward through the A-horizon, it dissolves minerals, leaving this horizon depleted in soluble material. The A-horizon is therefore known as the zone of leaching. Minerals left behind in this horizon are therefore mostly clays and relatively insoluble minerals like quartz.

B horizon : is not as rich in organic matter as the A-horizon, but is much richer in soluble material, which is deposited by the same solutions that leached horizon A as they percolated downward, or by solutions rising upward by capillary action as a result of evaporation on the surface (in horizon A). The B-horizon is therefore known as the zone of accumulation.

C horizon : consists of slightly altered but fragmented bedrock, mixed with a small amount of clay.

D horizon : is the zone of unaltered bedrock.

These four horizons are not always well developed in all soils. Where these zones are not well developed, the soils are generally described as immature, reflecting the short time over which they have formed. Such is the case in areas of steep slopes where the soil formed is continually being eroded.

The factors which affect the amount and composition (or type) of a soil include :

- *Type of bedrock being weathered :* A rock consisting of only quartz (a quartzite or a mature sandstone) is unlikely to produce soil, as the weathered material under all conditions of weathering will be predominated by quartz and/or the elements Si and O, which are insufficient to sustain plant life. Moreover, quartz is one of the minerals that are most resistant minerals to weathering, so the supply of debris will be slow; possibly slower than the rate at which such debris is eroded.

- *The climate :* In warm wet regions, where chemical weathering is predominant, soils may be very rich in Fe-oxides, and Al-oxides and hydroxides, the most stable minerals under conditions of intense chemical weathering. These soils are known as lateritic soils, and although the climate and availability of water may be suitable for plant life, such soils are not very useful for growing crops (as they, once more, do not contain the important elements Na, Ca and K, which have been carried away in solution (*i.e.* leached) since chemical weathering is so strong). On the other hand, in temperate zones that receive enough water from rainfall, but where the rates of chemical weathering are not excessive, good quality soils may form.

- *Slope and topography :* Even where the climate and bedrock are both suitable for the development of soil, the topography plays a major role in controlling where these soils will be found in this area. For example, soils will not develop on steep slopes, as whatever mineral fragments are produced by weathering, they are rapidly removed by erosion. The process of removal of soils from slopes is known as "sheet wash".

- *Time :* The longer the time for weathering under conditions suitable for the formation of soils, the better developed the soil will be (with well-marked horizons).

Rates of Weathering and Soil Formation

Rates of weathering will vary from one place to another, and also over time. Nevertheless, in broad general terms :

- The average rate of weathering is between 5 and 40 cm/1000 years.
- The rate of soil formation is < 6 cm/1000 years.

WEATHERING AND SOILS

Rock weathering can produce spectacular landscapes. In Monument Valley, Arizona differential rock weathering and erosion of sandstone formations results in stair-stepped topography and distinctive towers of rock. Weathering etches less resistant layers so that bedding becomes visible, and the pattern of jointing and overall rock strength varies with rock type (lithology). Talus, a blanket of material fallen from cliff walls, covers some of the lower slopes.

The thin layer of mechanically broken and chemically altered rock at the ground surface provides the substrate in which plants root, terrestrial life derives sustenance, and erosion acts to shape landscapes. This thin skin of the Earth houses the sphere of life — the biosphere. From a geomorphologist's perspective, erosional processes are much more effective at shaping landscapes once solid rock breaks down into transportable sediment and decomposes into unconsolidated secondary minerals or dissolved ions. Weathering is the chemical or physical alteration of parent material (rock or sediment), and the weathering products that mantle fresh rock (or sediment) are called soil. Weathering processes influence the physical and chemical properties of weathered rock and soil, which in turn influence geomorphological processes and landforms.

Soils form Earth's outer skin, the transition from its rocky interior to a gaseous atmosphere. Soils not only harbour and sustain life, they are themselves partly composed of organic matter. Plants and animals both depend on soils, and in turn influence rates and styles of soil formation. Weathering and soil formation processes create the thin layer of regolith (soil and weathered rock) that provides the foundation for terrestrial life. Physical, chemical, and biological processes all act as weathering agents pretty much everywhere, but their rates and relative importance vary dramatically among different landscapes around the world. But not all regolith was produced *in situ* — the glacial debris covering New England was imported by glaciers from Canada.

Weathering and erosion do not necessarily progress at the same pace. Where weathering outpaces erosion, landscapes develop thick mantles of saprolite, highly decomposed rock that lost mass but not volume during weathering. Saprolite can extend hundreds of meters deep in flat-lying, slowly eroding areas of tropical Africa and South America. It can erode like loose sand yet still retain structures or features of the original parent material, like igneous dikes or sedimentary bedding planes even though its density can be half that of intact rock due to mass loss. This chapter reviews the processes that act to weather rocks, sediment, and minerals at and near Earth's surface, and introduces the physical, chemical, and biological processes that produce soils. We will explore the dominant controls on the transformation of primary rock-forming minerals into secondary minerals and consider global and lithologic controls on weathering and the resulting influences on landforms. We will also address how soil scientists recognise diagnostic characteristics and classify soil types.

Physical Processes of Weathering

Physical processes mechanically break rocks into smaller pieces (disaggregation), and chemical processes alter their mineral composition (decomposition). During physical weathering, rocks and rock-forming minerals break apart without changes in composition. In contrast, chemical weathering involves reactions that change primary rock-forming minerals into secondary minerals, such as clays. In the process some elements are lost in solution to surface and groundwater.

The two types of weathering (physical and chemical) are related. The greater surface area that results from breaking rocks into smaller fragments accelerates chemical weathering, and changes in mass or volume accompanying chemical decomposition can promote physical weathering. So even though different environmental conditions tend to favour physical versus chemical weathering, the processes are complementary and often strongly influence each other. Biological activity catalyses both physical and chemical weathering, and depends on the active species, substrate, and weathering processes in question. Tree roots, for example, slowly pry rocks apart as they grow and the organic acids they secrete promote chemical weathering. And in many environments the soil is a nano-scale jungle of microbial activity that promotes weathering.

Physical Weathering

Physical weathering reduces parent material into smaller pieces, and in some cases even fragments of individual mineral grains. Zones of weakness in the original material, like cleavage or bedding planes, metamorphic foliation, and even mineral boundaries often determine the size and shape of rock fragments. Cracks form where stresses imparted by expansion, contraction or shearing exceed the strength of rocks or minerals. The most pronounced physical weathering in bedrock occurs where there is a strong directional contrast in pressure, as is the case at and near the land surface where overburden is minimal and open fractures are more abundant than at depth where confining pressures are greater and more uniform. Important general mechanisms of physical weathering include release of confining pressure that allows rock to expand and fracture, thermal expansion from heating by insolation or forest fires, and cyclic expansion and contraction from freeze/thaw action in cold environments. Wetting and drying is important in rocks with minerals susceptible to shrinking and swelling. And expansion by growth of salt crystals can crack rocks in arid environments.

Erosion of overlying material unloads rocks, allowing them to expand upward and crack as they are exhumed. Because vertical confining stresses lessen as the weight of overlying rock decreases, rocks that are brought up from depth by tectonic uplift and unburdened by erosion tend to break into thin, rind-like sheets that are oriented perpendicular to the direction of stress release (normal to the minimum principle stress). These exfoliation sheets are typically 1 to 10 m thick and generally follow the shape of the land surface, resulting in onion-like fracture patterns. Exfoliation sheets are most apparent on bare rock surfaces, but slope-parallel fractures also develop within soil-mantled slopes, creating networks of near-surface discontinuities that can greatly influence near-surface hydrology and slope stability. Because exfoliation sheets are more readily eroded than underlying unfractured rock, slope-parallel fractures tend to mimic topographic forms as subsequent sheets form and erode. The process of exfoliation commonly produces dome-shaped surfaces, like Yosemite Valley's famous granite Half Dome.

Exfoliation is typically better developed in igneous and metamorphic rocks that formed deep within Earth's crust than in sedimentary and volcanic rocks formed at shallow depths. Sedimentary rocks form within a few kilometers of Earth's surface. In contrast, igneous and metamorphic rocks usually form at depths exceeding ten kilometers, and are commensurately stronger due to higher temperatures and pressures of formation. As erosion exhumes once deeply buried rocks, bringing them closer to Earth's surface, it increases the contrast between the stresses locked in during crystallisation or metamorphism and the stress imposed by adjacent rock. When the difference becomes greater than the strength of the rock, it cracks and sets the stage for exfoliation. Even in sedimentary rocks, upward expansion from unloading separates bedding planes, allowing outward expansion where bedding tilt matches the topographic slope. Fracturing due to stress release is an important mechanism by which water, oxygen, and plant roots penetrate into a rock mass.

Rock masses subject to tectonic stresses or that have shrunk upon cooling develop patterns of joints, fractures along which no significant movement has occurred, but that provide planes of weakness along which weathering and erosion can penetrate and preferentially remove material. Because the tensile (extensional) strength of rock is typically many times less than its compressive or shear strength, jointing is often well developed even in relatively strong, erosion resistant rocks. Parallel sets of extensional joints develop orthogonal to the direction of maximum stress as a result of either crustal extension or cooling of igneous rock. Joint systems can have strong topographic expression in arid and semi-arid landscapes where bedrock properties dominate hillslope morphology. Joints provide avenues for water and plant roots to promote weathering that focus erosion and facilitate infiltration and groundwater flow into rock that, in turn, promote more aggressive weathering. Erosion along intersecting sets of joints can produce isolated columns of rock separated by weathered out joints. Joints also form parallel to the strike of bedding as a result of bending or folding of brittle rocks. Lithology greatly influences the degree of jointing, with joints typically being better developed in more brittle rocks such as sandstone and granite and less well developed in more flexible rocks like shale. In addition, joint patterns developed in igneous rocks tend to be less linear and more irregularly spaced than those developed in sedimentary rocks.

The expansion of water in fractures and pore spaces as it freezes into ice is particularly effective at breaking rocks apart. Consequently, frost shattering is a primary weathering process in alpine and polar environments that are subject to frequent freeze-thaw cycles. Water expands by almost ten per cent when it freezes, so rock generally will not shatter unless about 80 per cent of the available pore space is saturated. Ice simply expands into partially saturated voids without generating high pressures on the surrounding rock. However, hydrofracturing where freesing of water proceeds from the outside in forces water into the tiny ends of fractures, producing an effect like a hydraulic jack. In addition, as ice lenses form, water and vapor flow towards them, imparting enough force that growing ice crystals crack rocks. Many alpine slopes in environments subject to frost shattering are covered by a blanket of rock blocks called felsenmeer (German for sea of rocks). The rubble-covered summit of Mt. Washington in New England is a great example.

Rocks and minerals expand when subjected to heat, but low thermal conductivity generally prevents the zone of high heating from penetrating more than a few centimeters into rocks over the course of a forest fire or a day of exposure to intense sunlight. Low thermal conductivity explains why rock surfaces retain heat at night better than do vegetated surfaces and why stone buildings stay relatively cool in summer and warm in winter without the need for additional insulation. Extreme temperature differences between a rock's surface and its cool interior can produce large differential stresses that result in spalling of a thin outer layer that can be up to several centimeters thick.

The role of daily temperature fluctuations due to heating by sunlight has long been debated, but recent work has documented convincing evidence that daily heating and cooling can lead to rock fracture. For example, it has been shown that the majority of cracks in desert rocks that are not related to rock heterogeneities like bedding are oriented north-south. That the majority of vertical cracks in rocks in deserts around the world are preferentially oriented north-south suggests that thermal stresses produce north-south oriented cracks due to differential heating when the sun crosses the sky from east to west. The coefficient of thermal expansion of certain minerals such as calcite can cause a rock to break apart along mineral boundaries as individual grains expand and contract with small changes in temperature. The weathering product of this type of granular disintegration, known as grus, is essentially loose sand made of individual mineral grains.

The addition and removal of water from minerals, processes known as hydration and dehydration, can cause swelling or contraction capable of fracturing rocks or disaggregating them into individual mineral grains. The volume increase in the anhydrite to gypsum transformation, in particular, has been ruinous to archaeological monuments. Most micas and clay minerals exhibit some shrink when they dry and swell when they absorb water. Some can expand to twice their original volume, and certain particularly susceptible clays, like smectite (montmorillonite), swell several-fold when they get wet. Expansive soils that are composed of minerals with shrink/swell behaviour cause substantial engineering problems and extensive damage through cracked foundations annually in the United States, primarily in the South and the western plains wherever such soils are common. Swelling of bentonite-rich soils in Montana, Wyoming, and Colorado can make travel on dirt roads nearly impossible after rainstorms as sticky mud bogs down even 4-wheel drive vehicles.

In some rocks, like granite, expansion of biotite micas during weathering to hydratable vermiculite clay can pry apart grain boundaries, and produces grus. Similarly, cycles of wetting and drying can lead to repeated expansion and contraction that disaggregates micaceous sandstone and turns hard, erosion-resistant rock into a loose pile of sand in a single season. In rock types susceptible to this process, exposure to seasonal cycles of wetting and drying leads to rapid rates of bedrock erosion through the annual formation of a loose crust or outer covering of highly erodible material easily removed by subsequent high flows.
Hydration and expansion of salts such as gypsum or halite within pore spaces also cause spalling and rock disintegration. Salts can expand several fold when hydrated. Repeated wetting and drying, as well as the growth of salt crystals due to evaporation of fluids in near-surface fractures and pore spaces, can gradually pry rocks apart in arid climates. Salt weathering may even be an important mechanism of rock disintegration in the dry desert landscapes on Mars.

Chemical Weathering

Chemical weathering occurs because the minerals in rocks form at deep Earth pressure and temperature conditions that are not in equilibrium with conditions at Earth's surface, and are thus vulnerable to chemical decomposition and transformation. A primary weathering agent is rainwater that percolates into the ground and promotes chemical weathering because it contains dissolved ions gained from the atmosphere and from the soils through which it moves. The bipolar water molecule is a potent solvent — given time. The metabolism of soil micro-organisms and decay of organic matter enhance weathering as they add organic acids to water moving through soils. Root respiration and microbial oxidation make a soil atmosphere rich in CO_2. The addition of water makes carbonic acid. Sulfuric and nitric-acid weathering are important in some areas and always present at some level. In addition, bio-chemical activity tends to increase rates of chemical reactions between soil fluids and minerals by lowering pH and increasing temperature.

Processes of chemical weathering include solution, oxidation and reduction, hydrolysis, ion exchange, and the formation of new, more stable secondary minerals, like clays and hydrous oxides. The end results of chemical weathering depend on a variety of interacting factors including the composition and texture of the parent material and the chemical, physical, and bio-chemical processes acting in a particular environment. The mobility and stability of the secondary minerals and solutions produced depend on environmental conditions like pH, Eh, and temperature. Chemical weathering is the breaking of chemical bonds — metallic, ionic, and covalent. The corresponding principle weathering processes are electron exchange (oxidation/reduction), ionisation (solution), and ion exchange (as in acid attack). As many rocks are dominated by a mix of ionic and covalent bonds, solution and acid attack are major weathering processes. Hydration and dehydration are also important weathering mechanisms for certain rock types and in certain environments. Chemical weathering is essential for the biosphere in the critical zone, where vegetation demand for Ca, Mg, K, NO_3, and P is extraordinary. Nutrients derived from minerals are cycled through ecological systems because of the slow pace of weathering or the depleted nature of surficial materials (especially in key limiting elements such as phosphorus).

The general susceptibility of rock minerals to weathering is the inverse of the sequence in which they form deep within the earth. Rocks that formed at the greatest temperatures and pressures are furthest from equilibrium at surface conditions and are therefore most susceptible to weathering when exposed to the elements. Among the common silicate minerals that account for the majority of rock forming minerals, olivine and pyroxene are most susceptible to weathering, followed in order of decreasing susceptibility to breakdown by amphibole, biotite, feldspar, muscovite and quartz. This progression, known as Goldich's Weathering Series, is the opposite of the order of crystallisation as magma cools, familiar to geologists as Bowen's Reaction Series. Under similar environmental conditions, rocks composed of more mafic iron- and magnesium-rich minerals (olivine, pyroxene, amphibole, and biotite) will weather faster than those composed of more felsic minerals (feldspar, muscovite, and quartz). However, it is not always as simple as this. Due to the importance of both covalent and ionic bonds, silicates with complicated mineral structures break down more readily than do those with simpler structures like quartz (SiO_2) or zircon ($ZrSiO_4$), a very stable mineral even though it has a very high melting temperature. So both complexity and formation conditions are central to mineral stability.

The mobility of cations in rock-forming minerals varies greatly, and influences the relative ease and order in which weathering strips cations from rocks and secondary minerals, with the sequence from most to least mobile proceeding as Ca^{2+}, Na^+, Mg^{2+} > K^+ > Fe^{2+} > Si^{3+} > Fe^{3+} > Al^{3+}. The most mobile cations (Ca^{2+}, Na^+, Mg^{2+}) are readily stripped from mineral surfaces, tend to remain in solution, and are the first to be lost from rocks as they weather. The least mobile cations (Si^{3+}, Fe^{3+}, and Al^{3+}) are relatively insoluble and become concentrated in residual soils over time as weathering strips away more mobile elements.

Oxidation is a process during which an element loses an electron to a receptor, often an oxygen ion, like when iron rusts. Conversely, reduction is defined as the gain of an electron. Free oxygen is rare at crustal depths where rocks form, but abundant at Earth's surface. Rocks and rock-forming minerals typically oxidise when they are exposed to well-oxygenated soil water, directly to the atmosphere, or to gases in soil pores. Reducing conditions in oxygen-poor waters with lots of organic matter, like swamps and peat bogs with high seasonal water tables, generally prevent oxidation, retard organic decay, and slow down weathering. When soils alternate between saturated and unsaturated conditions, a speckled colour pattern known as mottling develops — with gray colours due to the lack of oxidised iron as well as reddish colours due to oxidation.

Oxidising potential is expressed in terms of redox potential (Eh), the availability of free oxygen, which is greatly influenced by the amount of dissolved organic matter in pore fluids. Soils typically have Eh high enough to oxidise most common elements, but iron, manganese and sulfur are especially prone to rapid oxidation and typically occur as red, black, and yellow coatings in soils. Redox potential exerts a substantial influence on ion mobility, and oxidation is often the first form of weathering to alter freshly exposed rock surfaces. Oxidised versions of elements are relatively immobile, whereas reduced versions are far more mobile. Over time, rinds of oxidised material form on the surfaces of outcrops or boulders as other material is leached away. When rocks containing common iron bearing carbonates, sulfides, and silicates (such as olivine and biotite) oxidise, they become susceptible to additional physical weathering. Oxidation produces relatively insoluble ferric oxides, like hematite (Fe_2O_3), or rust, and oxyhdroxides like goethite ($FeO(OH)$) that colour soils and weathered rock various shades of reddish or yellowish brown. By the same token, oxygen starved conditions (such as those under stagnant water rich in decomposing organic matter) turns iron and manganese back into reduced forms, which allows them to be dissolved and leached if the water drains or is flushed from the soil.

The flow rate and acidity of pore water are two of the most important factors influencing the amount of disolution from soil, sediment, or rock in a given weathering environment. In addition, pH strongly affects the solubility of most elements. Rainwater is slightly acidic from dissolved atmospheric CO_2, and chemical and biologic weathering processes often act to lower the pH of water moving through a weathering zone. In weathering zones with active groundwater circulation, fresh water comes in contact with parent material, and weathering continues as leaching removes dissolved material. Slowly circulating pore waters retard dissolution as the number of dissolved ions in solution approaches an equilibrium concentration.

Natural groundwater tends to be slightly acidic due to dissolution of carbon dioxide in water to produce carbonic acid ($H_2CO_3^-$). Carbonic acid is not a strong acid but it is extremely abundant because it forms wherever water encounters CO_2 through the carbonation reaction :

$$H_2O + CO_2 \longleftrightarrow 2H_2CO_3^-$$

Decay of organic matter together with respiration of soil invertebrates, bacteria, and root systems can elevate CO_2 concentrations in soil pores so that they are 10 to 100 times greater than atmospheric concentrations. This makes carbonation a particularly important factor in heavily vegetated areas. Cold temperatures also favour formation of carbonic acid in soil water, because the solubility of CO_2 is inversely proportional to temperature, as is true of most gases.

In solution, carbonic acid readily disassociates into hydrogen (H^+) and bicarbonate ions (HCO_3^-) :

$$2H_2CO_3^- \longleftrightarrow H^+ + HCO_3^-$$

Consequently, bicarbonate is the most common cation in natural groundwater.

Atoms exposed on the mineral surfaces of rock and soil particles are electrically charged ions that react with dissociated hydrogen (H^+) and hydroxide (OH^-) ions in water. This interaction breaks bonds, effectively disassociating individual mineral molecules and causing exchanges that release cations from the mineral surface into solution. Mineral structures become unstable and vulnerable to further weathering when they lose cations, so a little weathering promotes more weathering. Congruent dissolution occurs when all the constituents of an individual molecule are separated and remain in solution. During incongruent dissolution, some of the released ions recombine to create new compounds and secondary minerals. Dissolved material may remain in solution and move along with flowing water and re-precipitate elsewhere, or it may emerge into streams and rivers, and eventually reach the ocean. The origin of the salinity of the oceans lies in the long-term delivery of dissolved material in stream water. Most common elements are soluble to some degree in both rainwater and soil water. Consequently, water circulation promotes solution by introducing fresh water that removes dissolved ions from mineral surfaces.

The dissolution of calcite ($CaCO_3$, calcium carbonate) is a particularly important chemical weathering reaction. This occurs in the presence of carbon dioxide (CO_2) and introduces bicarbonate ions (HCO_3^-) into solution. The resulting reaction is expressed as

$$CaCO_3 + H_2O + CO_2 \longleftrightarrow Ca^{++} + 2HCO_3^-$$

The carbonate dissolution reaction is reversible. An increase in CO_2 concentration within soil gasses, a decrease in pH, or dilution will drive the reaction to the right (as written above); the rate of carbonate dissolution will increase and the bicarbonate concentration in groundwater will go up. This effect helps percolating water erode fractures and form extensive cave systems typical of regions underlain by carbonate rocks (limestone or dolomite). Conversely, decreased CO_2 concentration, increased pH, or evaporation will drive the reaction to the left and favour precipitation of calcium carbonate ($CaCO_3$). It is this reaction that deposits stalagmites and stalactites in caves, as well as calcite in desert soils.   When carbonic acid dissociates to form an "acid" of protons, the resulting weathering of aluminosilicate minerals consumes CO_2 and thus helps to cool global climate through the general reaction

aluminosilicate + H_2O + CO_2 → clay mineral + cations + OH^- + HCO_3^- + H_4SiO_4

Earth's long-term climate is thus mediated by organic matter burial and carbonate formation (both of which sequester carbon in the geologic record) and silicate weathering, which consumes CO_2 (producing bicarbonate). Over the long run, glaciations and the anthropogenic contribution to atmospheric CO_2 are short-term perturbations of this geologic control on global climate through carbonate burial and the influence of weathering on the concentration of CO_2 in atmosphere. Calcite and salts are readily dissolved in water, so carbonate rocks and evaporites are particularly susceptible to dissolution, especially in regions with abundant precipitation. In contrast, quartz and most other rock-forming silicate minerals are not very soluble at typical Earth surface conditions, leading to slow rates of dissolution in most environments. Solubility varies greatly between minerals, but even the least soluble minerals will dissolve over time if exposed to solutions refreshed by water circulation. Iron and aluminum oxides, however, are virtually insoluble under oxygenated soil water conditions, so these compounds are typically left behind as more soluble, mobile material is depleted. Consequently, the abundance of iron and aluminum oxides increases as rocks and sediment are exposed to weathering. The bright red soils of the south-eastern United States are an example of how these oxides can accumulate through time. Dissolution can play an important role in increasing pore space and thus increasing the percolation of water, soil acids, oxygen, and bacteria into the regolith.

Hydrolysis is a chemical reaction in which water molecules (H_2O) are split into protons (H^+) and hydroxide anions (OH^-) that react with primary rock-forming minerals to form new compounds (secondary minerals). Hydrolysis is an important chemcial weathering process that acts to break rocks apart and transform silicate minerals into weathering products. In hydrolysis reactions, mineral cations are released into solution and replaced by hydrogen (H^+), producing a new mineral. This process results in the transformation of aluminosilicate minerals, like feldspars and micas, into various clay minerals. *The conversion of potassium feldspar (orthoclase) into illite or kaolinite clay are examples of hydrolysis :*

Orthoclase (feldspar) illite (clay)

$$3KAlSi_3O_8 + 2H^+ + 12H_2O \rightarrow 2K^+ + KAl_3Si_3O_{10}(OH)_2 + 6H_4SiO_4$$

Orthoclase (feldspar) kaolinite (clay)

$$2KAlSi_3O_8 + 2H^+ + 9H_2O \rightarrow 2K^+ + H_4Al_2Si_2O_9 + 4H_4SiO_4$$

Hydrolysis is not reversible. Once secondary minerals are formed, further weathering can strip additional cations and can convert secondary aluminosilicates like illite into other, more cation-depleted clays. Upon more intensive weathering, each step in the weathering of clay minerals strips additional cations from the mineral structures. Eventually, intensive weathering can leave kaolinite, which consists of just hydrogen, aluminum, silica, and oxygen, and has no additional cations left to exchange. Progressive alteration of silicate minerals due to weathering reduces the complexity of mineral structures.

Hydration describes the process in which silicate minerals combine with water or hydroxide ions (OH^-) to form hydrated compounds. Hydration is another way that primary minerals are converted to secondary minerals. Common forms of hydration reactions include the conversion of anhydrite ($CaSO_4$) to gypsum ($CaSO_4 \bullet 2H_2O$), and the formation of relatively insoluble iron and aluminum hydrous oxides, like limonite ($FeO(OH) \bullet nH_2O$) and gibbsite ($AlOH_3$) in regions of intense tropical weathering like the Amazon basin.

Clay minerals are both a product and a player in processes of hydrolysis and hydration. Unlike most primary minerals, with the exception of quartz, secondary minerals like clays and hydrous oxides are chemically stable under earth surface conditions. They become a major constituent in soils because their relative stability and immobility leaves them as common in situ by-products of weathering.

Most clay minerals are layer silicates composed of sheets of alumina octahedra (an atom of aluminum bonded to six atoms of oxygen) or silica tetrahedra (an atom of silica bonded to four atoms of oxygen). Both tetrahedral (T) and octahedral (O) layers are organised around central Al, Fe, or Mg cations. These sheets generally are bonded together in either a 1 : 1 structure (TO) in which each layer of alumina octahedra is paired with a layer of silica tetrahedra, or in a 2 : 1 structure (TOT) in which each octahedral layer is sandwiched between two tetrahedral layers. These building blocks are themselves interlayered and bound together by shared ions between the sheets. Layer architecture (1 : 1 vs. 2 : 1) and the ions between the sheets determine the physical properties of different clay minerals.

Adjacent layers in kaolinite, a clay mineral with a 1 : 1 structure, are held together by ionic bonds that are strong enough to prevent cations or water from entering the spaces between the sheets. Because it has few exchangeable cations held between its layers, kaolinite does not swell much when wetted, and has low plasticity (and thus little capacity to be molded). Clays with 2 : 1 layer structures exhibit much more variability in the chemical composition of their octahedral sheets (typically due to substitution of Fe^{2+} and Mg^{2+} for Al^{3+}) and in the abundance and type of ions present between layers. Smectite clays, like montmorillonite, have weak bonds between the silicate layers, which allows water and ions to readily penetrate the crystal structure. Also known as swelling clays, smectites expand readily upon wetting and are a main component of expansive soils. In expanding clays there are layers of water in the interlayer position, which explains why they can so readily take on or lose water and why they are so weak when expanded. Illite, the most common clay mineral in soils, has a strongly bonded 2 : 1 structure and fewer exchangeable cations between its layers, so it has less swelling potential than smectite.

Weathering of secondary minerals involves stripping off layers of silicate structure. The modification of muscovite (mica) to illite (clay), both of which consist of TOT "sandwiches" involves removal of the interlayer cations. Extreme weathering conditions can go beyond leaching of the intermediary cations and remove one of the two T layers, leaving a TO sequence of silicate layers, which is kaolinite clay. Stripping the remaining tetrahedral layer leaves the basic octahedral layer of gibbsite. In general, smectites weather to illites, and ultimately become kaolinite. Deeply weathered soils generally have high concentrations of kaolinite. It is worth noting, however, that kaolinite can form directly from primary minerals, depending on the parent material, climate, and intensity of weathering.

Chelation is the process through which relatively immobile metal ions, like iron and aluminum, are rendered mobile by soluble organic compounds that form ring structures around metal ions, making them susceptible to solution and transport. Chelation is particularly important in moving iron and aluminum, which are otherwise immobile in most soils. Chelation is facilitated by organic acids (particularly fulvic acid) produced by the breakdown of soil organic matter and by lichens, which produce chelating agents that accelerate weathering and liberate nutrients that help sustain the lichen. Iron and aluminum mobilised by chelating agents may be carried along in solution with soilwater flow until concentration changes or microbial actions break the chelating agent, causing the metal to reprecipitate. Incomplete conifer needle decay in cool, moist environments is a common source of chelating agents, leading to the stripping of Fe and other elements from forest soils.

An important outcome of chemical weathering is the ability of secondary minerals to exchange cations with soil water, thereby making nutrients available to plants. Clay minerals loosely hold exchangeable cations adsorbed on their surfaces. In many soils, the exchangable cations are associated mainly with organic matter. Ion exchange is the process by which ions in solution substitute for ions on mineral surfaces. Exchangeable cations are readily taken up in soil fluids, and they provide the dominant source of mineral nutrients for plants.

Clays and organic compounds vary in their ability to adsorb and release cations, a property called cation exchange capacity. Ion exchange is controlled by cation exchange capacity as well as by the ionic composition and pH of soil water. Strongly acidic (low pH) pore fluids allow H+ to substitute for and replace metal cations. As hydrogen ions exchange places with nutrient cations held on a clay surface, the number of potentially exchangeable cations decreases. The degree to which the exchange sites are occupied by exchangeable cations other than H^+ and Al^{3+} is called base saturation. Cation exchange progressively lowers base saturation in clay minerals by removing cations from between clay sheets. A clay with a high cation exchange capacity but a low base saturation has had its balancing cations stripped out and replaced by hydrogen ions through substantial chemical weathering. Such clays are typically found in tropical regions with high temperature and rainfall, such as Hawaii. The progressive loss of exchangeable cations reduces soil fertility — older, more intensively weathered soils are less fertile.

Biological activity can greatly influence both physical and chemical weathering. Organisms directly affect mineral weathering by mechanically breaking down rocks and indirectly by catalysing chemical reactions and causing changes in environmental conditions (such as changing soil pH). The growth of plant roots helps disintegrate rocks by gradually prying apart cracks, fractures, and other openings. Burrowing animals, like gophers, ants, and termites excavate rock fragments and mix them into overlying soils. The process of biologically mediated mixing, called bioturbation, disrupts original structures or fabric in the parent material. Biological activity also increases the potential for chemical weathering by increasing surface area and creating and enlarging pathways for water flow. Because oxygenated water is reactive, the more water that moves through a soil the more weathering happens. Decaying organic matter and respiration of soil micro-organisms and plant roots indirectly influence chemical weathering by increasing the concentration of CO_2 in soil gases, thereby promoting acidification of soil water and enhancing chemical weathering in the fractures and pores through which soil water flows. Plant roots release organic acids that enter soil water and either attack fresh mineral surfaces directly or exchange hydrogen ions for nutrient cations that plant roots absorb as they take up water.

Decaying organic matter also releases humic acids that facilitate further weathering. Bacteria can mediate weathering reactions and lichen colonisation is the first step in weathering of many rocky surfaces. Evapotranspiration by plants can also strongly affect the amount of soil water available for weathering, an effect that is particularly important in desert climates. In short, more life leads to more breakdown and thus more weathering.

Global Patterns of Weathering

The presence, amount, and phase of water (*i.e.*, vapour, liquid, or ice) strongly influence both chemical and physical weathering. It is not surprising, then, that global patterns in the style and intensity of weathering generally track regional climate. Differences in mean annual temperature and precipitation control the magnitude and relative importance of physical and chemical weathering processes. Chemical weathering is strongest in regions with high year-round temperatures and abundant precipitation, namely the equatorial tropics, and weakest in cold, dry areas, like the poles. Physical weathering is least active in dry environments and most active in regions where temperatures drop below freesing for part of the year. Consequently, tropical regions are generally dominated by chemical weathering and high latitudes by physical weathering. Similarly, topography also greatly influences weathering, as high mountains promote mechanical weathering and low-relief environments favour the dominance of chemical weathering. Rates of physical and chemical weathering, however, are correlated because high rates of mechanical breakdown promote higher rates of chemical decomposition by exposing greater surface area to chemical attack and *vice-versa* (through reduced material strength). Consequently, the highest rates of rock weathering typically occur in environments conducive to both physical and chemical weathering.

Styles and rates of weathering vary among rock types because of differences in chemical (mineralogical) characteristics and physical attributes (strength and fracturing). In particular, the ease with which water is able to penetrate rocks strongly influences weathering rates. Mineral composition is a primary factor in chemical weathering. Some rock types, like limestone and marble, are particularly vulnerable to chemical weathering in humid climates due to the fact that they can be dissolved by weakly acidic soil water or rainwater. But these rock types can be quite erosion resistant in arid regions where water is not sufficiently abundant to carry away dissolved ions. In contrast, quartz is far more resistant to both chemical and physical weathering, and rocks like quartz sandstone (quartz arenite) and quartzite weather more slowly than most other rocks regardless of environmental conditions.

SOILS

To a geomorphologist, soil is a layered residue of decomposed rock and organic matter left by weathering over an extended period of time. Soil will have mineralogical, biological, and morphological characteristics that are distinct from the material from which it weathered, its parent material. Other disciplines have their own definitions of soil — engineers generally consider soil to be any loose material above bedrock (if it can be dug it is soil), and agronomists often consider it anything in which one can grow a plant.

A tremendous variety of soils has resulted from the wide array of parent materials, climates, and weathering environments around the world. But due to differences in soil-forming processes, patterns in the distribution of soil types mirror different climate zones and geologic settings. Soils also reflect differences in landscape history, and in some cases their characteristics can be used to infer ancient climate changes, periods of landscape stability, or the nature of ancient environments. Soils are generally unconsolidated, but calcite ($CaCO_3$), silica (SiO_2), and iron oxide (Fe_2O_3) cements that develop in some soils add strength and allow formation of distinct erosion-resistant landforms.

Factors of Soil Formation

The factors that control soil development are climate, organisms (biological activity), topography (slope steepness), parent material, and time. Climate and time are the predominant factors in soil development at a regional scale, but geologic factors greatly influence local soil characteristics. Other than time, soil-forming factors are to some degree interdependent. Climate, for example, influences vegetation, and differences in parent material influence topography. Still, considering these factors independently helps to organise thinking about the things that favour soil development.

Climate, the long-term expression of weather, characterises general averages and variability of temperature, humidity, wind, and precipitation. As discussed above, temperature and moisture conditions strongly influence rates and styles of weathering and hence soil formation, with differences most pronounced on level, stable slopes. Weathering reactions proceed faster at higher temperatures. In hot, wet tropical regions, intense weathering creates thick soils depleted of most of their original constituents, leaving behind relatively stable weathering products like kaolinite clay and aluminum and iron oxides. In contrast, soils formed slowly in cold, dry polar regions tend to be relatively shallow (thin) and have only minimal chemical alteration.

In mid-latitude temperate regions that have moderate temperatures, precipitation, and evaporation, the resulting soils typically have intermediate depths and weathering intensities. In arid regions, low precipitation rates and high evaporation rates cause development of carbonate- or salt-enriched soils because dissolved constituents (usually derived from weathered dust) percolate into the soil, reprecipitate and then accumulate over time. Systematic differences in weathering intensity among different climate zones lead to a general association of different soil types with different climates. Altitudinal effects on temperature and precipitation can also greatly influence weathering processes, with mechanical weathering dominating above timberline in mountain environments. Soils formed from limestone in temperate and humid climates usually consist of fine-grained residual clays and dust left behind after carbonate dissolution. In contrast, soils formed on limestone sediment in deserts form thick accumulations of secondary calcium carbonate that can persist for hundreds of thousands of years.

Organisms from microbes to plants and animals affect rates and patterns of soil formation. Vegetation influences weathering and erosion directly when the growth of roots and tree throw break up bedrock and mix it into the soil. Vegetation influences weathering indirectly when fallen leaves add a seasonal source of soil organic matter that decays to form organic acids. The type and amount of vegetation growing in a soil influences erosion resistance and the ratio of surface water run-off to infiltration after precipitation events. In arid regions, relatively bare, unvegetated slopes do little to slow run-off; during storms a substantial proportion of the precipitation simply runs off over the ground surface and does not contribute to weathering.

Conversely, well-vegetated slopes promote infiltration, chemical weathering, and soil formation. This happens both by directly impeding flow and due to the layer of vegetation and organic debris covering the ground surface. Vegetation further promotes soil formation because both roots and decomposing vegetation add organic acids to soil. Root size, density, and depth can greatly influence soil formation through the location of organic inputs, whether at the surface or at depth.

Vegetation also helps keep soil in place and shades the soil, helping it retain moisture and increasing chemical weathering. Animals contribute to soil formation and mixing through burrowing activity. In tropical regions ants and termites can build great mounds that move soil vertically. In more temperate regions, burrowing mammals, such as gophers, can churn the regolith and mix weathered rock fragments into the soil. Microbes, fungi, and other decomposers are central to the breakdown of organic matter, and the cycling and recycling of key nutrients derived from rock weathering.

Topography greatly influences rates and styles of soil formation through rates of soil erosion and patterns of soil moisture. Gentler slopes usually foster greater plant growth and accelerated soil formation because they retain soil moisture better than steeper slopes. In upland regions, soils are better developed on gentle slopes, while shallow soils and even bare rock characterise steep slopes. Steeper slopes also promote faster rates of downslope soil movement, leading to higher soil turnover rates and therefore younger, less well-developed soils. Moderate relief promotes the greatest weathering as slopes are not too steep to prevent infiltration, but not too flat to prevent flushing of soil and groundwater. In general, soils developing at the base of slopes are better developed and thicker due to runon of water and deposition of sediment from upslope. In addition, slope aspect (the direction a slope faces) can affect micro-climate and local moisture. In some landscapes, different types and densities of vegetation and soils develop on north *versus* south-facing slopes because systematic differences in the amount of direct sunlight translate into differences in soil moisture. North-facing slopes in the northern hemisphere have lower evaporation rates and can retain snow cover longer in the spring and tend to hold soil moisture longer into the summer growing season. Such differences may result in local differences in soil thickness, moisture, pH, and organic matter content. Topography also has a large effect on drainage, including the water table position and its seasonal variation, such as in the location of saturated zones and wetlands.

Rocks or unconsolidated sediments provide the raw material for soil formation, and the mineralogy, porosity, and permeability (*e.g.*, fracturing) of the parent material greatly influences the style of weathering and types of secondary minerals that form from weathering. In particular, the structure, hydraulic conductivity, and fracturing of rock and sediments greatly influence the movement of soil water and groundwater and thus the progress of weathering into parent material. Differences in parent material generally exert the greatest influence during initial soil formation, and determine clay mineralogy, chemistry and certain physical characteristics of the soils that develop from them. For example, young desert soils are strongly influenced by parent material, particularly the contrast between soils developing in carbonates (limestone) versus aluminosilicates (like granite or basalt). Likewise, in the Piedmont of the eastern United States smectite clays in soils developed from Ca and Mg-rich rocks like gabbros and limestones tend to have thicker, more organic-rich topsoil and redder, clay-rich subsoil than well-drained soils developed on granite due to the expansive clays sealing up soil structures and impeding soil drainage. In tropical regions like Brazil, deeply weathered residual soils developed on silica-rich granite tend to consist mainly of kaolinite and quartz (SiO_2), whereas those formed from silica-poor basalt typically consist of aluminum oxides, like gibbsite ($Al(OH)_3$).

Soils develop and change over time as weathering processes mechanically and chemically transform parent material. Soils developing in volcanic deposits have many unique characteristics early in their development that become less distinct through time. Young topographic surfaces have poorly developed soils. Old surfaces can have quite well-developed soils. On relatively flat-lying surfaces, where erosion and deposition rates are low, soils generally remain in place and chemically evolve over long periods of time (tens of thousands to hundreds of thousands of years). Given enough time, a soil will progressively mature as weathering proceeds until it approaches an equilibrium between the rate at which weathering products are removed and the rate at which they are produced. The time scale to reach an equilibium soil thickness varies greatly, with the vast majority of desert soils likely never attaining such a state. Rates of weathering and ion loss from soils tend to decrease over time because the supply of fresh minerals declines as weathering progresses. Soil formation can be interrupted and equilibrium conditions reset by a changing regional climate or environmental changes like glaciation that scrapes away the soil.

On steep slopes with high erosion rates, soils typically have less time to develop than on gentler slopes because minerals have shorter average residence time, the amount of time between initial weathering from the underlying rock and erosion from the soil. In sloping, actively eroding landscapes where soil production and erosion are in equilibrium, soil residence time is equal to the ratio of the soil thickness to the rate of erosion, so the degree to which soils accumulate on slopes depends in part upon the rate at which the landscape is eroding. One meter of soil developed on a hillside that is eroding at an average rate of 0.1 mm yr^{-1} thus has a residence time of 10,000 years, whereas soil 0.1 m thick developed on a slope that is eroding at 1 mm yr^{-1} has an average residence time of just a century. In general, soils with longer residence times will be better developed than those with shorter residence times.

Soil production involves the breakdown of parent material and its mixing with organic matter. Because soil erosion occurs on most landforms, soil can only be maintained if it is replaced at least as fast as it erodes. Soil production is critically important for maintaining soil and soil fertility in the form of fresh minerals. It is also globally important for producing substrate for weathering to work on and for sequestering carbon. Globally, the average rate of soil production is <0.1 mm yr^{-1}, and human activity is thought to have increased erosion over soil production by more than 10 fold.

Geomorphologists have proposed two general models for the relationship between soil depth and soil production from unweathered bedrock. In one model, soil production declines with increasing soil depth because soil-forming processes have less access to fresh bedrock beneath deeper soils. For example, burrowing activity in thick soils breaks down less rock to produce fresh soil than does burrowing in thin soils where organisms are more likely to entrain weathered rock fragments in their digging. Likewise, once soils become deeper than the average depth of tree rooting, the roots of falling trees simply rework soil rather than rip up bedrock. For soils produced by such processes, soil production rates are highest in shallow soils and decline with increasing soil thickness. In this case, feedback between soil production and soil thickness can match soil production to soil erosion because soil production rates increase as a soil thins, while soil production slows as soils thicken. Thus, given time, soil thickness will approach a steady-state that reflects local environmental conditions in which soil production and soil erosion are matched.

In the other model, soils promote rock weathering because they hold more water and support more vegetation than bare rock outcrops. This model predicts that soil production rates are highest at some intermediate soil thickness, and lower in both thin soils that hold less water and support less vegetation, and in thick soils that act to protect the underlying bedrock. In landscapes that have this relationship between soil production and thickness, soil that thins to the point that soil production declines substantially will erode to expose bare rock. In this case, bedrock weathers and erodes slower than surrounding terrain, giving rise to isolated, high-standing bedrock knobs.

Processes of Soil Formation

Soil production occurs by a variety of processes through the addition, loss, transformation, and translocation (movement) of material within a soil profile. These processes include the accumulation of organic matter at the ground surface and its subsequent integration down into the soil, decomposition of primary minerals into secondary minerals, the leaching of particular organic and mineral constituents from the parent material, and the movement of material leached or washed from upper soil layers deeper into the soil.

New material may be added to a soil profile from sources below, within, or above the ground surface. The breakdown of bedrock adds mineral matter to soil from below. Organic matter is added to soil profiles from above when leaves and other organic debris fall to the ground and the remains of once-living plants and animals become mixed into the soil, or roots die within the soil. Water infiltrating into the ground deposits dissolved matter within the soil profile in response to significant changes in pH or solute concentration (such as through evaporation). In addition, floods can deposit fluvial sediment on floodplains and wind can add dust to the top of a soil profile. Soil profiles lose material by either erosion or dissolution. Erosion by surface water flow and wind removes material from the top of the soil profile, and dissolution removes mineral material from within the profile. Soil-forming material is transformed by chemical weathering processes that convert primary rock-forming minerals into secondary minerals, and the breakdown of organic matter into humus.

Soil scientists refer to the movement of material from one part of a soil profile to another as translocation. Dissolved matter may move within the soil or leave the soil in groundwater, a process known as leaching. Water percolating down through the upper soil may also physically entrain clays, silt, and colloids and re-deposit them deeper in the soil. Precipitation infiltrating into a soil dissolves material from the upper layers. Soil water flow carries it downward and may concentrate it in lower layers. The process of transporting clays and mobile ions out of a soil horizon is called eluviation; washing of material into a soil horizon is called illuviation. The formation of caliche, a hard carbonate-cemented layer, is an important process in arid region soils where infiltrating water evaporates within the soil and deposits the dissolved material it was carrying. Over time such deposition can build thick layers of erosion resistant carbonate hardpan. Cumulic soils are those that accumulate slowly enough, or episodically enough, that material added at the top of the soil profile weathers to some degree before being buried. Examples of such inputs include wind-blown dust or volcanic ash and flood-deposited silts on floodplains. Fine-grained cumulic soils are particularly vulnerable to wind erosion if stripped of their vegetation cover, such as happened to the fertile silt (loess) originally deposited by wind across the Great Plains of North America in glacial times that blew away when plowed during the Dust Bowl era drought in the 1930s. Addition of wind deposited dust may also significantly alter soil properties. For example, incorporation of dust infiltrating down into a permeable soil may eventually clog pore spaces and decrease permeability enough to alter the intensity and pace of chemical weathering.

Soil Profiles

The cumulative action of soil-forming processes acts over time to differentiate vertical soil profiles into distinct, recognisable horizons at different depths. Soils exhibit tremendous variability, but almost all soils have an upper zone of leaching that is composed of inorganic primary minerals mixed with organic matter in temperate climates and silt-sised dust in dry climates. Most soils also have a lower zone of accumulation, the nature of which is determined by climate. In temperate and tropical climates, clays and iron oxides dominate the accumulation zone. In dry climates, carbonate ($CaCO_3$) and salts dominate. The upper zone, where losses dominate, is called the zone of eluviation, and the lower zone where particles accumulate and chemical compounds precipitate out of solution is called the zone of illuviation. The ways in which and degree to which soil profiles develop reflect environmental conditions and the cumulative operation of soil-forming processes.

Biological activity can disrupt the development of soil horizons and mix soil by moving material upward or downward. Bio-turbation by plant roots and burrowing animals acts to disrupt development of soil horizons. In particular, the burrowing activity of worms, termites, and gophers may thoroughly mix soil profiles, dragging rock fragments to the surface and organic matter down into the soil. Some bio-turbation can augment layering, such as when ants move coarse material to the ground surface, or termites preferentially move fines to the ground surface. Significant material mixing and net downslope transport also occurs when a tree falls and the material pulled up by roots gradually settles back into or near the resulting hole.

As soils develop, they become differentiated into distinct soil horizons with compositions that reflect the action and interaction of different soil forming processes. Soil scientists use an alphabetical convention to designate distinct master horizons within soil profiles. Lettered horizons are used to identify major differences in the physical and chemical characteristics and soil-forming processes among soil layers. Soil profiles typically consist of certain associations and combinations of soil horizons that are typical of common climatic zones with distinctive environmental conditions. The layer of decomposing organic matter at the ground surface is known as the O horizon. Organic matter (leaves, twigs, pine needles, fallen logs, animal remains, etc...) makes up more than half the volume of solids in an O horizon, and inorganic minerals make up less than half. In some soil profiles, the O horizon is a thick layer that acts as organic mulch in various stages of decomposition. In others it is thin or missing altogether. The next layer down, the A horizon, is a mixture of decomposed organic matter and mineral grains. It typically has a dark brown to black colour because of high concentration of decomposed organic matter. Humic acids and high soil CO_2 concentrations promote decomposition of fresh mineral grains within the A horizon, and percolating water leaches material downward in the soil profile. Many A horizons have a relatively loose, friable (crumbly) texture because of soil aggregates, relatively high organic matter concentration, and pervasive disruption by plant roots. In soils developed in dry climates, the A horizons are often light tan in colour and have accumulated dust rather than organic matter. In cool, moist, acidic soils, extreme leaching removes organic matter and iron oxides, and a nutrient-poor, bleached gray E horizon develops below either the O or A horizon.

The B horizon lies below the O, A, or E horizon, and consists of inorganic minerals in which leached material has accumulated after percolating down from the overlying horizons. Consequently, clay content is usually higher in B horizons than in the overlying or underlying horizon(s). The majority of soil transformations occur in the B horizon where water tends to sit for longer periods of time. Depending on the environment, these accumulations may consist of clay minerals, carbonates, salts, or iron and aluminum oxides. B horizons have textures that are distinct from overlying soil horizons as well as from underlying parent material, and often are redder than the overlying A horizon because they contain less organic matter and more iron oxides. B horizons retain little to no evidence of original rock or sediment structure such as bedding or foliation. Clay films deposited by water seeping down into the soil often coat and line fractures and voids within the B horizon. In arid and semi-arid regions, evaporation of pore water concentrates calcium carbonate ($CaCO_3$) or gypsum ($CaSO_4 \cdot 2H_2O$) within the soil profile, and leads to development of distinctive Bk horizons in which whitish evaporites coat mineral grains. Extreme cases in which carbonate makes up more than half the soil horizon are called K or petrocalcic horizons ("caliche") by geomorphologists working in desert environments.

The C horizon consists of unweathered parent material and weathered material that retains some original rock or depositional structure, including sedimentary bedding planes, depositional fabrics, and core stones surrounded by altered weathering rinds. Thoroughly weathered saprolite is generally noted as a Cr horizon and consists of material that has weathered in place. Although it is not really part of the soil profile, bedrock is sometimes referred to as the R horizon. However, the parent material for a soil can be lithified rock or unconsolidated sediment like dune sand or river gravel. Deeply weathered soils and saprolite extending down hundreds of meters into weathered rock are common in flat, hot, wet tropical landscapes in the cratons of Africa and South America.

Soil Classification

Soils are classified based on field-observable characteristics. Different soils and soil horizons are distinguishable by differences in one or more soil properties that reflect the degree of soil development and provide both indications of process and measures of time and landform stability. The most important soil properties observable in the field — colour, texture, and structure — reflect a soil's clay and organic matter content and its ability to hold moisture.

Soil colour indicates both mineral composition and organic matter content. Soils with high organic matter content are typically dark brown to black. Iron-rich soils are red to yellow-brown in colour in oxidising environments and iron is blue to black or gray in reducing environments such as wetlands. A mottled pattern of red and bluish colours often indicates seasonally saturated soils that alternate between oxidising and reducing conditions. Light colours, like white, gray and beige are generally associated with accumulations of calcite ($CaCO_3$) or leached horizons consisting mostly of residual silica (SiO_2). In the field, soil scientists use a book of colour charts resembling those used at paint stores to identify soil colours based on a palette of standardised colours.

Soil texture is measured in terms of the proportions of different particle sizes in the soil. The U.S. Department of Agriculture defines twelve standard soil textures based on the relative proportions of clay, silt, and sand (*e.g.*, clay loam, silt loam, silty clay, etc…). Although soils may contain larger clasts (like gravel), soil properties measured in the field generally are described based on particles less than 2 mm in diameter (*i.e.*, sand and smaller). Soil texture greatly influences the stickiness of a soil as well as its ability to resist deformation (plasticity). A simple field test for soil composition is to moisten a sample of soil and try to make a ball in one's hand. Clay can be rolled into a "worm"; silt and sand cannot. Another simple field test — one that is not always advisable to employ — is that silt feels gritty between one's teeth, while clay has a smooth, creamy texture. Sand grains are visible to the naked eye.

Soil structure describes the shapes in which soil particles cluster together. Most A horizons have granular or crumb structures in which individual clumps, or aggregates, are more or less spheroidal. In contrast, B horizons often have structures, called peds, that are shaped like blocks, prisms, or plates with rounded or angular edges. Peds can range from pea-size to fist size and reflect the deposition of clay along soil partings.

Soil organic matter includes undecomposed material like leaves, branches and bones (litter) as well as amorphous decomposed material (humus). Soil micro-organisms that decompose litter into humus become active at temperatures slightly above freesing (generally >5°C), and rates of microbial decomposition increase as temperatures rise. At temperatures above about 25°C, little humus accumulates because it decomposes as fast as it is produced. Consequently, soil organic matter content tends to be greatest in mid-latitude regions with mean annual temperatures between 5 and 25°C (41 to 77°F).

Soil orders are characterised by certain kinds of soil profiles. The U.S. Natural Resource Conservation Service (formerly the Soil Conservation Service) uses a system of soil taxonomy that recognises twelve soil orders that are distinguished by diagnostic horizons and characteristics that reflect different environments, processes, and duration of formation. Soils can be further subdivided into 64 common sub-orders and then into specific soil types, but the major soil orders are sufficient to characterise soil characteristics, parent material, and degree of weathering at a regional scale. Some soils are formed by factors that change over short distances, such as from differences in parent material or topography. Others reflect factors that vary over long distances (climate and vegetation types). Yet others reflect the influence of particular parent materials or time on the degree of soil development.

Immature soils and wetland soils can look the same in widely varying envrionments. Young, incompletely developed soils that are essentially raw sediment are either entisols that have a faint A horizon, or inceptisols with a weak B horizon, but not with enough horizon development to quality as one of the more developed soil orders.

Soils formed in seasonally sub-merged or frozen environments have distinctive characteristics. Histosols are organic-rich soils that form in wetlands where reducing conditions restrict rates of organic decomposition. Histosols are identified by their black colour, high organic matter content (more than 25 per cent), and lack of evidence for oxidation. Gelisols form in polar and subpolar environments where the ground stays frozen all year. These cold climate soils include, by definition, a layer of permanently frozen soil (permafrost) within two meters of the land surface. They typically show little evidence of weathering and have an A horizon developed over permafrost. Gelisols are often structurally disrupted and mixed by seasonal freeze/thaw processes called cryoturbation, which complicates foundation engineering in periglacial environments.

Several types of soil reflect distinctive parent materials. Andisols are formed from volcanic parent material. Vertisols are soils that contain high concentrations of swelling clay (smectite) and exhibit significant shrink/swell behaviour. Vertisols do not have distinctive horizons; they exhibit large desiccation cracks when dry, and tend to become annoyingly sticky when moist. The other six soil orders are associated with particular climate or environmental settings, and form under certain combinations of temperature and relative wetness.

Mollisols typically develop in temperate regions with grassland vegetation, like the Great Plains of North America. Mollisols have well-developed, black, organic-rich A horizons that are often more than a meter deep and B horizons that are enriched in clay. Grassland soils accumulate abundant organic matter because of the high seasonal production of plant litter and because grass species are generally alkaline and promote low soil acidity and therefore slow organic decomposition. Environments that permit incomplete decay of organics produce organic colloids that can leach down into the soil, making fine black A & B horizons. This commonly happens in grasslands where the soils are too cold for decomposition in the winter, but are too dry for much decomposition in the summer. Mollisols typically contain abundant organic matter and clays with high base saturation and thus generally make fertile agricultural soils. The fertility of the grassland soils of the American Midwest is sustained by A horizons so thick and rich that they remain productive even after more than a century of agricultural erosion.

Surface litter produced by forests tends to be more acidic than organic matter produced by grasses or shrubs. Consequently, forest soils are generally more acidic and have A horizons that are more leached of soluble ions (notably Ca, Na, and Mg) because of extensive chelation by organic compounds. Soils that develop beneath forests are called alfisols, ultisols, and spodosols, in order of driest to wettest environment of formation. Alfisols and ultisols have thin A horizons, but the A horizon is absent in most spodosols. Alfisols form under deciduous forests due to acidity and leaching by decaying leaves, and as such are common in the eastern United States. Alfisols are distinguished by higher base saturation and a fertile C horizon. Ultisols tend to form in old and highly weathered areas that are warm and wet and were not glaciated, reflecting more prolonged or intense weathering. Spodosols have even more extreme leaching, often with extensive E and B horizon development, mostly reflecting acidic conditions under coniferous forests due to organic acids produced by decaying needles. All three soil orders (alfisols, ultisols, and spodosols) are subject to iron, aluminum, and organic matter leaching from the A horizon and accumulating in the B horizon due to production of organic acids and prevalence of chelation. Alfisols and ultisols form in a wide range of temperature zones, but ultisols typically characterise wetter regions and are more weathered. In ultisols, colloidal Fe and Al oxides and hydroxides may dominate as Fe and Al bearing primary minerals are nearly entirely broken down and leached. The B horizons of alfisols and ultisols are enriched in clay, whereas B horizons of spodosols are enriched in organic matter and iron and aluminum oxides. Spodosols best develop in cold, wet environments and in parent materials with high permeability, such as sands on a coastal plain. Many extensively leached forest soils have a distinctive, light coloured E horizon at the top of the B horizon that consists mostly of silica that has been leached of organic matter, Al^{3+}, and Fe^{3+}.

Aridisols are soils that develop in dry regions. They usually have low organic matter content and an immature A horizon overlying a weak B horizon. In subtropical and arid climates, infiltrating precipitation typically evaporates within the soil profile before reaching the water table. When this occurs, ions and particles in percolating rainwater or mobilised from the A horizon precipitate or deposit in the B horizon. Calcification occurs when calcium carbonate ($CaCO_3$), gypsum, or other salts dissolved in groundwater precipitate out and coat mineral surfaces to form a distinctive white layer. The depth to this zone of evaporative accumulation reflects the amount of annual precipitation. Increasing aridity generally results in less well developed A horizons and shallower zones of carbonate accumulation. In extremely arid settings like the Negev Desert in southern Israel, salt-rich (salic) or gypsum-rich (gypsic) horizons develop because soil water is insufficient to leach away even these very soluble minerals. Given enough time, however, red, clay-rich B horizons can develop in some of Earth's driest deserts. Dustfall is also important in the development of arid and semi-arid soils, which can have thick A horizons that develop as dust infiltrates down through a surface layer of coarse clasts known as a desert pavement.

Oxisols are the most deeply weathered of all soil orders and are characterised by highly oxidised soil horizons that form in hot, wet (tropical) environments. They generally have distinctive red colouration and little organic matter accumulation because plant and animal remains decay rapidly in such conditions. Red-coloured oxisols have an abundance of iron oxides (hematite and goethite) and tend to be depleted in exchangeable nutrient cations. The most heavily weathered tropical soils (laterites) consist of little more than residual iron and aluminum oxides such as gibbsite ($Al(OH_3)$), an aluminum oxide that forms when silica has been leached from kaolinite. The components, once heated and dried by the sun, are basically brick. Organic matter and nutrients in oxisols are in the thin O or A horizon or the vegetation, posing impressive challenges to sustaining agricultural productivity when these surficial horizons are removed by erosion.

Soils and Landscapes

Soil properties and orders vary among climate zones and physiographic regions. Soil orders and the degree of soil development — particularly soil thickness and organic matter content — generally track latitudinal patterns in temperature and precipitation. Deeply weathered oxisols are typical of the equatorial tropics. Aridisols are typical of the mid-latitutde desert belts. Organic-rich mollisols and forest soils are typical of temperate latitudes. However, soils also differ within single landscapes because of variations in the factors of soil formation, erosion, and landscape history.

Soils mature over time. They gradually lose the physical and chemical characteristics of their parent materials, and take on characteristics that reflect soil-forming processes. Consequently, differences in parent material are better expressed in immature soils. More mature soils have characteristics determined by the dominant climate and vegetation. Soil properties like the amount of organic matter, the degree and depth of oxidation, the removal of cations or minerals, development of clay minerals, and clay or $CaCO_3$ content in the B horizon change as the soil weathers and matures, so the degree of weathering can serve as a proxy for the stage of development and relative age of a soil.

The parts of a soil profile also mature at different rates. As a soil ages, the A horizon generally achieves maturity (*i.e.*, steady state or equilibrium) first, followed by the B horizon, and then such features as carbonate accumulation and oxidation. Different soil orders take different times to reach maturity too. Mature spodosols and mollisols are capable of forming in millennia, while mature ultisols may take tens of thousands of years to form and mature oxisols may require hundreds of thousands of years. A chronosequence is a series of soils of different ages that formed on the same parent material under constant conditions of vegetation, topography, and climate, such as for example a sequence of soils formed on alluvial terraces at different heights above an incising river. Chronosequences of soils that have been dated by radiometric or other means are used to evaluate rates of change in soil properties. Calibrated soil development rates can then be used to infer the ages of landforms from soil properties, or to estimate the amount of time that a soil was exposed at the surface before being buried.

Different kinds of soils can develop on different parts of a single slope because of local variability in soil forming processes, erosion, and deposition. A soil catena is a suite of soils with profiles that grade into one another in different landscape positions because of variations in soil moisture, sediment transport rates, slope steepness and chemical weathering. Catenas reflect local differences in soil-forming processes that result from topographic position and the hydrological and geomorphological processes that govern run-off, infiltration, and soil erosion. For example, soils at the top of slopes where water readily washes over or through the soil profile may be well drained, oxidised, and reddish in colour, whereas those at the base of slopes tend to have a higher water table and reducing conditions that produce more gray to blue colours. Soils developed on the steep slopes ringing a plateau may be quite different from those developed on the flat plateau surface. In many environments, sloping surfaces are the rule and different soils develop at the top, mid-slope, and base of slopes. Soils in a catena reflect different records of the same processes acting on different parts of a landscape.

Most soils are geologically young, and actively evolving at the land surface. Paleosols, however, are ancient soils in which active soil processes have ceased. Some paleosols are buried soils preserved in the geologic record and some are relict soils that formed under different conditions but are still exposed at the surface. Buried soils are readily recognised, typically by an abrupt change in colour, and can be found in rocks of almost any geologic age. When found in the geologic record, paleosols indicate a land surface on which a soil was able to form. They thus represent a period during which soil formation outpaced both erosion and burial by deposition. Because the characteristics of paleosols reflect the soil forming factors of the landscape in which they developed, they can be useful indicators of past climatic or environmental conditions. Some soil features, such as organic matter, do not survive extended burial and are poorly preserved in the rock record, whereas other characteristics, like soil structure and texture, are geologically robust.

Weathering-Dominated Landforms

Weathering processes produce a variety of distinctive small- and large-scale landforms. At small scales, physical and chemical processes wear away the edges and corners of rock outcrops and boulders fastest, promoting spheroidal weathering that converts angular blocks of rock that result from tectonic and unloading fractures into rounded cobbles, boulders, and monoliths. Outward expansion of clay minerals on chemically altered rock surfaces enhances spheroidal weathering, and produces weathering rinds with an onion ring texture. Wetting and drying or freesing of water causes localised spalling and granular disintegration that creates weathering pits on bare rock surfaces of certain rock types. Extreme cases of weathering pit development result in cavernous or honeycomb textures known as tafoni. Dissolution pits and cavities often characterise highly soluble carbonate outcrops. At larger scales, weathering-dominated landforms include karst and thermokarst topography, inselbergs, tors, and duricrusts. Differential weathering in the valley and ridge province of the eastern United States results in quartzite ridgetops and limestone valley bottoms.

Inselbergs (island mountains in German) and tors are high standing bodies of exposed rock that rise above surrounding terrain and result from spatial variability in weathering and erosion. Inselbergs are large residual rock masses still attatched to bedrock after episodes of deep weathering and erosion removed the surrounding material. Tors form the same way, but they are smaller features consisting of multiple, smaller exhumed corestones not necessarily still attached to bedrock. What have today become inselbergs and tors weathered more slowly in the sub-surface than surrounding areas because of differences in either mineral composition or fracture density that rendered them less susceptible to chemical weathering than the surrounding rock. Deep weathering produces an uneven weathering front, and inselbergs and tors represent exhumation of less weathered core stones protruding above an advancing weathering front. Australia's famous Uluru (also known as Ayers Rock), a massive sandstone outcrop that rises hundreds of meters above the desert plain of the central outback, is a prime example of a large inselberg. Another is Rio de Janiero's famous Sugarloaf, a granitic inselberg towering above deeply weathered surroundings. Tors are common landforms in weathered landscapes around the world, but the rock towers of south-west England and Wales are well known examples. Inselbergs and tors record the changing balance between weathering and erosion or the change from an erosion-limited landscape to one where the pace of weathering limits erosion rates.

Duricrusts are erosion-resistant cemented soils turned back into rock by cementation within the pedogenic zone. In arid regions, enough calcium carbonate ($CaCO_3$) may accumulate within a soil profile to form a cement-like layer of caliche, or calcrete, that is erosion resistant when it is exposed at the land surface. Silcrete, a hard silica-rich (SiO_2) layer, behaves similarly. Both calcrete and silcrete form in arid regions. Duricrusts also form by evaporation of solute-rich groundwater along stream valleys. These protective shells can even be tough enough to cause topographic inversions, wherein formerly low-lying terrain becomes more resistant to erosion than the neighbouring uplands and eventually becomes elevated as the surrounding terrain erodes away. Extreme oxidation, intense weathering and leaching produce iron-rich residual soils known as laterite and ferricrete. These erosion resistant crusts form where intense weathering removes all but the least mobile elements, leaving behind just aluminum and iron oxides. Ferricrete is an iron-rich duricrust made of highly concentrated iron oxides as a result of intensive weathering. Laterite and ferricrete are often found on slowly eroding topographic highs in the tropics, and form the caprock for mesas in central Australia and the Amazon. Evaporation of water within the soil or at the ground surface can coat mineral surfaces with precipitates. In arid and semi-arid regions like the Owens Valley in south-east California, soil water containing silica dissolved from fine dust and rock moves towards the surface and reprecipitates a hardened shell where it evaporates on the rock surface.

Weathering is of fundamental geomorphological importance because almost nothing happens on slopes without weathering to make slope-forming material transportable. The physical breakdown of rock into smaller pieces and chemical transformation into secondary minerals influence the types and rates of geomorphological processes that shape topography.

Soils define the frontier between geology and biology. The thin skin of weathered rock and decomposing organic matter provides the nutrients that nourish the plants on which all terrestrial life depends. Cation exchange capacity and base saturation of soils are the basis for soil fertility. Plants can readily extract nutrients from soils with high base saturation. Sustained cultivation without replenishing soil nutrients leads to declining crop yields. This is one reason why floodplains that receive annual deposits of fresh minerals and volcanic soils that are periodically replenished by ash fall are prised and highly productive agricultural lands around the world. Modern industrial agriculture uses tremendous amounts of chemical fertilizers (principally nitrogen, phosphorus, and potassium) to supplement native soil fertility in place of traditional crop rotations, applications of manure, and organic farming techniques that are based on soil ecology. Understanding soil-forming processes and soil fertility are important for evaluating options for maintaining soil fertility and agricultural productivity in a post-petroleum (and potentially post-cheap fertilizer) world.

Maintenance of soil fertility depends not only upon maintaining soil nutrient levels, but upon conserving the soil itself. Extensive soil loss to erosion can follow deforestation, tillage, and destruction of vegetative cover by fire or overgrasing. On some now barren Caribbean islands, sugar cane cultivation on steep slopes sent most of the topsoil into the ocean within several generations. Recent earth history is rife with truncated and thinned soil profiles that record examples of ancient societies (such as Classical Greece, Rome, and Easter Island) that failed to prevent soil erosion from exceeding soil formation. Globally, the average rate of net soil loss from agricultural fields has been estimated to have increased by 10 to 20 fold as a result of tillage and exposure of bare soil to the effects of wind, rainfall, and run-off.

Farming practices have also reduced soil organic matter across vast areas of the continents, particularly in mollisols. About a third of the carbon dioxide added to the atmosphere by human activity since the Industrial Revolution came from the decay of soil organic matter resulting from plowing up fertile grassland soils, rather than from burning fossil fuels. But people can improve as well as degrade soil fertility. The recently discovered organic-rich, incredibly fertile "terra preta" soils in the Amazon jungle formed over millennia as indigenous people burned their trash and broken pottery in their fields. Today these soils form islands of fertility in otherwise infertile tropical soils. Soils are among the most diverse features of Earth's surface, reflecting both regional environmental and local factors. Basic training in geomorphology can help tailor sustainable agricultural practices to particular soils and landforms.

Weathering also provides minerals critical for modern life. Deep weathering on ancient land surfaces produced iron and aluminum ores through pervasive leaching and removal of other common elements that concentrated relatively immobile elements in residual soils. Aluminum is a common element in terms of its distribution in Earth's crust, but it is dispersed at such low concentrations in most rocks that it is not economically extractable. It takes millions of years to dissolve away everything else and make laterite soils that are enriched enough in aluminum that they constitute aluminum ores. Few people realise that we warp our soda in ancient soils.

Chapter 5

CHEMICAL EQUILIBRIUM

INTRODUCTION TO CHEMICAL EQUILIBRIUM

Chemical equilibrium applies to reactions that can occur in both directions. In a reaction such as :

$$CH_4(g) + H_2O(g) <\text{--}> CO(g) + 3H_2(g)$$

The reaction can happen both ways. So after some of the products are created the products begin to react to form the reactants. At the beginning of the reaction, the rate that the reactants are changing into the products is higher than the rate that the products are changing into the reactants. Therefore, the net change is a higher number of products.

Even though the reactants are constantly forming products and *vice-versa* the amount of reactants and products does become steady. When the net change of the products and reactants is zero the reaction has reached equilibrium. The equilibrium is a dynamic equilibrium. The definition for a dynamic equilibrium is when the amount of products and reactants are constant. (They are not equal but constant. Also, both reactions are still occurring.)

Some crystals of iodine, $I_2(s)$, are placed in an open gas jar (an **open system**) maintained at a constant room temperature. A purple vapour is seen to form above the crystals, the crystals gradually disappear, and eventually so does the purple iodine vapour.

The experiment is repeated but this time a lid is placed on the gas jar (a **closed system**). After a short time, a uniform purple vapour forms and some crystals of iodine remain. At the constant temperature the system remains like this indefinitely.

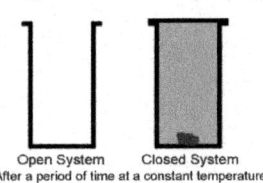

Open System   Closed System
After a period of time at a constant temperature

In this **closed system** an **equilibrium** has established. The equilibrium is between the iodine solid and the iodine vapour. Here we **do not** have a static condition but a **dynamic condition**. What is happening all of the time is that I_2 molecules are leaving the solid and entering the vapour phase at the same rate as they are leaving the vapour and entering the solid state. The forward and reverse rates have become the same and the system is at equilibrium. However, this is not a chemical equilibrium but a physical equilibrium.

$$I_2(s) \rightleftharpoons I_2(g)$$

Incidentally, in the closed system above, the pressure of the iodine vapour is equal to its vapour pressure. This does depend on temperature, but is independent of the volume of the vessel and of the amount of iodine solid, provided some solid is always present.

A Chemical Reaction can also Reach a State of Equilibrium

At the start of an experiment you have two sealed flasks maintained at 720K, one containing a mixture of $H_2(g)$ and $I_2(g)$, and the other HI(g) only.

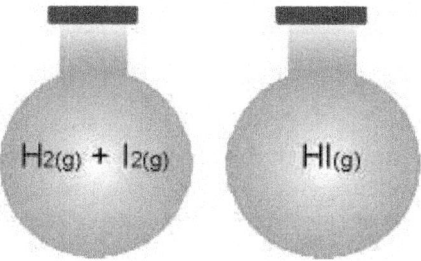

At intervals of time during the experiment the amount of HI(g) in each is measured (and expressed as the percent of all H and I atoms present as HI).

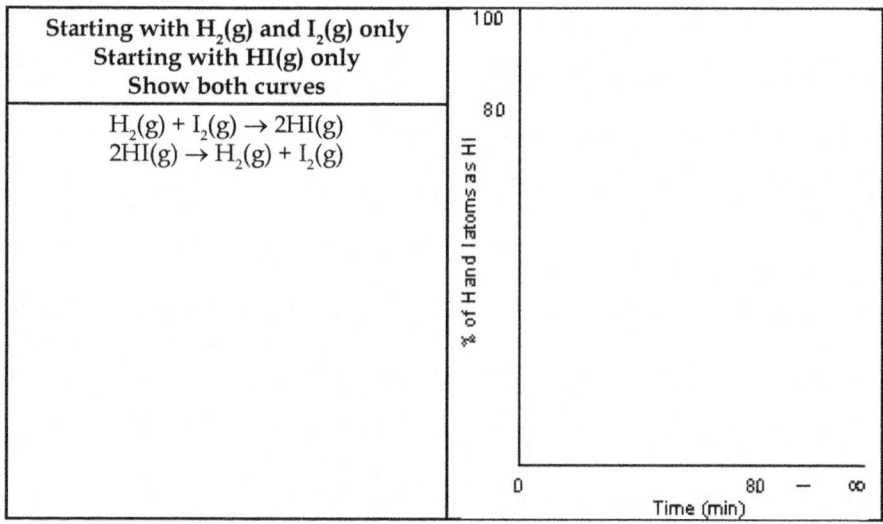

Starting with $H_2(g)$ and $I_2(g)$ only Starting with HI(g) only Show both curves
$H_2(g) + I_2(g) \rightarrow 2HI(g)$ $2HI(g) \rightarrow H_2(g) + I_2(g)$

Starting with $H_2(g)$ and $I_2(g)$ only, the two react producing $HI(g)$ in a **forward reaction**. The rate of this reaction decreases as it proceeds owing to the concentrations of $H_2(g)$ and $I_2(g)$ diminishing. As $HI(g)$ is formed, its molecules begin to react to reform $H_2(g)$ and $I_2(g)$ in a **reverse reaction**. As the concentration of $HI(g)$ increases as a result of the forward reaction, the rate of the reverse reaction increases. Eventually, the concentrations of all the chemical components will attain values such that the forward and reverse reactions are occuring at the same rate.

We now have a **reversible reaction** that is said to have attained **chemical equilibrium**.

In an equilibrium reaction, all chemical components are both reactants and products, though by tradition those shown on the left of the chemical equation are called reactants and those on the right products.

$$H_2(g) + I_2(g) \rightleftharpoons 2HI(g)$$

Remember, although at equilibrium the forward and reverse reactions may appear to have stopped they are, in fact, taking place all of the time and at the same rate - equilibrium is a dynamic not a static condition.

What Quantities of Reactants and Products are Present at Equilibrium

Equilibrium refers to equality of forward and reverse reaction rates, not of quantities of reactants and products. In the above chemical equilibrium, 78 per cent of the mixture is in the form of $HI(g)$, so appreciable quantities of both reactants and products are present. Some reactions establish equilibrium after the formation of only minute amounts of products. In others, only minute amounts of reactants remain at equilibrium. The composition of the equilibrium mixture is sometimes referred to in terms of **equilibrium position**.

How Long Does it Take for Equilibrium to be Reached

In the case of $HI(g)$, the mixture reaches equilibrium in less than 2 hours at 720 K. Other reactions may take only a tiny fraction of a second, and there are many more that would be nowhere close to equilibrium after a million years.

The Equilibrium Constant

Now consider we have a mixture of reactants and products that is not yet at equilibrium, or one that is at equilibrium.

Which way will the reaction go in order to reach equilibrium?

How far will the reaction go, that is, what will be the amounts of reactants and products at equilibrium?

If we make changes to the system, such as to its volume, or add more of a reactant, how will the equilibrium composition be affected?

To be able to answer these questions we have to know about the **Equilibrium Constant**. Now back to the equilibrium involving $H_2(g)$, $I_2(g)$ and $HI(g)$ at 720K.

$$H_2(g) + I_2(g) \rightleftharpoons 2HI(g)$$

In four separate experiments at 720 K, chemical equilibrium was allowed to establish. Two were started with different concentrations of $H_2(g)$ and $I_2(g)$ only in each, and two with differing initial concentrations of HI(g) only. In each of the four experiments, the equilibrium concentration of $H_2(g)$, $I_2(g)$ and HI(g) were measured. These are given in the table below :

Equilibrium concentrations (mol dm⁻³)		
$[H_2(g)]_{eqm}$	$[I_2(g)]_{eqm}$	$[HI(g)]_{eqm}$
1.14×10^{-2}	0.12×10^{-2}	2.52×10^{-2}
0.92×10^{-2}	0.20×10^{-2}	2.96×10^{-2}
0.34×10^{-2}	0.34×10^{-2}	2.35×10^{-2}
0.86×10^{-2}	0.86×10^{-2}	5.86×10^{-2}

If we substitute these equilbrium concentrations into the expression below :

$$\frac{[HI(g)]^2_{eqm}}{[H_2(g)]_{eqm} \; [I_2(g)]_{eqm}}$$

then a **constant value**, within experimental error, is obtained. The square brackets represent concentrations in mol dm⁻³.

The value of this ratio of equilibrium concentrations (in mol dm⁻³) is known as the **equilibrium constant**, represented by the symbol K_c.

$$K_c = \frac{[HI(g)]^2_{eqm}}{[H_2(g)]_{eqm} \; [I_2(g)]_{eqm}} \qquad (= 46.8 \text{ at } 720 \text{ K})$$

Many other chemical equilibrium reactions have been studied and in each case an equilibrium constant relating to the stoichiometric chemical equation has been calculated. This observation can be universally applied, in what is sometimes called the *Law of Chemical Equilibrium*, as illustrated by the general chemical equilibrium :

$$aA_{(g)} + bB_{(g)} \rightleftharpoons cC_{(g)} + dD_{(g)}$$

$$K_c = \frac{[C(g)]^c_{eqm} \; [D(g)]^d_{eqm}}{[A(g)]^a_{eqm} \; [B(g)]^b_{eqm}}$$

Can the Equilbrium Constant be Expressed in Other Terms

The Ideal Gas Equation shows that the pressure of a gas is proprtional to its concentration.

$$pV = nRT$$

where p is pressure of a particular gas (its partial pressure) in an equilibrium mixture, V is the total volume, n is the number of moles of the particular gas, R is the general gas constant, and T is the absolute temperature.

In the above equation where the temperature is also constant,

$$P \alpha \, n \, / \, V$$

The equilibrium constant can therefore also be expressed in terms of partial pressures, and is denoted as K_p :

$$K_p = \frac{pC(g)^c_{eqm} \quad pD(g)^d_{eqm}}{pA(g)^a_{eqm} \quad pB(g)^b_{eqm}}$$

Does the equilbrium Constant have Units

The equilibrium constant (K_c or K_p) may or may not have units; this depends precisely upon the equilibrium expression.

What Values can the Equilibrium Constant Have

Equilibrium constants come in all sizes, from almost zero (like 10^{-50}) to extremely large (like 10^{50}). When K (K_c or K_p) is much greater than 1, the partial pressures or concentrations of the products are large in comparison to those of the reactants.

These are ways in which we describe this situation...

"...the reaction goes nearly to completion..." or " ...goes mostly to the right..." or "...the equilibrium favours the products...".

Does the Equilibrium Constant Depend on Temperature

Yes. An equilibrium constant is constant only at a constant temperature. Changing the temperature of a chemical system at equilibrium will change the value of the equilibrium constant, and also the position of the equilibrium.

Are K_c and K_p Related

Yes. The relationship depends on the reaction and how it is written. Specifically, it depends on the number of moles of *gaseous* reactants and products.

The equation is :

$$K_p = K_c (RT)^{\Delta n}$$

where Δn is the number of moles of gaseous products minus the number of moles of gaseous reactants, R is the gas constant, and T is the absolute temperature at which the equilibrium exists.

Now let's make some use of the equilibrium constant...

Return to : The Equilibrium Constant

Which way will the reaction go in order to reach equilibrium?

We have a mixture of $SO_2(g)$, $O_2(g)$, and $SO_3(g)$ at 1000K which has still to reach equilibrium. The partial pressures are p_{SO2} = 0.48 atm, p_{O2} = 0.18 atm, and p_{SO3} = 0.72 atm. K_p = 3.40 atm^{-1}. The reaction is :

$$2SO_2(g) + O_2(g) \rightleftharpoons 2SO_3(g)$$

Which way must the reaction go to reach equilibrium?

answer...

How far will the reaction go, that is, what will be the amounts of reactants and products at equilibrium?

A sample of air is compressed so that $[N_2(g)]$ = 0.80 mol dm^{-3} and $[O_2(g)]$ = 0.20 mol dm^{-3}. (It might help to imagine the container has a volume of 1 dm^3.) It is then heated to 1500 K and allowed to reach equilibrium.

$$N_2(g) + O_2(g) \rightleftharpoons 2NO(g)$$

How much of each reactant and product will be present at equilibrium? K_c = 1.0 x 10^{-5} at 1500K.

answer...

If we make changes to the system, such as to its volume, or add more of a reactant, how will the equilibrium composition be affected?

It is of great practical importance to be able to predict, and to control, the equilibrium composition of a chemical reaction. In an industrial process it may be desirable to convert as much as possible of reactants to prodcuts. At equilibrium the product yield may not be satisfactory. But, the composition of the equilibrium mixture (equilibrium position) is dependent on certain conditions, such as temperature and pressure, and it might be possible to change these to achieve a better yield. **Le Chatelier's Principle** helps us deal with this.

Now consider another chemical reaction in a state of equilibrium at a constant temperature.

$$N_2(g) + 3H_2(g) \rightleftharpoons 2NH_3(g) \quad \Delta H°_{f, 298} = -46.0 \text{ kJ mol}^{-1}$$

In a closed system and at a pressure of 250 atmospheres in the presence of a finely divided iron catalyst at 500 °C, about 15 per cent of the gases are converted into ammonia at equilibrium.

We can impose the following changes [both an increase and decrease] upon this chemical system at equilibrium :

1. Amount of a substance / concentration of a reactant or product.

2. **Pressure / Volume (pressure and volume are inversely related).**

3. **Temperature.**

In general terms, this is what Le Chatelier's Principle says...

If a change is imposed upon a chemical system at equilibrium then it will respond in such a way as to undo, in part, the effect of the change imposed upon it.

Here are some examples :

Example 1

The reaction is at equilibrium in a closed vessel of fixed volume. The temperature is constant. Suddenly some nitrogen gas is added. The concentration of nitrogen has increased. The partial pressure of nitrogen is raised. The system is no longer at equilibrium. Le Chatelier's Principle says that the equilibrium will adjust so as to reduce the concentration of the nitrogen. This can only be achieved if the nitrogen reacts. The nitrogen can only react with hydrogen forming ammonia. The equilibrium therefore shifts from left to right. A new equilibrium position is established. Since the system responds 'to undo, in part, the effect of the change', at the new equilibrium position there will be slightly more nitrogen, less hydrogen, and more ammonia present.

Example 2

The reaction is at equilibrium in a closed vessel. The temperature is constant. The volume of the vessel is suddenly reduced. This is equivalent to increasing the total pressure of the system. The system is no longer at equilibrium. Le Chatelier's Principle says that the equilibrium will adjust so as to reduce the pressure. It can only achieve this by shifting in the direction of fewer moles of gas (fewer gaseous molecules). There are 2 moles of gas on the right of the equilibrium and 4 on the left. The equilibrium therefore shifts from left to right. A new equilibrium position is established. The yield of ammonia is increased.

Example 3

The reaction is at equilibrium in a closed vessel of fixed volume. The temperature of the system is suddenly reduced. The system is no longer at equilibrium. Le Chatelier's Principle says that the equilibrium will adjust so as to bring about a rise in temperature of the system. A reaction needs to occur so as to produce heat for this to happen. The equilibrium must shift in the exothermic direction, that is, from left to right. A new equilibrium position is established. The yield of ammonia is increased.

Getting the Best Yield of Ammonia

It seems that by raising the pressure of the system and at the same time reducing its temperature we can maximise the yield of ammonia at equilibrium. This is not without its problems. Maintaining a higher pressure is expensive; more robust plant (thicker pipes and stronger joints, for example) is required, and more electricity would have to be used to drive pumps to maintain this pressure. Lowering the temperature slows the rate at which chemical reactions take place resulting in a longer wait for equilibrium to be achieved. A 'compromise' has to be settled upon. The conditions selected approximate to those given above.

Does a Catalyst Affect the Equilibrium Position

No. The addition or removal of a catalyst does not cause a shift in equilibrium. A **catalyst** *is a substance that affects the rate of a reaction, but is not consumed in the* *reaction*. It does this by providing an alternative mechanism for the reaction but of lower activation enthalpy. It therefore changes the rate of approach of equilibrium and so affects the forward and reverse reaction rates in the same way. The composition of the equilibrium mixture is unchanged.

Heterogeneous Equilibrium

So far we have considered only chemical equilibria involving all gases, that is, **homogeneous equilibria**. An equilibrium involving more than one phase — gas and solid, for example, or liquid and solid — is said to be **heterogeneous**. For example :

$$H_2(g) + I_2(s) \rightleftharpoons 2HI(g)$$

In the above example, the equilibrium expressions for this reaction are :

$$K_c = \frac{[HI(g)]^2_{eqm}}{[H_2(g)]_{eqm}} \qquad K_p = \frac{p_{HI(g)}{}^2_{eqm}}{p_{H2(g)}{}_{eqm}}$$

Notice that I_2 is missing from the equilibrium expressions. This is because, at room temperature, iodine is a solid. The composition of the equilibrium mixture of $H_2(g)$ and $HI(g)$ is independent of the amount of solid iodine present, *as long as some of it is always present*. [The partial pressure of $I_2(g)$ (which is its vapour pressure) in equilibrium with $I_2(s)$ is constant at a constant temperature, and is independent of the volume of the container and of the quantity of solid.]

Equilibrium in Solution

Here we will consider homogeneous and heterogeneous chemical equilibria in which the solvent is much more abundant than all the other components put together. The solvent may also be a reactant or product itself.

First of all, let's look at an equilibrium that is **not in solution** :

$$CH_3COOH(l) + CH_3CH_2OH(l) \rightleftharpoons CH_3COOCH_2CH_3(l) + H_2O(l)$$

$$K_c = \frac{[CH_3COOCH_2CH_3(l)]_{eqm}\ [H_2O(l)]_{eqm}}{[CH_3CH_2OH(l)]_{eqm}\ [CH_3COOH(l)]_{eqm}}$$

In this case, water is not a solvent but a product only.

Now consider the same equilibrium **in very dilute aqueous solution**. The chemical equation is :

$$CH_3COOH(aq) + CH_3CH_2OH(aq) \rightleftharpoons CH_3COOCH_2CH_3(aq) + H_2O(l)$$

And the equilibrium expression is :

$$K_c = \frac{[CH_3COOCH_2CH_3(aq)]_{eqm}}{[CH_3CH_2OH(aq)]_{eqm}\ [CH_3COOH(aq)]_{eqm}}$$

Water is now both a product and a solvent. In reaching a state of equilibrium the concentration of the water changes so little that it is effectively constant. For this reason its value is taken along with the equilibrium constant, K_c.

Here are some more examples of chemical equilibria in aqueous solution :

$$CH_3COOH_{(aq)} + H_2O_{(l)} \rightleftharpoons CH_3COO^-_{(aq)} + H_3O^+_{(aq)}$$

$$K_a = \frac{[H_3O^+(aq)]_{eqm}\ [CH_3COO^-(aq)]_{eqm}}{[CH_3COOH\ (aq)]_{eqm}}$$

K_a is called the **acid dissociation constant**. Again, the concentration of water changes negligibly on reaching equilibrium and so its value is taken along with the equilibrium constant (K_c) to form the 'modified' equilibrium constant, K_a.

$$NH_{3(aq)} + H_2O_{(l)} \rightleftharpoons NH_4^+{}_{(aq)} + OH^-{}_{(aq)}$$

$$K_b = \frac{[NH_4^+(aq)]_{eqm}\ [OH^-(aq)]_{eqm}}{[NH_3\ (aq)]_{eqm}}$$

K_b is called the **base dissociation constant**. Again, the concentration of water changes negligibly on reaching equilibrium and so its value is taken along with the equilibrium constant (K_c) to form the 'modified' equilibrium constant, K_b.

$$AgCl_{(s)} \rightleftharpoons Ag^+{}_{(aq)} + Cl^-{}_{(aq)}$$

$$K_s = [Ag^+(aq)]_{eqm}\ [Cl^-(aq)]_{eqm}$$

K_s is the **Solubility Product**. It is the equilibrium constant for the equilibrium that exists between a slightly soluble salt and its ions in saturated solution. The solid, AgCl, is omitted from the equilibrium expression. As long as there is some solid present the concentration of the saturated solution of ions remains constant at a given temperature.

Chemical change is one of the two central concepts of chemical science, the other being *structure*. The very origins of Chemistry itself are rooted in the observations of transformations such as the combustion of wood, the freezing of water, and the winning of metals from their ores that have always been a part of human experience. It was, after all, the quest for some kind of constancy underlying change that led the Greek thinkers of around 200 BCE to the idea of elements and later to that of the atom.

It would take almost 2000 years for the scientific study of matter to pick up these concepts and incorporate them into what would emerge, in the latter part of the 19th century, as a modern view of chemical change.

1 Chemical Change : How Far, How Fast?

Chemical change occurs when the atoms that make up one or more substances rearrange themselves in such a way that new substances are formed. These substances are the *components* of the *chemical reaction system*; those components which decrease in quantity are called *reactants*, while those that increase are *products*.

A given chemical reaction system is defined by a *balanced net chemical equation* which is conventionally written as

$$reactants \rightarrow products$$

The first thing we need to know about a chemical reaction represented by a balanced equation is whether it can actually take place. If the reactants and products are all substances capable of an independent existence, then in principle, the answer is always "yes". This answer must be qualified, however, by the following considerations :

How Complete is the Reaction?

That is, what fraction of the reactants are converted into products? Some reactions convert essentially 100 per cent of reactants to products, while for others the quantity of products may be undetectable. Many are somewhere in between, meaning that significant quantities of all components remain at the end. Later on, in another part of the course, you will learn that the tendency of a reaction to occur can be predicted entirely from the properties of the reactants and products through the laws of thermodynamics.

How Fast Does the Reaction Occur?

Some reactions are over in micro-seconds; others take years. The speed of any one reaction can vary over a huge range depending on the temperature, the state of matter (gas, liquid, solid) and the presence of a catalyst. Unlike the question of completeness, there is no simple way of predicting reaction speed.

What is the Mechanism of the Reaction?

What happens, at the atomic or molecular level, when reactants are transformed into products? What intermediate species (those that are produced but later consumed so that they do not appear in the net reaction equation) are involved? This is the *microscopic*, or kinetic view of chemical change, and cannot be predicted by theory as it is presently developed and must be inferred from the results of experiments.

A reaction that is thermodynamically possible but for which no reasonably rapid mechanism is available is said to be *kinetically limited*. Conversely, one that occurs rapidly but only to a small extent is *thermodynamically limited*. As you will see later, there are often ways of getting around both kinds of limitations, and their discovery and practical applications constitute an important area of industrial chemistry.

2 What is Equilibrium?

Textile artist Hilary Rice offers this view of equilibrium which nicely evokes its dual character of internal dynamics contained within a static exterior.

Basically, the term refers to what we might call a "balance of forces". In the case of mechanical equilibrium, this is its literal definition. A book sitting on a table top remains at rest because the downward force exerted by the earth's gravity acting on the book's mass (this is what is meant by the "weight" of the book) is exactly balanced by the repulsive force between atoms that prevents two objects from simultaneously occupying the same space, acting in this case between the table surface and the book. If you pick up the book and raise it above the table top, the additional upward force exerted by your arm destroys the state of equilibrium as the book moves upward. If you wish to hold the book at rest above the table, you adjust the upward force to exactly balance the weight of the book, thus restoring equilibrium.

An object is in a state of **mechanical equilibrium** when it is either static (motionless) or in a state of unchanging motion. From the relation $f = ma$, it is apparent that if the net force on the object is zero, its acceleration must also be zero, so if we can see that an object is not undergoing a change in its motion, we know that it is in mechanical equilibrium.

Thermal Equilibrium

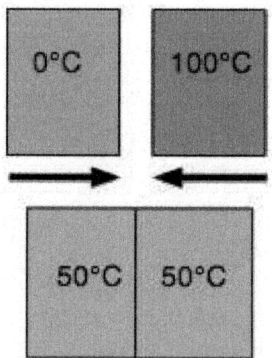

Another kind of equilibrium we all experience is **thermal equilibrium**. When two objects are brought into contact, heat will flow from the warmer object to the cooler one until their temperatures become identical. Thermal equilibrium arises from the tendency of thermal energy to become as dispersed or "diluted" as possible.

A metallic object at room temperature will feel cool to your hand when you first pick it up because the thermal sensors in your skin detect a flow of heat from your hand into the metal, but as the metal approaches the temperature of your hand, this sensation diminishes. The time it takes to achieve thermal equilibrium depends on how readily heat is conducted within and between the objects; thus a wooden object will feel warmer than a metallic object even if both are at room temperature because wood is a relatively poor thermal conductor and will therefore remove heat from your hand more slowly.

Thermal equilibrium is something we often want to avoid, or at least postpone; this is why we insulate buildings, perspire in the summer and wear heavier clothing in the winter.

Chemical Equilibrium

When a chemical reaction takes place in a container which prevents the entry or escape of any of the substances involved in the reaction, the quantities of these components change as some are consumed and others are formed. Eventually this change will come to an end, after which the composition will remain unchanged as long as the system remains undisturbed. The system is then said to be in its *equilibrium state*, or more simply, "at equilibrium".

Why Reactions Go Towards Equilibrium

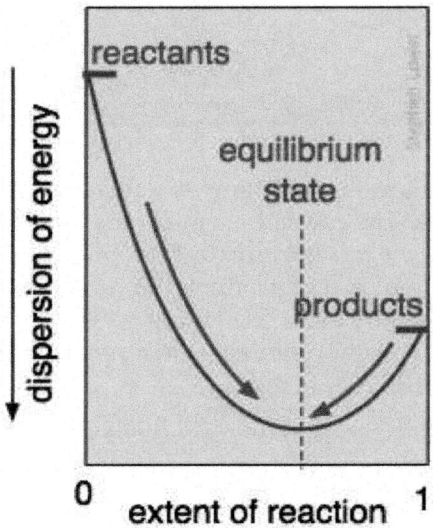

What is the nature of the "balance of forces" that drives a reaction towards chemical equilibrium? It is essentially the balance struck between the tendency of energy to reside within the chemical bonds of stable molecules, and its tendency to become dispersed and diluted. Exothermic reactions are particularly effective in this, because the heat released gets dispersed in the infinitely wider world of the surroundings.

In the reaction represented here, this balance point occurs when about 60 per cent of the reactants have been converted to products. Once this equilibrium state has been reached, no further net change will occur. (The only spontaneous changes that are allowed follow the arrows pointing towards maximum dispersal of energy.)

A more complete explanation of this must be deferred until the discussion of thermodynamics in a later chapter.

Chemical equilibrium is something you definitely want to avoid for yourself as long as possible. The myriad chemical reactions in living organisms are constantly moving *toward* equilibrium, but are prevented from getting there by input of reactants and removal of products. So rather than being in equilibrium, we try to maintain a "steady-state" condition which physiologists call *homeostasis* — maintenance of a constant internal environment. For more on this, see this Wikipedia article.**Equilibrium is death !**

For the time being, it's very important that you know this definition :

A chemical reaction is in equilibrium when there is no tendency for the quantities of reactants and products to change.

The **direction** in which we write a chemical reaction (and thus which components are considered reactants and which are products) is arbitrary. Thus the two equations

$H_2 + I_2 \rightarrow 2\,HI$	"synthesis of hydrogen iodide"
$2\,HI \rightarrow H_2 + I_2$	"dissociation of hydrogen iodide"

represent the same chemical reaction system in which the roles of the components are reversed, *and both yield the same mixture of components when the change is completed.*

This last point is central to the concept of chemical equilibrium. It makes no difference whether we start with two moles of HI or one mole each of H_2 and I_2; once the reaction has run to completion, the quantities of these two components will be the same. In general, then, we can say that the **composition of a chemical reaction system will tend to change in a direction that brings it closer to its equilibrium composition.** Once this equilibrium composition has been attained, no further change in the quantities of the components will occur as long as the system remains undisturbed.

It's the same both ways

The two diagrams below show how the concentrations of the three components of this chemical reaction change with time. Examine the two sets of plots carefully, noting which substances have zero initial concentrations, and are thus "products" of the reaction equations shown. Satisfy yourself that these two sets represent the same *chemical reaction system*, but with the reactions occurring in opposite directions. Most importantly, note how the final (equilibrium) concentrations of the components are the same in the two cases.

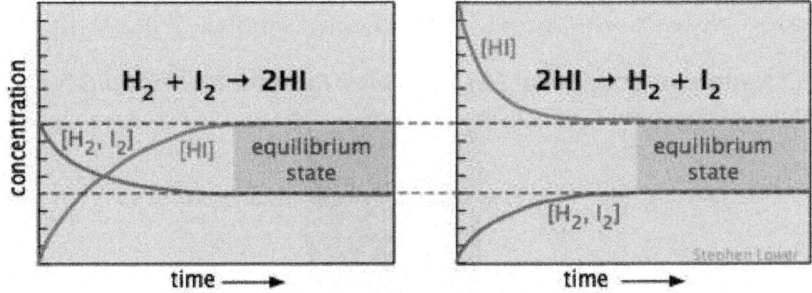

Whether we start with an equimolar mixture of H_2 and I_2 (left) or a pure sample of hydrogen iodide (shown on the right, using twice the initial concentration of HI to keep the number of atoms the same), the composition after equilibrium is attained (shaded regions on the right) will be the same.

The equilibrium composition is independent of the direction from which it is approached.

3 What is a Reversible Reaction?

A chemical equation of the form A $\rightarrow$ B represents the transformation of A into B, but it does not imply that *all* of the reactants will be converted into products, or that the *reverse* reaction B $\rightarrow$ A cannot also occur.

In general, *both* processes (forward and reverse) can be expected to occur, resulting in an *equilibrium mixture* containing finite amounts of *all* of the components of the reaction system. (We use the word *components* when we do not wish to distinguish between reactants and products.)

If the equilibrium state is one in which significant quantities of both reactants and products are present (as in the hydrogen iodide example given above), then the reaction is said to *incomplete* or *reversible*.

The latter term is preferable because it avoids confusion with "complete" in its other sense of being completed or finished, implying that the reaction has run its course and is now at equilibrium.

- If it is desired to emphasise the reversibility of a reaction, the single arrow in the equation is replaced with a pair of hooked lines pointing in opposite directions, as in A $\rightleftharpoons$ B.

  Note that there is no fundamental difference between the meanings of A $\rightarrow$ B and A $\rightleftharpoons$ B. Some older text-books just use A = B.

- A reaction is said to be *complete* or *quantitative* when the equilibrium composition contains no significant amount of the reactants. However, a reaction that is complete when written in one direction is said "not to occur" when written in the reverse direction.

In principle, all chemical reactions are reversible, but this reversibility may not be observable if the fraction of products in the equilibrium mixture is very small, or if the reverse reaction is very slow (the chemist's term is "*kinetically inhibited*")

How did Napoleon Bonaparte *Help Discover Reversible Reactions?*

We can thank Napoleon for bringing the concept of reaction reversibility to Chemistry.

Napoleon recruited the eminent French chemist Claude Louis Berthollet (1748-1822) to accompany him as scientific advisor on the most far-flung of his campaigns, the expedition into Egypt in 1798. Once in Egypt, Berthollet noticed deposits of sodium carbonate around the edges of some the salt lakes found there. He was already familiar with the reaction

$$Na_2CO_3 + CaCl_2 \rightarrow CaCO_3 + 2\,NaCl$$

which was known to proceed to completion in the laboratory. He immediately realised that the Na_2CO_3 must have been formed by the reverse of this process brought about by the very high concentration of salt in the slowly-evaporating waters. This led Berthollet to question the belief of the time that a reaction could only proceed in a single direction. His famous text-book *Essai de statique chimique* (1803) presented his speculations on chemical affinity and his discovery that an excess of the product of a reaction could drive it in the reverse direction.

Unfortunately, Berthollet got a bit carried away by the idea that a reaction could be influenced by the amounts of substances present, and maintained that the same should be true for the compositions of individual compounds. This brought him into conflict with the recently accepted Law of Definite Proportions (that a compound is made up of fixed numbers of its constituent atoms), so his ideas (the good along with the bad) were promptly discredited and remained largely forgotten for 50 years. (Ironically, it is now known that certain classes of compounds do in fact exhibit variable composition of the kind that Berthollet envisioned.)

4 What is the Law of Mass Action?

Berthollet's ideas about reversible reactions were finally vindicated by experiments carried out by others, most notably the Norwegian chemists (and brothers-in-law) Cato Guldberg and Peter Waage. During the period 1864-1879 they showed that an equilibrium can be approached from either direction (see the hydrogen iodide illustration above), implying that any reaction $aA + bB \rightarrow cC + dD$ is really a competition between a "forward" and a "reverse" reaction. When a reaction is at equilibrium, the rates of these two reactions are identical, so no *net* (macroscopic) change is observed, although individual components are actively being transformed at the microscopic level.

Equilibrium is Dynamic !

The "active masses" are essentially the *concentrations* of the reactants and products that combine directly in the manner represented by the reaction equation.

The meaning of the terms on the right sides of the equations is simply that the rate is proportional to the concentrations of the components.

Guldberg and Waage showed that for a reaction $aA + bB \rightarrow cC + dD$, the rate (speed) of the reaction in either direction is proportional to what they called the "active masses" of the various components :

$$\text{rate of forward reaction} = k_f[A]^a[B]^b$$

$$\text{rate of reverse reaction} = k_r[C]^c[D]^d$$

in which the proportionality constants k are called *rate constants* and the quantities in square brackets represent concentrations. If we combine the two reactants A and B, the forward reaction starts immediately; then, as the products C and D begin to build up, the reverse process gets underway. As the reaction proceeds, the rate of the forward reaction diminishes while that of the reverse reaction increases.

Eventually the two processes are proceeding at the same rate, and the reaction is at equilibrium :

rate of forward reaction = rate of reverse reaction

$$k_f[A]^a[B]^b = k_r [C]^c [D]^d$$

It is very important that you understand the significance of this relation. The equilibrium state is one in which there is no **net** change in the quantities of reactants and products. But do not confuse this with a state of "no change"; at equilibrium, the forward and reverse reactions continue, but *at identical rates*, essentially cancelling each other out.

Equilibrium is macroscopically static, but is microscopically dynamic!

To further illustrate the dynamic character of chemical equilibrium, suppose that we now change the composition of the system previously at equilibrium by adding some C or withdrawing some A (thus changing their "active masses"). The reverse rate will temporarily exceed the forward rate and a change in composition ("a shift in the equilibrium") will occur until a new equilibrium composition is achieved.

The Composition of the Equilibrium State Depends on the Ratio of the Forward- and Reverse Rate Constants

Be sure you understand the difference between the *rate* of a reaction and a *rate constant*. The latter, usually designated by k, relates the reaction rate to the concentration of one or more of the reaction components — for example, $rate = k [A]$.

At equilibrium the rates of the forward and reverse processes are identical, but the rate *constants* are generally different. To see how this works, consider the simplified reaction $A \rightarrow B$ in the following three scenarios.

$k_f >> k_r$

If the rate constants are greatly different (by many orders of magnitude), then this requires that the equilibrium concentrations of products exceed those of the reactants by the same ratio. Thus the equilibrium composition will lie strongly on the "right"; the reaction can be said to be "complete" or "quantitative".

$k_f << k_r$

The rates can only be identical (equilibrium achieved) if the concentrations of the products are very small. We describe the resulting equilibrium as strongly favouring the left; very little product is formed. In the most extreme cases, we might even say that "the reaction does not take place".

$k_f \approx k_r$

If k_f and k_r have comparable values (within, say, several orders of magnitude), then signficant concentrations of products and reactants are present at equilibrium; we say the the reaction is "incomplete" and "reversible".

The images shown below offer yet another way of looking at these three cases. The plots show how the relative concentrations of the reactant and product change during the course of the reaction.

The plots differ in the assumptions we make about the ratio of k_f to k_r. The equilibrium composition of the system is illustrated by the proportions of A and B in the horizontal parts of each plot where the composition remains unchanged. In each case, the two rate constants are sufficiently close in magnitude that each reaction can be considered "incomplete".

See here for a very nice online simulation that allows you to explore the effects of your own set of starting concentrations and rate constants on this kind of plot.

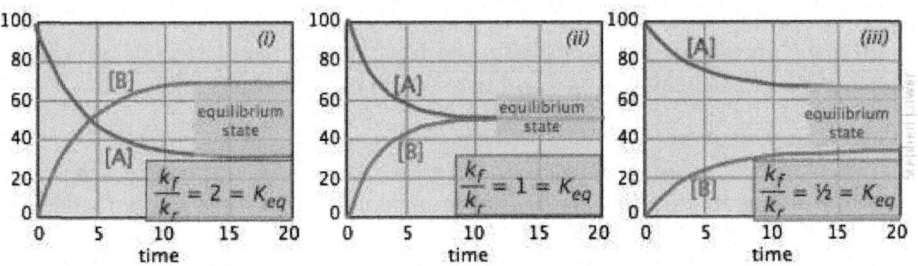

- In plot *(i)* the forward rate constant is twice as large as the reverse rate constant, so product (B) is favoured, but there is sufficient reverse reaction to maintain a significant quantity of A.
- In *(ii)*, the forward and reverse rate constants have identical magnitudes. Not surprisingly, so are the equilibrium values of [A] and [B].
- In *(iii)*, the reverse rate constant exceeds the forward rate constant, so the eqilibrium composition is definitely "on the left".

The Law of Mass Action is thus essentially the statement that the equilibrium composition of a reaction mixture can vary according to the quantities of components that are present. This of course is just what Berthollet observed in his Egyptian salt ponds, but we now understand it to be a consequence of the dynamic nature of chemical equilibrium.

5 How do We Know When a Reaction is at Equilibrium?

Clearly, if we observe some change taking place—a change in colour, the release of gas bubbles, the appearance of a precipitate, or the release of heat, we know the reaction is not yet at equilibrium.

But the absence of any apparent change does not by itself establish that the reaction is at equilibrium. The equilibrium state is one in which not only no change in composition take place, but also one in which no energetic tendency for further change is present. Unfortunately, "tendency" is not a property that is directly observable! Consider, for example, the reaction representing the synthesis of water from its elements :

$$2\ H_2(g) + O_2(g) \rightarrow 2\ H_2O(g)$$

You can store the two gaseous reactants in the same container indefinitely without any observable change occurring. But if you create an electrical spark in the container or introduce a flame, *bang* ! After you pick yourself up off the floor and remove the shrapnel from what's left of your body, you will know very well that the system was not initially at equilibrium ! It happens that this particular reaction has a tremendous tendency to take place, but for reasons that we will discuss in a later chapter, nothing can happen until we "set it off" in some way — in this case by exposing the mixture to a flame or spark, or (in a more gentle way) by introducing a platinum wire, which acts as a *catalyst*.

A reaction of this kind is said to be highly favoured thermodynamically, but inhibited kinetically. The similar reaction of hydrogen and iodine

$$H_2(g) + I_2(g) \rightarrow 2\ HI(g)$$

by contrast is only moderately favoured thermodynamically (and is thus incomplete), but its kinetics are both unspectacular and reasonably facile.

Some Simple Tests for the Equilibrium State

- As we explained above in the context of the law of mass action, addition or removal of one component of the reaction will affect the amounts of all the others. For example, if we add more of a reactant, we would expect to see the concentration of a product change. If this does not happen, then it is likely that the reaction is kinetically inhibited and that the system is unable to attain equilibrium.

- It is almost always the case, however, that once a reaction actually starts, it will continue on its own until it reaches equilibrium, so if we can observe the change as it occurs and see it slow down and stop, we can be reasonably certain that the system is in equilibrium. This is by far the chemist's most common criterion.

- There is one other experimental test for equilibrium in a chemical reaction, although it is really only applicable to the kind of reactions we described above as being reversible. As we shall see later, the equilibrium state of a system is always sensitive to the temperature, and often to the pressure, so any changes in these variables, however, small, will temporarily disrupt the equilibrium, resulting in an observable change in the composition of the system as it moves towards its new equilibrium state.

Summary

Make sure you thoroughly understand the following essential ideas which have been presented above. It is especially imortant that you know the precise meanings of all the highlighted terms in the context of this topic.

- Any reaction that can be represented by a balanced chemical equation can take place, at least in principle. However, there are two important qualifications :
  - o The tendency for the change to occur may be so small that the quantity of products formed may be very low, and perhaps negligible. A reaction of this kind is said to be *thermodynamically inhibited*. The tendency for chemical change is governed solely by the properties of the reactants and products, and can be predicted by applying the laws of thermodynamics.
  - o The rate at which the reaction proceeds may be very small, or even zero, in which case we say the reaction is *kinetically inhibited*. Reaction rates depend on the *mechanism* of the reaction — that is, on what actually happens to the atoms as reactants are transformed into products. Reaction mechanisms cannot generally be predicted, and must be worked out experimentally. Also, the same reaction may have different mechansims under different conditions.
- As a chemical change proceeds, the quantities of the components on one side of the reaction equation will decrease, and those on the other side will increase. Eventually the reaction slows down and the composition of the system stops changing. At this point the reaction is in its *equilibrium state*, and no further change in composition will occur as long as the system is left undisturbed.
- For many reactions, the equilibrium state is one in which components on both sides of the equation (that is, both reactants and products) are present in significant amounts. Such a reaction is said to be *incomplete* or *reversible*.
- **The equilibrium composition is independent of the direction from which it is approached**; the labelling of substances as "reactants" or "products" is entirely a matter of convenience. (See the hydrogen iodide reaction plots above.)
- The *law of mass action* states that any chemical change is a competition between a *forward* reaction (left-to-right in the chemical equation) and a *reverse* reaction. The rate of each of these processes is governed by the concentrations of the substances reacting; as the reaction proceeds, these rates approach each other and at equilibrium they become identical.
- From the above, it follows that **equilibrium is a *dynamic process*** in which *microscopic change* (the forward and reverse reactions) continues to occur, but *macroscopic change* (changes in the quantities of substances) is absent.

- When a chemical reaction is at equilibrium, any disturbance of the system, such as a change in temperature, or addition or removal of one of the reaction components, will "shift" the composition to a new equilibrium state. This is the only unambiguous way of verifying that a reaction is at equilibrium. The fact that the composition remains static does not in itself prove that a reaction is at equilibrium, because the change may be kinetically inhibited.

What You Should be Able to Do

- Define "*the **equilibrium state** of a chemical reaction system*". What is its practical significance?
- State the meaning and significance of the following terms :
  - *reversible reaction*
  - *quantitative reaction*
  - *kinetically inhibited reaction.*
- Explain the meaning of the statement "*equilibrium is macroscopically static, but microscopically dynamic*". Very important !
- Explain how the relatve magnitudes of the forward and reverse ***reaction rate constants*** in the Mass Action expression affect the equilibrium composition of a reaction system.
- Describe several things you might look for during an experiment that would help determine if a reaction system is in its equilibrium state.

Concept Map

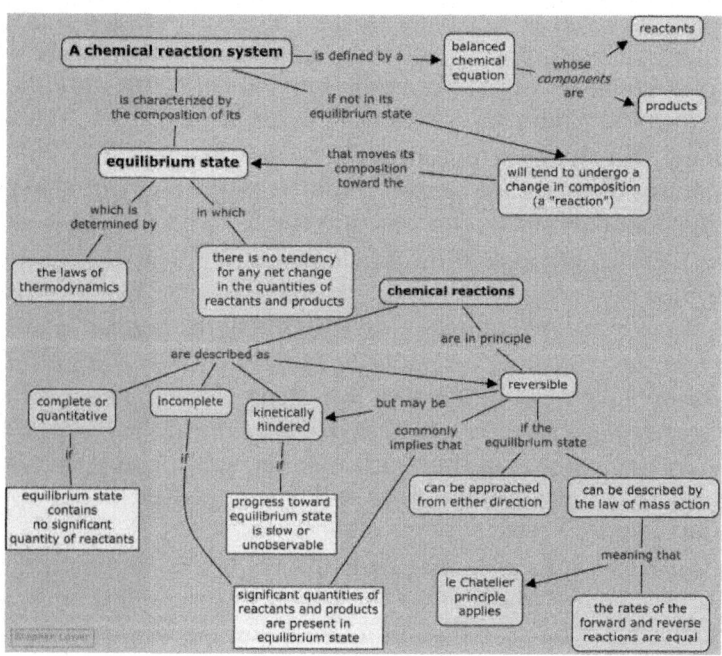

EQUILIBRIUM CONSTANT

To determine the amount of each compound that will be present at equilibrium you must know the equilibrium constant. To determine the equilibrium constant you must consider the generic equation :

$$aA + bB <\text{-->} cC + dD$$

The upper case letters are the molar concentrations of the reactants and products. The lower case letters are the coefficients that balance the equation. Use the following equation to determine the equilibrium constant (K_c).

$$K_c = \frac{[C]^c[D]^d}{[A]^a[B]^b}$$

For example, determining the equilibrium constant of the following equation can be accomplished by using the K_c equation.

Using the following equation, calculate the equilibrium constant.

$$N_2(g) + 3H_2(g) <\text{-->} 2NH_3(g)$$

A one-liter vessel contains 1.60 moles NH_3, .800 moles N_2, and 1.20 moles of H_2. What is the equilibrium constant?

$$K_c = \frac{[1.60]^2}{[.800][1.20]^3} = 29.6$$

Answer : 1.85

For a general *elementary* chemical reaction,

$$aA + bB \leftrightarrow cC + dD$$

The concentrations of the reactants and products are related to each other according to

$$K_c = \frac{[C]^c[D]^d}{[A]^a[B]^b}$$

The number K_c is called the *equilibrium constant*, and is a function of temperature only (*i.e.*, its numerical value doesn't change unless the temperature changes – we'll hold the temperature constant for now). Note the word *elementary*, more on that later; for now, all reactions are elementary. The stoichiometric coefficients a, b, c and d show up as powers of the corresponding reactants (and products – just call them all reactants from now on). For our example reaction,

$$A + B \leftrightarrow C$$

the equilibrium constant is defined

$$K_c = \frac{[C]}{[A][B]}$$

Notice that the units of K_c in this case are $M^{-1} = L/mol$.

There is another useful definition of the equilibrium constant based on pressure rather than concentration. The ideal gas law reads :

$$PV = nRT$$

Here P is the total pressure. In the case of several components, each has a partial pressure, all of which sum up to the total pressure :

$$P = P_A + P_B + P_C$$

For each component, we can write the ideal gas law (putting a subscript where applicable)

$$P_A V = n_A RT \Rightarrow \frac{n_A}{V} = [A] = \frac{P_A}{RT}$$

This works for the other components, too, and gives us the relation between the concentrations and the partial pressures. If we plug in the partial pressures in the definition of K_c above, we get :

$$K_c = \frac{[C]}{[A][B]} = \frac{P_C}{RT}\frac{RT}{P_A}\frac{RT}{P_B} = \frac{P_C}{P_A P_B}(RT) = K_P(RT)$$

In the more general case where the reaction is

$$aA + bB \leftrightarrow cC + dD$$

And K_c is

$$K_c = \frac{[C]^c[D]^d}{[A]^a[B]^b}$$

Then the new equilibrium constant becomes

$$K_P = K_C(RT)^{(c+d)-(a+b)} = K_C(RT)^{\Delta n}$$

Where :

$$\Delta n = (c + d) - (a + b)$$

Is the change in number of moles in the gas phase. These numbers come from the balanced equation and are basically the sum of the stoichiometric coefficients of the products minus the sum of the stoichiometric coefficients of the reactants.

Problems using K_p instead of K_c are solved the same way, except we are looking for the partial pressures P_x instead of the concentrations [X], X = A, B, C, etc. is particularly useful for gas phase problems since the pressure is usually more convenient to measure than the concentration. Also, there is a direct relationship between K_p and the Gibbs energy of the reaction, ΔG^0_{rxn} .

LE CHATELIER'S PRINCIPLE

Le Chatelier's principle states that when a system in chemical equilibrium is disturbed by a change of temperature, pressure, or a concentration, the system shifts in equilibrium composition in a way that tends to counteract this change of variable. The three ways that Le Chatelier's principle says you can affect the outcome of the equilibrium are as follows :

- Changing concentrations by adding or removing products or reactants to the reaction vessel.
- Changing partial pressure of gaseous reactants and products.
- Changing the temperature.

These actions change each equilibrium differently, therefore you must determine what needs to happen for the reaction to get back in equilibrium.

Example involving change of concentration :

In the equation

$$2NO_{(g)} + O_{2(g)} <--> 2NO_{2(g)}$$

If you add more $NO_{(g)}$ the equilibrium shifts to the right producing more $NO_{2(g)}$

If you add more $O_{2(g)}$ the equilibrium shifts to the right producing more $NO_{2(g)}$

If you add more $NO_{2(g)}$ the equilibrium shifts to the left producing more $NO_{(g)}$ and $O_{2(g)}$

Example involving pressure change :

In the equation

$$2SO_{2(g)} + O_{2(g)} <--> 2SO_{3(g)}$$

an increase in pressure will cause the reaction to shift in the direction that reduces pressure, that is the side with the fewer number of gas molecules. Therefore an increase in pressure will cause a shift to the right, producing more product. (A decrease in volume is one way of increasing pressure.)

Example involving temperature change :

In the equation

$$N_{2(g)} + 3H_{2(g)} <--> 2NH_3 + 91.8 \text{ kJ,}$$

an increase in temperature will cause a shift to the left because the reverse reaction uses the excess heat. An increase in forward reaction would produce even more heat since the forward reaction is exothermic. Therefore the shift caused by a change in temperature depends upon whether the reaction is exothermic or endothermic.

If a reaction is at equilibrium and we alter the conditions so as to create a new *equilibrium state*, then the composition of the system will tend to change until that new equilibrium state is attained. (We say "tend to change" because if the reaction is *kinetically inhibited*, the change may be too slow to observe or it may never take place.) In 1884, the French chemical engineer and teacher **Henri Le Châtelier** (1850-1936) showed that in every such case, the new equilibrium state is one that partially reduces the effect of the change that brought it about.

This law is known to every Chemistry student as the *Le Châtelier principle*. His original formulation was somewhat complicated, but a reasonably useful paraphrase of it reads as follows :

To see how this works (and you *must* do so, as this is of such fundamental importance that you simply cannot do any meaningful chemistry without a thorough working understanding of this principle), look again at the hydrogen iodide dissociation reaction

$$2\ HI \rightarrow H_2 + I_2$$

Consider an arbitrary mixture of these three components at equilibrium, and assume that we inject more hydrogen gas into the container. Because the H_2 concentration now exceeds its new equilibrium value, the system is no longer in its equilibrium state, so a net reaction now ensues as the system moves to the new state.

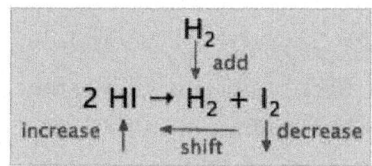

The Le Châtelier principle states that the net reaction will be in a direction that tends to reduce the effect of the added H_2. This can occur if some of the H_2 is consumed by reacting with I_2 to form more HI; in other words, a net reaction occurs in the reverse direction. Chemists usually simply say that "the equilibrium shifts to the left".

To get a better idea of how this works, carefully examine the diagram below which follows the concentrations of the three components of this reaction as they might change in time (the time scale here will typically be about an hour) :

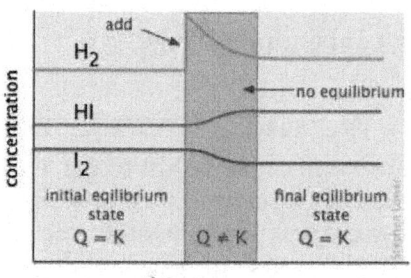

Disruption and Restoration of Equilibrium

At the left, the concentrations of the three components do not change with time because the system is at equilibrium. We then add more hydrogen to the system, disrupting the equilibrium. A net reaction then ensues that moves the system to a new equilibrium state (right) in which the quantity of hydrogen iodide has increased; in the process, some of the I_2 and H_2 are consumed. Notice that the new equilibrium state contains more hydrogen than did the initial state, but not as much as was added; as the le Châtelier principle predicts, the change we made (addition of H_2) has been partially counteracted by the "shift to the right".

The following tabLe Contains several examples showing how changing the quantity of a reaction component can shift an established equilibrium.

system	change	result
$CO_2 + H_2 \rightarrow H_2O(g) +$ CO	a drying agent is added to absorb H_2O	**Shift to the right**. Continuous removal of a product will force any reaction to the right
$H_2(g) + I_2(g) \rightarrow 2HI(g)$	Some nitrogen gas is added	**No change**; N_2 is not a component of this reaction system.
$NaCl(s) + H_2SO_4(l) \rightarrow$ $Na_2SO_4(s) + HCl(g)$	reaction is carried out in an open container	Because HCl is a gas that can escape from the system, the reaction is forced to the **right**. This is the basis for the commercial production of hydrochloric acid.
$H_2O(l) \rightarrow H_2O(g)$	water evaporates from an open container	Continuous removal of water vapour forces the reaction to the **right**, so equilibrium is never achieved.
$HCN(aq) \rightarrow H^+(aq) +$ $CN^-(aq)$	the solution is diluted	Shift to **right**; the product $[H^+][CN^-]$ diminishes more rapidly than does [HCN].
$AgCl(s) \rightarrow Ag^+(aq) +$ $Cl^-(aq)$	some NaCl is added to the solution	Shift to **left** due to increase in Cl^- concentration. This is known as the common ion effect on solubility.
$N_2 + 3 H_2 \rightarrow 2 NH_3$	a catalyst is added to speed up this reaction	**No change**. Catalysts affect only the rate of a reaction; the have no effect at all on the composition of the equilibrium state.

How Do Changes in Temperature Affect Equilibria

Virtually all chemical reactions are accompanied by the liberation or uptake of heat. If we regard heat as a "reactant" or "product" in an endothermic or exothermic reaction respectively, we can use the Le Châtelier principle to predict the direction in which an increase or decrease in temperature will shift the equilibrium state. Thus for the oxidation of nitrogen, an endothermic process, we can write

$$[heat] + N_2 + O_2 \rightarrow 2\ NO$$

raise temperature

$$N_2(g) + O_2(g) \rightarrow 2\ NO_2(g)$$

0 kJ ↓ 0 kJ   66 kJ ↑
decrease ——shift——→ increase

Suppose this reaction is at equilibrium at some temperature T_1 and we raise the temperature to T_2. The Le Châtelier principle tells us that a net reaction will occur in the direction that will partially counteract this change. Since the reaction is endothermic, a shift of the equilibrium to the right will take place.

Nitric oxide, the product of this reaction, is a major air pollutant which initiates a sequence of steps leading to the formation of atmospheric smog. Its formation is an unwanted side reaction which occurs when the air (which is introduced into the combustion chamber of an engine to supply oxygen) gets heated to a high temperature. Designers of internal combustion engines now try, by various means, to limit the temperature in the combustion region, or to restrict its highest-temperature part to a small volume within the combustion chamber.

How Do Changes in Pressure Affect Equilibria

You will recall that if the pressure of a gas is reduced, its volume will increase; pressure and volume are inversely proportional. With this in mind, suppose that the reaction :

$$2\,NO_2(g) \rightarrow N_2O_4(g)$$

is in equilibrium at some arbitrary temperature and pressure, and that we double the pressure, perhaps by compressing the mixture to a smaller volume. From the Le Châtelier principle we know that the equilibrium state will change to one that tends to counteract the increase in pressure. This can occur if some of the NO_2 reacts to form more of the dinitrogen tetroxide, since two moles of gas are being removed from the system for every mole of N_2O_4 formed, thereby decreasing the total volume of the system. Thus increasing the pressure will shift this equilibrium to the right.

It is important to understand that changing the pressure will have a significant effect only on reactions in which there is a change in the number of moles of gas.

For the above reaction, this change :

$$\Delta n_g = (n_{products} - n_{reactants}) = 1 - 2 = -1.$$

In the case of the nitrogen oxidation reaction

$N_2 + O_2 \rightarrow 2\,NO$, $\Delta n_g = 0$ and pressure will have no effect.

The volumes of solids and liquids are hardly affected by the pressure at all, so for reactions that do not involve gaseous substances, the effects of pressure changes are ordinarily negligible. Exceptions arise under conditions of very high pressure such as exist in the interior of the Earth or near the bottom of the ocean. A good example is the dissolution of calcium carbonate $CaCO_3(s) \rightarrow Ca^{2+} + CO_3^{2-}$. There is a slight decrease in the volume when this reaction takes place, so an increase in the pressure will shift the equilibrium to the right, with the results that calcium carbonate becomes more soluble at higher pressures.

The skeletons of several varieties of microscopic organisms that inhabit the top of the ocean are made of $CaCO_3$, so there is a continual rain of this substance towards the bottom of the ocean as these organisms die. As a consequence, the floor of the Atlantic ocean is covered with a blanket of calcium carbonate. This is not true for the Pacific ocean, which is deeper; once the skeletons fall below a certain depth, the higher pressure causes them to dissolve. Some of the seamounts (undersea mountains) in the Pacific extend above the solubility boundary so that their upper parts are covered with $CaCO_3$ sediments.

The effect of pressure on a reaction involving substances whose boiling points fall within the range of commonly encountered temperature will be sensitive to the states of these substances at the temperature of interest.

For reactions involving gases, only changes in the partial pressures of those gases directly involved in the reaction are important; the presence of other gases has no effect.

Problem Example

The commercial production of hydrogen is carried out by treating natural gas with steam at high temperatures and in the presence of a catalyst ("steam reforming of methane") :

$$CH_4 + H_2O \rightarrow CH_3OH + H_2$$

Given the following boiling points : CH_4 (methane) = –161°C, H_2O = 100°C, CH_3OH = 65°, H_2 = –253°C, predict the effects of an increase in the total pressure on this equilibrium at 50°, 75° and 120°C.

Solution : Calculate the change in the moles of gas for each process :

temp	equation	Δn_g	shift
50°	$CH_4(g) + H_2O(l) \rightarrow CH_3OH(l) + H_2(g)$	0	none
75°	$CH_4(g) + H_2O(l) \rightarrow CH_3OH(g) + H_2(g)$	+1	to left
120°	$CH_4(g) + H_2O(g) \rightarrow CH_3OH(g) + H_2(g)$	0	none

What is the Haber Process and Why is it Important

The Haber process for the synthesis of ammonia is based on the exothermic reaction :

$$N_2(g) + 3\,H_2(g) \rightarrow 2\,NH_3(g) \qquad \Delta H = -92 \text{ kJ/mol}$$

The Le Châtelier principle tells us that in order to maximise the amount of product in the reaction mixture, it should be carried out at high pressure and low temperature. However, the lower the temperature, the slower the reaction (this is true of virtually all chemical reactions.) As long as the choice had to be made between a low yield of ammonia quickly or a high yield over a long period of time, this reaction was infeasible economically.

Nitrogen is available for free, being the major component of air, but the strong triple bond in N_2 makes it extremely difficult to incorporate this element into species such as NO_3^- and NH_4^+ which serve as the starting points for the wide variety of nitrogen-containing compounds that are essential for modern industry. This conversion is known as *nitrogen fixation*, and because nitrogen is an essential plant nutrient, modern intensive agriculture is utterly dependent on huge amounts of fixed nitrogen in the form of fertilizer. Until around 1900, the major source of fixed nitrogen was the $NaNO_3$ found in extensive deposits in South America. Several chemical processes for obtaining nitrogen compounds were developed in the early 1900's, but they proved too inefficient to meet the increasing demand.

$CH_4 + H_2O \rightarrow CO + 3\ H_2$ formation of synthesis gas from methane

$CO + H_2O \rightarrow CO_2 + H_2$ shift reaction carried out in reformer

The Haber-Bosch process is considered the most important chemical synthesis developed in the 20th century. Besides its scientific importance as the first large-scale application of the laws of chemical equilibrium, it has had tremendous economic and social impact; without an inexpensive source of fixed nitrogen, the intensive crop production required to feed the world's growing population would have been impossible. Haber was awarded the 1918 Nobel Prize in Chemistry in recognition of his.

The Le Châtelier Principle in Physiology

Many of the chemical reactions that occur in living organisms are regulated through the Le Châtelier principle.

Oxygen Transport by the Blood

Few of these are more important to warm-blooded organisms than those that relate to aerobic respiration, in which oxygen is transported to the cells where it is combined with glucose and metabolised to carbon dioxide, which then moves back to the lungs from which it is expelled.

$$hemoglobin + O_2 \rightleftharpoons oxyhemoglobin$$

The partial pressure of O_2 in the air is 0.2 atm, sufficient to allow these molecules to be taken up by hemoglobin (the red pigment of blood) in which it becomes loosely bound in a complex known as oxyhemoglobin. At the ends of the capillaries which deliver the blood to the tissues, the O_2 concentration is reduced by about 50 per cent owing to its consumption by the cells. This shifts the equilibrium to the left, releasing the oxygen so it can diffuse into the cells.

Much more detail about the mechanism by which hemoglobin transfers oxygen from the lungs to tissues, and then carries carbon dioxide and hydrogen ions back to the lungs, can be found on this U. of Virginia page.

Maintence of Blood Ph

Carbon dioxide reacts with water to form a weak acid H_2CO_3 which would cause the blood pH to fall to dangerous levels if it were not promptly removed as it is excreted by the cells. This is accomplished by combining it with carbonate ion through the reaction :

$$H_2CO_3 + CO_3^{2-} \rightleftharpoons 2\,HCO_3^-$$

which is forced to the right by the high local CO_2 concentration within the tissues. Once the hydrogen carbonate (bicarbonate) ions reach the lung tissues where the CO_2 partial pressure is much smaller, the reaction reverses and the CO_2 is expelled.

Carbon Monoxide Poisoning

Carbon monoxide, a product of incomplete combustion that is present in automotive exhaust and cigarette smoke, binds to hemoglobin 200 times more tightly than does O_2. This blocks the uptake and transport of oxygen by setting up a competing equilibrium

$$O_2\text{-hemoglobin} \rightleftharpoons \text{hemoglobin} \rightleftharpoons \text{CO-hemoglobin}$$

Air that contains as little as 0.1 per cent carbon monoxide can tie up about half of the hemoglobin binding sites, reducing the amount of O_2 reaching the tissues to fatal levels. Carbon monoxide poisoning is treated by administration of pure O_2 which promotes the shift of the above equilibrium to the left. This can be made even more effective by placing the victim in a hyperbaric chamber in which the pressure of O_2 can be made greater than 1 atm.

Make sure you thoroughly understand the following essential ideas which have been presented above. It is especially imortant that you know the precise meanings of all the highlighted terms in the context of this topic.

- A system in its *equilibrium state* will remain in that state indefinitely as long as it is undisturbed. If the equilibrium is destroyed by subjecting the system to a change of pressure, temperature, or the number of moles of a substance, then a net reaction will tend to take place that moves the system to a new equilibrium state. *Le Châtelier's principle* says that this net reaction will occur in a direction that partially offsets the change.
- The Le Châtelier Principle has practical effect only for reactions in which signficant quantities of both reactants and products are present at equilibrium — that is, for reactions that are *thermodynamically reversible*.
- Addition of more product substances to an equilibrium mixture will shift the equilibrium to the left; addition of more reactant substances will shift it to the right. These effects are easily explained in terms of competing forward — and reverse reactions — that is, by the *law of mass action*.
- If a reaction is *exothermic* (releases heat), an increase in the **temperature** will force the equilibrium to the left, causing the system to absorb heat and thus partially ofsetting the rise in temperature. The opposite effect occurs for *endothermic* reactions, which are shifted to the right by rising temperature.
- The effect of **pressure** on an equilibrium is significant only for reactions which involve different numbers of moles of gases on the two sides of the equation. If the number of moles of gases increases, than an increase in the total pres-

sure will tend to initiate a reverse reaction that consumes some the products, partially reducing the effect of the pressure increase.

- The classic example of the practical use of the Le Châtelier principle is the *Haber-Bosch process* for the synthesis of ammonia, in which a balance between low temperature and high pressure must be found.

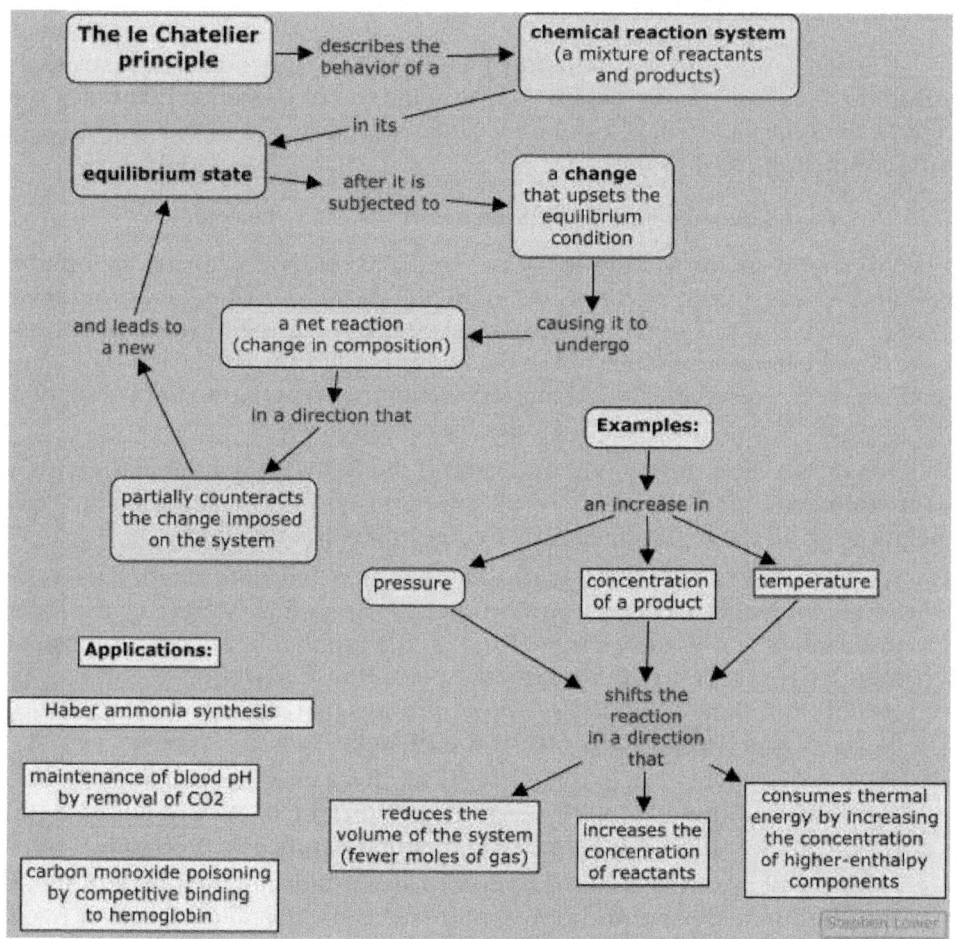

Q AND K : WHAT'S THE DIFFERENCE

Given a chemical reaction system and its equilibrium constant, together with the concentrations or pressures of the reactants and products actually present, we often need to predict whether, and in which direction, a net change in composition will tend to take place. The concept of the reaction quotient makes this very easy to do, but the similarity of the algebraic forms of K and Q can easily lead to confusion. It is hoped that the more detailed treatment offered below will help.

What is the Equilibrium Quotient

In the previous section we defined the *equilibrium expression* for the reaction

$$a\,A + b\,B \rightarrow c\,C + d\,D \quad \text{as} \quad \frac{[C]^{c}[D]^{d}}{[A]^{a}[B]^{b}}$$

In the general case in which the concentrations can have any arbitrary values (including zero), this expression is called the *reaction quotient* (the term *equilibrium quotient* is also commonly used.) and its value is denoted by Q (or Q_c or Q_p if we wish to emphasise that the terms represent molar concentrations or partial pressures.) If the terms correspond to *equilibrium* concentrations, then the above expression is called the *equilibrium constant* and its value is denoted by K (or K_c or K_p.)

K is thus the special value that Q has when the reaction is at equilibrium.

The value of Q in relation to K serves as an index how the composition of the reaction system compares to that of the equilibrium state, and thus it indicates the direction in which any net reaction must proceed.

For example, if we combine the two reactants A and B at concentrations of 1 mol L^{-1} each, the value of Q will be 0÷1=0. The only possible change is the conversion of some of these reacants into products. If instead our mixture consists only of the two products C and D, Q will be indeterminately large (1÷0) and the only possible change will be in the reverse direction.

It is easy to see (by simple application of the le Châtelier principle) that the ratio of Q/K immediately tells us whether, and in which direction, a net reaction will occur as the system moves towards its equilibrium state. A schematic view of this relationship is shown below :

$$Q = \frac{[products]}{[reactants]} \qquad K = Q = \frac{[products]}{[reactants]} \qquad Q = \frac{[products]}{[reactants]}$$

$Q > K$	$Q = K$	$Q < K$
net reaction to left	no net reaction	net reaction to right

More formally, we simply look at which of the following describes the relative values of Q and K :

Q/K	
>1	Product concentration too high for equilibrium; net reaction proceeds to **left**.
=1	System is at equilibrium; **no net change** will occur.
<1	Product concentration too low for equilibrium; net reaction proceeds to **right**.

It is very important that you be able to work out these relations for yourself, not by memorising them, but from the definitions of Q and K.

Problem Example

The equilibrium constant for the oxidation of sulfur dioxide is $K_p = 0.14$ at 900 K.

$$2\,SO_2(g) + O_2(g) \rightarrow 2\,SO_3(g)$$

If a reaction vessel is filled with SO_3 at a partial pressure of 0.10 atm and with O_2 and SO_2 each at a partial pressure of 0.20 atm, what can you conclude about whether, and in which direction, any net change in composition will take place?

Solution :

The value of the equilibrium quotient Q for the initial conditions is

$$Q = \frac{(p_{SO_3})^2}{(p_{O_2})\,(p_{SO_2})^2} = \frac{(0.10\ \text{atm})^2}{(0.20\ \text{atm})\,(0.20\ \text{atm})^2} = 1.25\ \text{atm}^{-1}$$

Since $Q > K$, the reaction is not at equilibrium, so a net change will occur in a direction that decreases Q. This can only occur if some of the SO_3 is converted back into products. In other words, the reaction will "shift to the left".

A Visual Way of Thinking about Q and K

The formal definitions of Q and K are quite simple, but they are of limited usefulness unless you are able to relate them to real chemical situations. The following diagrams illustrate the relation between Q and K from various standpoints. Take some time to study each one carefully, making sure that you are able to relate the description to the illustration.

Example : Dissociation of Dinitrogen Tetroxide

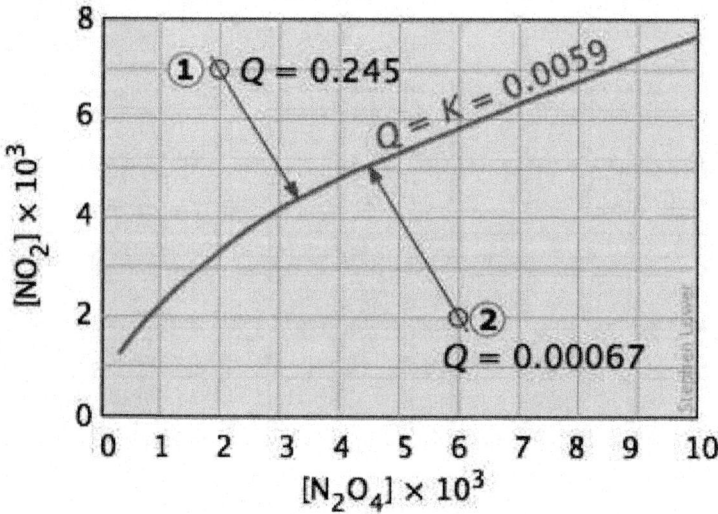

For the reaction $N_2O_4 \rightarrow 2\,NO_2$, $K_c = 0.0059$ at 298K.

This equilibrium condition is represented by the red curve that passes through all points on the graph that satisfy the requirement that —

$$Q = [NO_2]^2 / [N_2O_4] = 0.0059\,.$$

There are of course an infinite number of possible Q's of this system within the concentration boundaries shown on the plot. Only those points that fall on the red line correspond to *equilibrium states* of this system (those for which $Q = K$). The line itself is a plot of $[NO_2]$ that we obtain by rearranging the equilibrium expression $[NO_2] = ([N_2O_4]K)^{0.5}$.

If the system is initially in a non-equilibrium state, its composition will tend to change in a direction that moves it to one that is on the line.

Two such non-equilibrium states are shown. The state indicated by (1) has $Q > K$, so we would expect a net reaction that reduces Q by converting some of the NO_2 into N_2O_4; in other words, the equilibrium "shifts to the left". Similarly, in state (2), $Q < K$, indicating that the forward reaction will occur.

The blue arrows in the above diagram indicate the successive values that Q assumes as the reaction moves closer to equilibrium. The slope of the line reflects the stoichiometry of the equation. In this case, one mole of reactant yields two moles of products, so the slopes have an absolute value of 2 : 1.

Example 2 : Phase-change Equilibrium

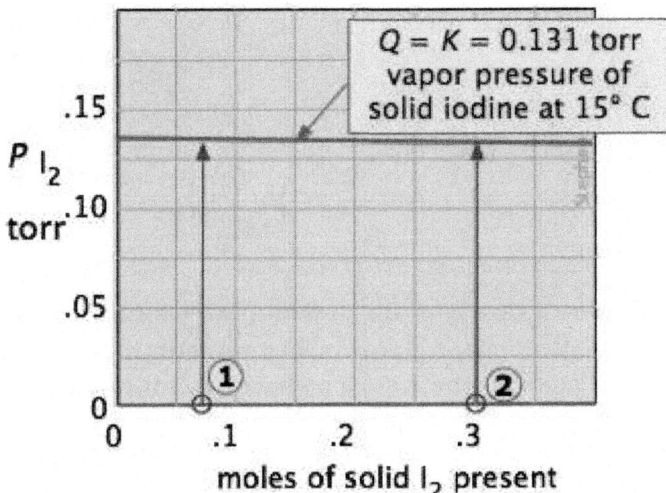

One of the simplest equilibria we can write is that between a solid and its vapour. In this case (as will be explained more fully in the next lesson), the equilibrium constant is just the vapour pressure of the solid. Thus for the process :

$$I_2(s) \rightarrow I_2(g)$$

all possible equilibrium states of the system lie on the horizontal red line and is independent of the quantity of solid present (as long as there is at least enough to supply the relative tiny quantity of vapour.) So adding various amounts of the solid to an empty closed vessel (states ① and ②) causes a gradual buidup of iodine vapour. Because the equilibrium pressure of the vapour is so small, the amount of solid consumed in the process is negligible, so the arrows go straight up and all lead to the same equilibrium vapour pressure.

Example : Heterogeneous Chemical Reaction

The decomposition of ammonium chloride is a common example of a heterogeneous (two-phase) equilibrium. Solid ammonium chloride has a substantial vapour pressure even at room temperature :

$$NH_4Cl(s) \rightarrow NH_3(g) + HCl(g)$$

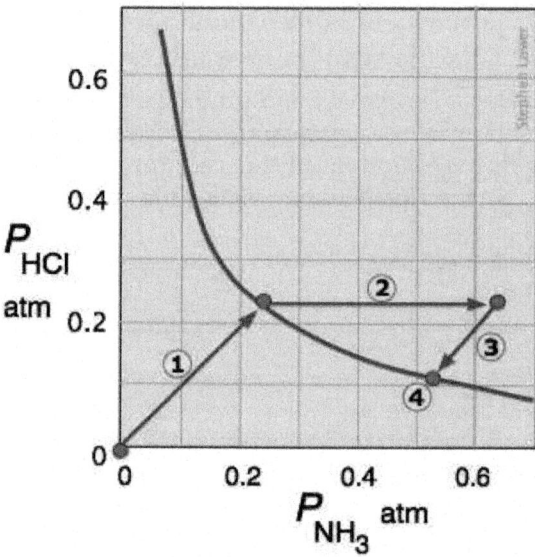

Arrow ① traces the states the system passes through when solid NH_4Cl is placed in a closed container. Arrow ② represents the addition of ammonia to the equilibrium mixture; the system responds by following the path ③ back to a new equilibrium state ④ which, as the Le Châtelier principle predicts, contains a smaller quantity of ammonia than was added. The unit slopes of the paths ① and ③ reflect the 1 : 1 stoichiometry of the gaseous products of the rection.

Make sure you thoroughly understand the following essential ideas which have been presented above. It is especially imortant that you know the precise meanings of all the highlighted terms in the context of this topic.

- When arbitrary quantities of the different *components* of a *chemical reaction system* are combined, the overall system composition will not likely correspond to the equilibrium composition. As a result, a net change in composition ("a shift to the right or left") will tend to take place until the *equilibrium state* is attained.

- The status of the reaction system in regard to its equilibrium state is characterised by the value of the *equilibrium expression* whose formulation is defined by the coefficients in the balanced reaction equation; it may be expressed in terms of concentrations, or in the case of gaseous components, as partial pressures.

- The various terms in the equilibrium expression can have any arbitrary value (including zero); the value of the equilibrium expression itself is called the *reaction quotient Q*.

- If the concentration or pressure terms in the equilibrium expression correspond to the equilibrium state of the system, then Q has the special value K, which we call the *equilibrium constant*.

- The ratio of Q/K (whether it is 1, >1 or <1) thus serves as an index of how far the system is from its equilibrium composition, and its value indicates the direction in which the net reaction must proceed in order to reach its equilibrium state.

- When $Q = K$, then the equilibrium state has been reached, and no further net change in composition will take place as long as the system remains undisturbed.

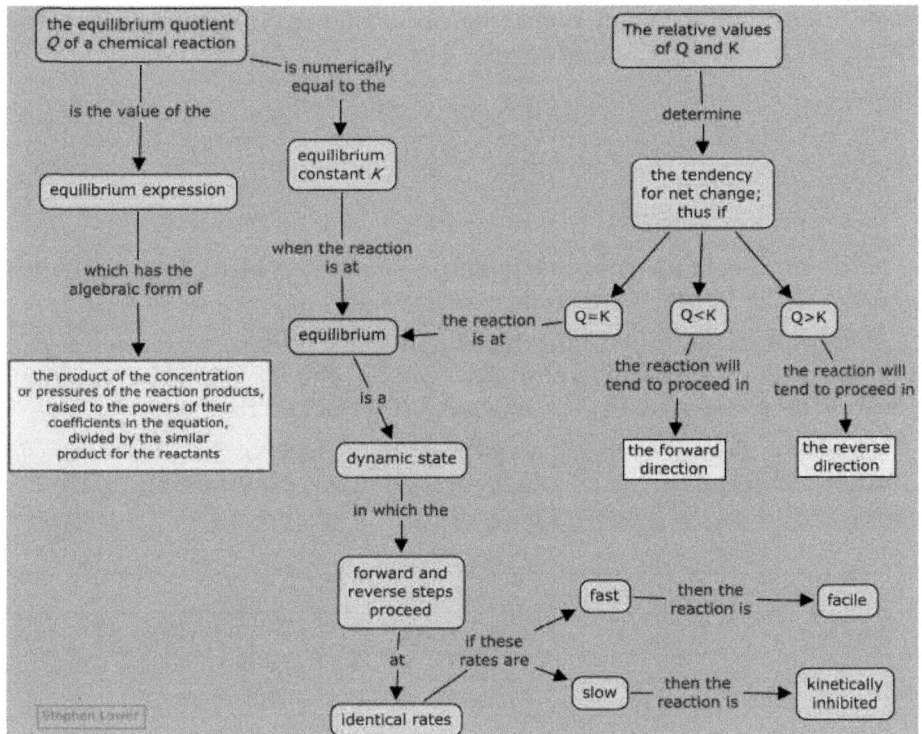

EQUILIBRIUM CALCULATIONS

Measuring and Calculating Equilibrium Constants

Clearly, if the concentrations or pressures of all the components of a reaction are known, then the value of K can be found by simple substitution. Observing individual concentrations or partial pressures directly may be not always be practical, however. If one of the components is coloured, the extent to which it absorbs light of an appropriate wavelength may serve as an index of its concentration. Pressure measurements are ordinarily able to measure only the total pressure of a gaseous mixture, so if two or more gaseous products are present in the equilibrium mixture, the partial pressure of one may need to be inferred from that of the other, taking into account the stoichiometry of the reaction.

Problem Example

In an experiment carried out by Taylor and Krist (*J. Am. Chem. Soc.* 1941: 1377), hydrogen iodide was found to be 22.3 per cent dissociated at 730.8°K. Calculate K_c for $2\,HI(g) \rightarrow H_2(g) + I_2$.

Solution:

No explicit molar concentrations are given, but we do know that for every n moles of HI, $0.223n$ moles of each product is formed and $(1-0.233)n = 0.777n$ moles of HI remains. For simplicity, we assume that $n=1$ and that the reaction is carried out in a 1.00-L vessel, so that we can substitute the required concentration terms directly into the equilibrium expression.

$$K_c = \frac{[H_2][I_2]}{[HI]^2} = \frac{(0.233)^2}{(0.777)^2} = 0.12$$

Problem Example

Ordinary white phosphorus, P_4, forms a vapour which dissociates into diatomic molecules at high temperatures: $P_4(g) \rightarrow 2\,P_2(g)$

A sample of white phosphorus, when heated to 1000°C, formed a vapour having a total pressure of 0.20 atm and a density of 0.152 g L^{-1}. Use this information to evaluate the equilibrium constant K_p for this reaction.

Solution: Before worrying about what the density of the gas mixture has to do with K_p, start out in the usual way by laying out the information required to express K_p in terms of an unknown x.

	P_4	$2\,P_2$	
initial moles :	1	$1 - x$	Since K is independent of the number of moles, assume the simplest case.

moles at equilibrium :	$1 - x$	$2x$	x is the fraction of P_4 that dissociates.
eq. mole fractions :	$\left(\dfrac{1-x}{1+x}\right)$	$\left(\dfrac{2x}{1+x}\right)$	The denominator is the total number of moles : $(1-x) + 2x = 1+x$.
eq. partial pressures :	$\left(\dfrac{1-x}{1+x}\right) \times 0.2$	$\left(\dfrac{2x}{1+x}\right) \times 0.2$	Partial pressure is the mole fraction times the total pressure.

Expressing the equilibrium constant in terms of x gives :

$$K_p = \frac{p_{P2}^2}{p_{P4}} = \frac{\left(\dfrac{1-x}{1+x}\right)^2}{\left(\dfrac{2x}{1+x}\right)} = \frac{(1-x)^2}{2x}$$

Now we need to find the dissociation fraction x of P_4, and at this point we hope you remember those gas laws that you were told you would be needing later in the course ! The density of a gas is directly proportional to its molecular weight, so you need to calculate the densities of pure P_4 and pure P_2 vapours under the conditions of the experiment. One of these densities will be greater than 0.152 gL^{-1} and the other will be smaller; all you need to do is to find where the measured density falls in between the two limits, and you will have the dissociation fraction.

The molecular weight of phosphorus is 31.97, giving a molar mass of 127.9 g for P_4. This mass must be divided by the volume to find the density; assuming ideal gas behaviour, the volume of 127.9 g (1 mole) of P_4 is given by RT/P, which works out to 522 L (remember to use the absolute temperature here.) The density of pure P_4 vapour under the conditions of the experiment is then

$$d = m/V = (128 \text{ g mol}^{-1}) \times = (522 \text{ L mol}^{-1}) = 0.245 \text{ g L}^{-1}.$$

The density of pure P_2 would be half this, or 0.122 g L^{-1}. The difference between these two limiting densities is 0.123 g L^{-1}, and the difference between the density of pure P_4 and that of the equilibrium mixture is $(.245 - .152)$ g L^{-1} or 0.093 g L^{-1}. The ratio 0.093   0.123 = 0.76 is therefore the fraction of P_4 that remains and its fractional dissociation is $(1 - 0.76) = 0.24$. Substituting into the equilibrium expression above gives $K_p = 1.2$.

Predicting Equilibrium Concentrations

This is by far the most common kind of equilibrium problem you will encounter : starting with an arbitrary number of moles of each component, how many moles of each will be present when the system comes to equilibrium?

The principal source of confusion and error for beginners relates to the need to determine the values of several unknowns (a concentration or pressure for each component) from a single equation, the equilibrium expression. The key to this is to make use of the stoichiometric relationships between the various components, which usually allow us to express the equilibrium composition in terms of a single variable. The easiest and most error-free way of doing this is adopt a systematic approach in which you create and fill in a small table as shown in the following problem example. You then substitute the equilibrium values into the equilibrium constant expression, and solve it for the unknown.

This very often involves solving a quadratic or higher-order equation. Quadratics can of course be solved by using the familiar quadratic formula, but it is often easier to use an algebraic or graphical approximation, and for higher-order equations this is the only practical approach. There is almost never any need to get an exact answer, since the equilibrium constants you start with are rarely known all that precisely anyway.

Problem Example

Phosgene ($COCl_2$) is a poisonous gas that dissociates at high temperature into two other poisonous gases, carbon monoxide and chlorine. The equilibrium constant $K_p = 0.0041$ at 600°K. Find the equilibrium composition of the system after 0.124 atm of $COCl_2$ is allowed to reach equilibrium at this temperature.

Solution : Start by drawing up a table showing the relationships between the components :

	$COCl_2$	CO	Cl_2
initial pressures :	0.124 atm	0	0
change :	$-x$	$+x$	$+x$
equilibrium pressures :	$0.124 - x$	x	x

Substitution of the equilibrium pressures into the equilibrium expression gives :

$$\frac{x^2}{0.124 - x} = 0.0041$$

This expression can be rearranged into standard polynomial form $x^2 + .0041\,x - 0.00054 = 0$ and solved by the quadratic formula, but we will simply obtain an approximate solution by iteration. Because the equilibrium constant is small, we know that x will be rather small compared to 0.124, so the above relation can be approximated by

$$\frac{x^2}{0.124} = 0.0041$$

which gives $x = 0.0225$. To see how good this is, substitute this value of x into the denominator of the original equation and solve again :

$$\frac{x^2}{0.124 - 0.0225} = \frac{x^2}{0.102} = 0.0041$$

This time, solving for x gives 0.0204. Iterating once more, we get :

$$\frac{x^2}{0.124-0.0204}=\frac{x^2}{0.104}=0.0041$$

and $x = 0.0206$ which is sufficiently close to the previous to be considered the final result. The final partial pressures are then 0.104 atm for $COCl_2$, and 0.0206 atm each for CO and Cl_2.

Problem Example

The gas-phase dissociation of phosphorus pentachloride to the trichloride has Kp = 3.60 at 540°C :

$$PCl_5 \rightarrow PCl_3 + Cl_2$$

What will be the partial pressures of all three components if 0.200 mole of PCl_5 and 3.00 moles of PCl_3 are combined and brought to equilibrium at this temperature and at a total pressure of 1.00 atm?

Solution : As always, set up a table showing what you know (first two rows) and then expressing the equilibrium quantities :

	PCl_5	PCl_3	Cl_2
initial moles :	0.200	3.00	0
change :	$-x$	$+x$	$+x$
equilibrium moles :	$0.200 - x$	$3.00 + x$	x
eq. partial pressures :	$\left(\dfrac{2.00-x}{3.20+x}\right)$	$\left(\dfrac{3.00+x}{3.20+x}\right)$	$\left(\dfrac{x}{3.20+x}\right)$

The partial pressures in the bottom row were found by multiplying the mole fraction of each gas by the total pressure : $P_i = X_i P_t$. The term in the denominator of each mole fraction is the total number of moles of gas present at equilibrium : $(0.200 - x) + (3.00 + x) + x = 3.20 + x$. Substituting the equilibrium partial pressures into the equilibrium expression, we have :

$$\frac{(3.00+x)(x)}{(0.200-x)(3.20+x)}=3.60$$

whose polynomial form is 4.60 x^2 + 13.80 x – 2.304 = 0.

Plotting this on a graphical calculator yields x = 0.159 as the positive root :

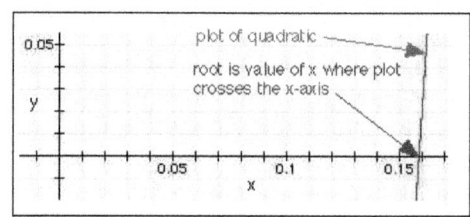

Substitution of this root into the expressions for the equilibrium partial pressures in the table yields the following values : $P(PCl_5) = 0.012$ atm, $P(PCl_3) = 0.94$ atm, $P(Cl_2) = 0.047$ atm.

Effects of Dilution on Equilibrium

In the section that introduced the Le Châtelier principle, it was mentioned that diluting a weak acid such as acetic acid CH_3COOH ("HAc") will shift the dissociation equilibrium to the right :

$$HAc + H_2O \rightarrow H_3O^+ + Ac^-$$

Thus a 0.10M solution of acetic acid is 1.3 per cent ionised, while in a 0.01M solution, 4.3 per cent of the HAc molecules will be dissociated. This is because as the solution becomes more dilute, the product $[H_3O^+][Ac^-]$ decreases more rapidly than does the [HAc] term. At the same time the concentration of H_2O becomes greater, but because it is so large to start with (about 55.5M), any effect this might have is negligible, which is why no $[H_2O]$ term appears in the equilibrium expression.

For a reaction such as $CH_3COOH(l) + C_2H_5OH(l) \rightarrow CH_3COOC_2H_5(l) + H_2O(l)$ (in which the water concentration does change), dilution will have no effect on the equilibrium; the situation is analogous to the way the pressure dependence of a gas-phase reaction depends on the number of moles of gaseous components on either side of the equation.

Problem Example

The bio-chemical formation of a disaccharide (double) sugar from two mono-saccharides is exemplified by the reaction

fructose + glucose-6-phosphate $\rightarrow$ sucrose-6-phosphate

(Sucrose is ordinary table sugar.) To what volume should a solution containing 0.050 mol of each mono-saccharide be diluted in order to bring about 5 per cent conversion to sucrose phosphate?

Solution : The initial and final numbers of moles are as follows :

td>

	fructose	glucose-6-P	sucrose-6-P
initial moles :	0.05	0.05	0
equilibrium moles :	0.0485	0.0485	0.0015

Substituting into the expression for K_c in which the solution volume is the unknown, we have :

$$\frac{[suc6P]}{[Fruc][gluc6P]} = \frac{\left(\dfrac{.0485}{V}\right)}{\left(\dfrac{.0485}{V}\right)^2} = 0.05$$

Solving for V gives a final solution volume of **78 mL.**

Phase Distribution Equilibria

It often happens that two immiscible liquid phases are in contact, one of which contains a solute. How will the solute tend to distribute itself between the two phases? One's first thought might be that some of the solute will migrate from one phase into the other until it is distributed equally between the two phases, since this would correspond to the maximum dispersion (randomness) of the solute. This, however, does not take into the account the differing solubilities the solute might have in the two liquids; if such a difference does exist, the solute will preferentially migrate into the phase in which it is more soluble.

For a solute S distributed between two phases a and b the process $S_a = S_b$ is defined by the distribution law

$$K_{a,b} = \frac{[S]_a}{[S]_b}$$

in which $K_{a,b}$ is the *distribution ratio* (also called the distribution coefficient) and $[S]_i$ is the solubility of the solute in the phase.

The transport of substances between different phases is of immense importance in such diverse fields as pharmacology and environmental science. For example, if a drug is to pass from the aqueous phase with the stomach into the bloodstream, it must pass through the lipid (oil-like) phase of the epithelial cells that line the digestive tract. Similarly, a pollutant such as a pesticide residue that is more soluble in oil than in water will be preferentially taken up and retained by marine organism, especially fish, whose bodies contain more oil-like substances; this is basically the mechanism whereby such residues as DDT can undergo *biomagnification* as they become more concentrated at higher levels within the food chain. For this reason, environmental regulations now require that oil-water distribution ratios be established for any new chemical likely to find its way into natural waters. The standard "oil" phase that is almost universally used is octanol, $C_8H_{17}OH$.

In preparative chemistry it is frequently necessary to recover a desired product present in a reaction mixture by extracting it into another liquid in which it is more soluble than the unwanted substances. On the laboratory scale this operation is carried out in a **separatory funnel** as shown here.

The two immiscible liquids are poured into the funnel through the opening at the top. The funnel is then shaken to bring the two phases into intimate contact, and then set aside to allow the two liquids to separate into layers, which are then separated by allowing the more dense liquid to exit through the stopcock at the bottom.

If the distribution ratio is too low to achieve efficient separation in a single step, it can be repeated; there are automated devices that can carry out hundreds of successive extractions, each yielding a product of higher purity. In these applications our goal is to exploit the Le Châtelier principle by repeatedly upsetting the phase distribution equilibrium that would result if two phases were to remain in permanent contact.

Problem Example

The distribution ratio for iodine between water and carbon disulfide is 650. Calculate the concentration of I_2 remaining in the aqueous phase after 50.0 mL of 0.10M I_2 in water is shaken with 10.0 mL of CS_2.

Solution:

The equilibrium constant is

$$K_d = \frac{C_{CS_2}}{C_{H_2O}} = 650$$

Let m_1 and m_2 represent the numbers of millimoles of solute in the water and CS_2 layers, respectively. K_d can then be written as $(m_2/10\ \text{mL}) \div (m_1/50\ \text{mL}) = 650$. The number of moles of solute is $(50\ \text{mL}) \times (0.10\ \text{mmol mL}^{-1}) = 5.00\ \text{mmol}$, and mass conservation requires that $m_1 + m_2 = 5.00$ mmol, so $m_2 = (5.00 - m_1)$ mmol and we now have only the single unknown m_1. The equilibrium constant then becomes :

$((5.00 - m_1)$ mmol $/ 10$ mL$) \div (m_1$ mmol $/ 50$ mL$) = 650.$

Simplifying and solving for m_1 yields $(0.50 - 0.1)m_1 / (0.02\, m_1) = 650$, with $m_1 = 0.0382$ mmol. The concentration of solute in the water layer is $(0.0382$ mmol$) / (50$ mL$) = \textbf{0.000763 M}$, showing that almost all of the iodine has moved into the CS_2 layer.

The six Problem Examples presented above were carefully selected to span the range of problem types that students enrolled in first-year college chemistry courses are expected to be able to deal with. If you are able to reproduce these solutions on your own, you should be well prepared on this topic.

The first step in the solution of all but the simplest equilibrium problems is to sketch out a table showing for each component the *initial* concentration or pressure, the *change* in this quantity (for example, $+2x$), and the *equilibrium* values (for example, $.0036 + 2x$). In doing so, the sequence of calculations required to get to the answer usually becomes apparent.

Equilibrium calculations often involve quadratic- or higher-order equations. Because concentrations, pressures, and equilibrium constants are seldom known to a precision of more than a few significant figures, there is no need to seek exact solutions. Iterative approximations are adequate and convenient.

Phase distribution equilibria play an important role in chemical separation processes on both laboratory and industrial scales. They are also involved in the movement of chemicals between different parts of the environment, and in the bioconcentration of pollutants in the food chain.

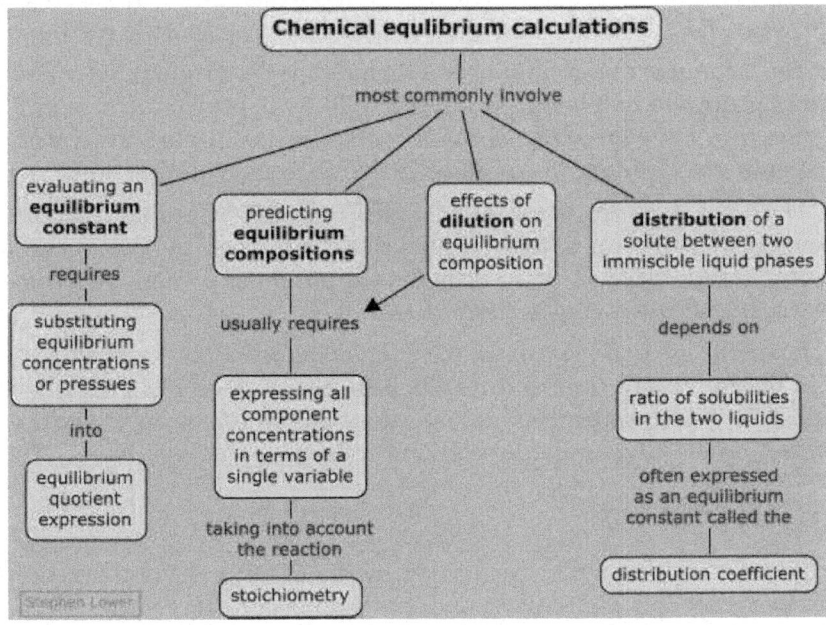

Chapter 6

SOLID PHASE PEPTIDE SYNTHESIS

INTRODUCTION

General Principles

The revolutionary principle behind solid phase peptide synthesis (SPPS) as conceived by Merrifield in 1959 is that if the peptide is bound to an insoluble support then any unreacted reagents left at the end of any synthetic step can be removed by a simple wash procedure, greatly decreasing the time required for synthesis. What is more, the arrangement is amenable to automation. This is only valid, however, if the individual synthetic steps occur with essentially quantitative yields. This latter requirement entails rigourous testing of the chemistry and tends to result in the use of a limited range of tried and tested methodologies.

The two most commonly used schemes are the original Merrifield method and that proposed by Sheppard in 1975. While the protection schemes used are different, they have the same basic concepts, a resin support, the use of an excess of reagents and building the peptide in a C→N terminal direction.

The resin supports used are designed to swell in commonly used solvents, expanding to many times their original size. Thus, the reactions occur not on the surface of a rigid particle, but within the support in a solvated gel which permits easy access to the growing peptide chain.

Reagents are used in excess, this allows the reactions to proceed to completion in the minimum time and results in faster synthesis of peptides with high purity. The couplings proceed in a C→N terminal direction to allow the use of racemisation limiting amine protection for the activated species.

Merrifield Synthesis

This methodology is characterised by the use of *tert*-butyl based temporary α-amino protection and benzyl, or substituted benzyl, groups for permanent side chain protection. There are over one hundred different substituted resins suitable for peptide synthesis generally based on polystyrene and polyethylene glycol. These resins allow introduction of an amino acid through either substitution, condensation or addition reactions. The traditional resin used for Merrifield synthesis was a chloromethylphenyl substituted resin. The first amino acid was attached to the resin through substitution of the chloride by the caesium salt of the BOC-amino acid, generating an equivalent to a benzyl ester.

Deprotection of the temporary BOC group uses a 20-50 per cent solution of trifluoroacetic acid (TFA) in dichloromethane and has to be followed by neutralisation of the resulting ammonium salt with a hindered tertiary base. Final cleavage from the resin as well as deprotection of benzyl based side chain protecting groups is achieved using strong acids, usually liquid hydrogen fluoride or trifluoromethane sulphonic acid. Such procedures require specialised apparatus and the highly acidic conditions catalyse several possible re-arrangements.

Fmoc Polyamide Synthesis

The fundamental differences between the Fmoc polyamide strategy when compared to the Merrifield approach are that the reactions are carried out under continuous flow and that the conditions for α-amino deprotection and cleavage from the resin are far more mild. This arises from the adoption of the base labile Fmoc protecting group for α-amino protection. The side chains are generally protected with *tert*-butyl based groups which, in common with the linkage to the resin, can be cleaved by TFA in the presence of scavengers.

As discussed above, a large number of resins are available. Traditionally, resins with 4-hydroxymethylphenoxy substitution were used. These allowed were esterified with the anhydride of the first amino acid. As a result of the mesomerically electron donating *para* oxygen atom stabilising the resultant carbocation, cleavage of the peptide from the resin occurs under more mild acid conditions, typically using trifluoroacetic acid with scavengers.

The use of continuous flow means that the reagents are passed through a reaction chamber containing the resin supported peptide. Having passed through this chamber, the reagents can be recirculated back into the chamber again or taken to a waste collection bottle. This allows the resin to be washed clean of excess reagents and unwanted reaction products, which in turn helps drive deprotection steps to completion following Le Chatelier's principle. In addition, the solution can be passed through a u.v. detector and monitored at a suitable wavelength for the Fmoc chromophore.

A typical cycle consists of :

1. Deprotection of the preceding residue with piperidine.
2. Wash to remove any remaining reagents from 1.
3. Acylation in a recirculatory mode.
4. Wash to remove excess reagents.

A typical uv trace indicates these steps.

In this way, the cycle can be qualitatively monitored. Such monitoring can be used in conjunction with automation. For example, a slow deprotection step suggests inaccessibility of the peptide amino terminal, indicating that the next coupling step may require a longer acylation time. If this is detected then the cycle can be interrupted for later manual intervention.

Since the turn of the century, there has been a constant drive to develop new and improved strategies for peptide synthesis. The areas of protection, deprotection, activation and coupling have all received attention. The field of solid phase peptide synthesis has rapidly grown since its inception by Merrifield but is dominated by use of BOC/benzyl and Fmoc/*tert*-butyl protection schemes.

SOLUTION PHASE PEPTIDE SYNTHESIS

A flow chart figure indicates the general method of solution phase synthesis. It should be noted that peptides are normally synthesised in a carboxyl to amino terminal direction. This allows the racemisation resistant urethane protected amino acid to be activated in preference to the racemisation prone carboxyl terminal of the peptide.

Care has to be taken in selection of protecting groups so that they can be removed with appropriate selectivity. Side chain and carboxyl protecting groups need to be selected so as not to be labile under the conditions required to deprotect the amino group. If a long peptide is to be made then it is common to synthesise it in smaller sections, which are later coupled to form the overall peptide—a process called fragment condensation. If a fragment condensation technique is being adopted then the carboxyl terming must also be deprotected without loss of side chain protection.

There are two ways of achieving this selective deprotection :

1. Chose protecting groups that are deprotected with completely different reagents (refered to as orthogonal protection) *e.g. tert butyl* (acid), fluorenyl methyl (base), benzyl (catalytic hydrogenolysis).

2. Chose protecting groups that are deprotected with the same type of reagent but under different conditions. *e.g. tert* butyl and benzyl which require increasingly strong acids for deprotection.

Chapter 7

THE IMPORTANCE OF SOIL ORGANIC MATTER

INTRODUCTION

On the basis of organic matter content, soils are characterised as mineral or organic. Mineral soils form most of the world's cultivated land and may contain from a trace to 30 per cent organic matter. Organic soils are naturally rich in organic matter principally for climatic reasons. Although they contain more than 30 per cent organic matter, it is precisely for this reason that they are not vital cropping soils.

This soils bulletin concentrates on the organic matter dynamics of cropping soils. In brief, it discusses circumstances that deplete organic matter and the negative outcomes of this. The bulletin then moves on to more proactive solutions. It reviews a "basket" of practices in order to show how they can increase organic matter content and discusses the land and cropping benefits that then accrue.

Soil organic matter is any material produced originally by living organisms (plant or animal) that is returned to the soil and goes through the decomposition

Plate
Crop residues added to the soil are decomposed by soil macro-fauna and micro-organisms,
increasing the organic matter content of the soil.
A.J. BOT

process (Plate 1). At any given time, it consists of a range of materials from the intact original tissues of plants and animals to the substantially decomposed mixture of materials known as humus.

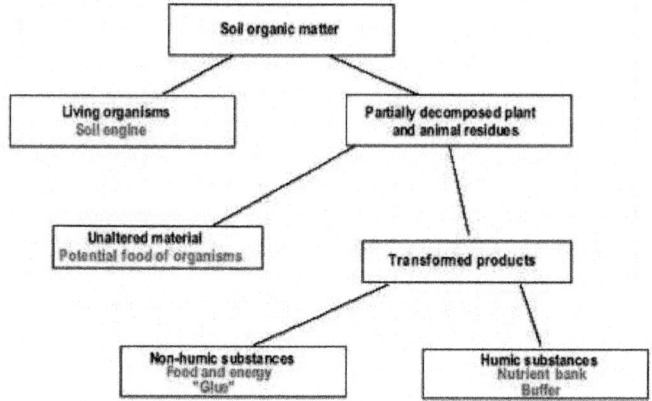

Fig. Components of soil organic matter and their functions.

Most soil organic matter originates from plant tissue. Plant residues contain 60-90 per cent moisture. The remaining dry matter consists of carbon (C), oxygen, hydrogen (H) and small amounts of sulphur (S), nitrogen (N), phosphorus (P), potassium (K), calcium (Ca) and magnesium (Mg). Although present in small amounts, these nutrients are very important from the viewpoint of soil fertility management.

Soil organic matter consists of a variety of components. These include, in varying proportions and many intermediate stages, an active organic fraction including micro-organisms (10-40 per cent), and resistant or stable organic matter (40-60 per cent), also referred to as humus.

Forms and classification of soil organic matter have been described by Tate (1987) and Theng (1987). For practical purposes, organic matter may be divided into aboveground and belowground fractions. Aboveground organic matter comprises plant residues and animal residues; belowground organic matter consists of living soil fauna and micro-flora, partially decomposed plant and animal residues, and humic substances. The C : N ratio is also used to indicate the type of material and ease of decomposition; hard woody materials with a high C : N ratio being more resilient than soft leafy materials with a low C : N ratio.

Although soil organic matter can be partitioned conveniently into different fractions, these do not represent static end products. Instead, the amounts present reflect a dynamic equilibrium. The total amount and partitioning of organic matter in the soil is influenced by soil properties and by the quantity of annual inputs of plant and animal residues to the ecosystem. For example, in a given soil ecosystem, the rate of decomposition and accumulation of soil organic matter is determined by such soil properties as texture, pH, temperature, moisture, aeration, clay mineralogy and soil biological activities. A complication is that soil organic matter in turn influences or modifies many of these same soil properties.

Organic matter existing on the soil surface as raw plant residues helps protect the soil from the effect of rainfall, wind and sun. Removal, incorporation or burning of residues exposes the soil to negative climatic impacts, and removal or burning deprives the soil organisms of their primary energy source.

Organic matter within the soil serves several functions. From a practical agricultural standpoint, it is important for two main reasons : (i) as a "revolving nutrient fund"; and (ii) as an agent to improve soil structure, maintain tilth and minimise erosion.

As a revolving nutrient fund, organic matter serves two main functions :

* As soil organic matter is derived mainly from plant residues, it contains all of the essential plant nutrients. Therefore, accumulated organic matter is a storehouse of plant nutrients.

* The stable organic fraction (humus) adsorbs and holds nutrients in a plant-available form.

Organic matter releases nutrients in a plant-available form upon decomposition. In order to maintain this nutrient cycling system, the rate of organic matter addition from crop residues, manure and any other sources must equal the rate of decomposition, and take into account the rate of uptake by plants and losses by leaching and erosion.

Where the rate of addition is less than the rate of decomposition, soil organic matter declines. Conversely, where the rate of addition is higher than the rate of decomposition, soil organic matter increases. The term steady state describes a condition where the rate of addition is equal to the rate of decomposition.

In terms of improving soil structure, the active and some of the resistant soil organic components, together with micro-organisms (especially fungi), are involved in binding soil particles into larger aggregates. Aggregation is important for good soil structure, aeration, water infiltration and resistance to erosion and crusting.

Traditionally, soil aggregation has been linked with either total C (Matson et. al., 1997) or organic C levels (Dalal and Mayer, 1986a, 1986b). More recently, techniques have developed to fractionate C on the basis of lability (ease of oxidation), recognising that these sub-pools of C may have greater effect on soil physical stability and be more sensitive indicators than total C values of carbon dynamics in agricultural systems (Lefroy, Blair and Strong, 1993; Blair, Lefroy and Lisle, 1995; Blair and Crocker, 2000). The labile carbon fraction has been shown to be an indicator of key soil chemical and physical properties. For example, this fraction has been shown to be the primary factor controlling aggregate breakdown in Ferrosols (non-cracking red clays), measured by the per centage of aggregates measuring less than 0.125 mm in the surface crust after simulated rain in the laboratory (Bell et. al., 1998, 1999).

The resistant or stable fraction of soil organic matter contributes mainly to nutrient holding capacity (cation exchange capacity [CEC]) and soil colour. This fraction of organic matter decomposes very slowly. Therefore, it has less influence on soil fertility than the active organic fraction.

Soil degradation has become a major concern in Canada. Erosion, salinisation, acidification and loss of organic matter are the main forms of soil deterioration. This publication deals with the role of organic matter in soil productivity and the effects of various management practices on soil organic matter.

What is Organic Matter

Soil organic matter consists of a variety of components. These include, in varying proportions and many Intermediate stages :

- **raw plant residues and micro-organisms** (1 to 10 per cent)
- **"active" organic traction** (10 to 40 per cent)
- resistant or stable organic matter (40 to 60 per cent) also referred to as humus.

Raw plant residues, on the surface, help reduce surface wind speed and water run-off. Removal, incorpration or burning of residues predisposes the soil to serious erosion.

The "active" and some of the resistant soil organic components, together with micro-organisms (especially fungi) are involved in binding small soil particles into larger aggregates. Aggregation is important for good soil structure, aeration, water infiltration and resistance to erosion and crusting.

The resistant or stable fraction of soil organic matter contributes mainly to **nutrient holding capacity** (cation exchange capacity) and **soil colour**. This fraction of organic matter decomposes very slowly and therefore has less influence on soil fertility than the "active" organic fraction.

Organic matter in soil serves several functions. From a practical agricultural standpoint, it is important for two main reasons. First as a **"revolving nutrient bank account"**; and second, as an agent to improve **soil structure, maintain tilth, and minimise erosion.**

As a **revolving nutrient bank account**, organic matter serves two main functions :

- Since soil organic matter is derived mainly from plant residues, it contains all of the essential plant nutrients. Accumulated organic matter, therefore, is a storehouse of plant nutrients. Upon decomposition, the nutrients are released in a plant-available form.

- The stable organic fraction (humus) adsorbs and holds nutrients in a plant available form.

Organic matter does not add any "new' plant nutrients but releases nutrients in a plant available form through the process of decomposition. In order to maintain this nutrient cycling system, the rate of addition from crop residues and manure must equal the rate of decomposition.

If the rate of addition is less than the rate of decomposition, soil organic matter will decline and, conversely if the rate of addition is greater than the rate of decomposition, soil organic matter will increase. The term steady state has been used to describe a condition where the rate of addition is equal to the rate of decomposition.

Fertilizer can contribute to the maintenance of this revolving nutrient bank account by increasing crop yields and consequently the amount of residues returned to the soil.

Organic Matter in Virgin and Cultivated Soils

Soils in Alberta are divided into soil groups (zones) based on the amount of organic matter they contain. They occur in geographic zones from the southeast to the north-west and are identified as the Brown, Dark Brown, and Black Chernozemic (prairie) soils.

The Brown soils have the least amount of organic matter because of the relatively small inputs of plant residues contributed by the short grass prairie vegetation under which these soils developed. Black soils developed under cooler and wetter conditions which allowed for more grass growth and thus a greater accumulation of organic matter.

Further north and west, trees became the dominant vegetation. Soils influenced by forest vegetation for a moderate length of time constitute the Dark Gray or transitional soils. Where the forest cover was established for a longer period, Luvisolic (forest) soils developed. Organic (peat) soils. occur in low lying areas throughout the Black, Dark Gray and Gray soil zones. These soils are saturated with water for much or all of the year thereby reducing the rate of organic matter decomposition. The amount of soil organic matter characteristic of virgin and cultivated soils in the various zones is shown in Table. Cultivation generally has resulted in a 30 to 50 per cent loss of organic matter.

Table. Organic matter in native and cultivated soils (per cent).

Soil zone	Virgin	Cultivated
Brown	3-4	2-3
Dark Brown	4-5	3-4
Black	6-10	4-6
Dark Gray	4-5	2-3
Gray	1-2	1-2

Before our soils were cultivated, they had achieved a "steady state". In most of our prairie soils, the increased rate of decomposition associated with cultivation, combined with the low rates of crop residue addition associated with crop-fallow rotations has caused a fairly rapid decline in soil organic matter. The rate of decline decreases with time as the amount of total soil organic matter decreases and particularly as the "active" organic fraction is depleted.

Cultivation of soils that are naturally high in organic matter will usually result in a decrease of organic matter. In the case of Luvisolic soils, their poor physical properties and low fertility have encouraged the use of forages, fertilizers, manure and judicious tillage. Such management practices have resulted in an increase in soil organic matter on Luvisolic soils, whereas excessive tillage, fallowing and minimal fertilization have lead to further depletion of the soil organic matter.

Effects of Organic Matter Decline

As stated in the introduction, soil degradation is becoming a major concern in Canada. Loss of organic matter is often identified as one of the main factors contributing to declining soil productivity, but it is misleading to equate a loss in soil organic matter with a loss in soil productivity.

Soil organic matter contributes to soil productivity in several ways, but there is no direct quantitative relationship between soil productivity and total soil organic matter. In fact, it has been the decline in organic matter that has contributed to the productivity of the crop-fallow system.

This decline in organic matter has resulted in the release of large amounts of plant nutrients, particularly nitrogen. For example, a decrease in soil organic matter of 2 per cent releases about 2,400 lb/ac of nitrogen. If this decline occurred over a 60 years period, an average of 40 lb/ac/yr of plant-available nitrogen has come from the soil organic matter. We therefore view prairie soils which had relatively high levels of organic matter as being nitrogen fertile, but this fertility could only be attained under a management system that allowed for organic matter to decline. Frequent fallowing has been a major factor contributing to this decline.

Insofar as organic matter contributes to improved soil physical properties (*e.g.*, tilth, aggregation, moisture holding capacity and resistance to erosion) increasing soil organic matter will generally result in increased soil productivity. But on many soils, suitable soil physical properties occur at relatively low levels of organic matter (2-4 per cent).

A level of organic matter higher than required to produce suitable physical properties is beneficial in that the soil has a greater buffering and nutrient holding capacity, but it does not contribute directly to soil productivity. If soils are managed so organic matter is not declining (steady-state), soils higher in organic matter (*e.g.*, 8 per cent) are not inherently more productive or fertile than those that have less organic matter (*e.g.*, 5 per cent).

To equate the ability to supply nutrients with total soil organic matter is not valid. The "active" fraction of organic matter is a more reliable indicator of soil fertility than is total soil organic matter. In cultivated soil, the "active" fraction is influence mainly by previous management.

Soil organic matter cannot be increased quickly even when management practices that conserve soil organic matter are adopted. The increased addition of organic matter associated with continuous cropping, and the production of higher crop yields, are accompanied by an increase in the rate of decomposition. Moreover, only a small fraction of crop residues added to soil remains as soil organic matter.

After an extended period of time, the return of all crop residues and the use of forages in rotations with cereals and oilseeds may significantly increase soil organic matter, particularly, the "active" fraction.

Managing Soil Organic Matter

There have been vast changes in the nature of agricultural production. In the past, farms were small, and much of what was produced was consumed on the farm. This system allowed for the limited removal of soil nutrients since there was an opportunity to return most of the nutrients back to the land.

The advent of the internal combustion engine, migration from rural to urban communities, increasing farm size and specialisation in production have resulted in a system of production where there is greater removal of plant nutrients from the soil and less opportunity for nutrient cycling.

Maintenance of organic matter for the sake of maintenance alone is not a practical approach to farming. It is more realistic to use a management system that will give sustained profitable production.

The greatest source of soil organic matter is the residue contributed by current crops. Consequently, crop yield and type, method of handling residues and frequency of fallow are all important factors. Ultimately, soil organic matter must be maintained at a level necessary to maintain soil tilth. The effects of specific management practices are discussed below.

Summerfallow

Summerfallowing accelerates the loss of organic matter. Aeration of the soil associated with tillage, and the increase in soil temperature and moisture results in increased organic matter decomposition. Since little In the way of residues are added to the soil, a net loss of organic matter occurs. Research has shown that as the frequency of fallow increases, the amount of soil organic matter decreases.

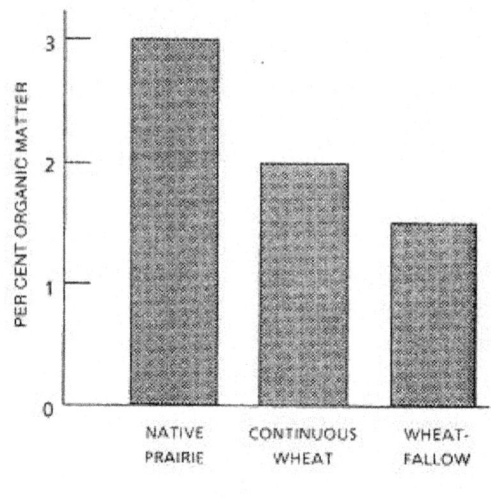

Fig. Effect of frequency of fallow on per cent organic matter.

Summerfallowing for moisture conservation may be a necessary practice in the Brown and Dark Brown soil zones. However, it must be questioned in the Black and the Gray soil zones. Periodic fallowing may be acceptable in the higher rainfall regions for control of persistent perennial weeds and volunteer grains in pedigreed seed production.

In the crop-fallow system common to the prairie region, the nitrogen removed has far exceeded that gained from crop residues, manure, legumes and fertilizer. The large reserves of nitrogen present in the organic matter of our prairie soils have been the major source of nitrogen in this cropping system. Continued reliance on soil organic matter reserves to supply the nitrogen requirements of crops will ultimately lead to a decline in soil productivity, and increased soil erosion.

When a change from a cropping system involving fallow to continuous cereal grain production, the nitrogen requirement increases. The nitrogen requirement is greatest in the first few years of continuous cropping as the nutrient cycling process adjusts to the new cropping system.

Crop Rotations

The value of forage crops in rotations with cereals and oilseeds has long been recognised, especially in the Luvisolic soils. Several long-term crop rotation studies conducted in Western Canada have shown that crop rotations involving perennial forages tend to stabilise soil organic matter at a higher level than crop rotations involving summerfallow.

Figure summarises data obtained by the University of Alberta from the Breton Plots and the University of Manitoba. The Breton Plots compared a two-year fallow-wheat rotation with a five-year rotation involving wheat, oats and barley followed by two years of hay production.

In research done by the University of Manitoba, the effects of various cultural practices on the level of organic matter are compared. It is interesting to note that the highest level of soil organic matter was maintained under continuous cropping.

The beneficial effects of perennial forages are the result of :

- a more extensive root system and crop aftermath contributing more organic matter to the soil,
- the fibrous nature of the root system of perennial grasses. These are particularly effective as a binding agent in soil aggregation,
- Nitrogen fertility enhancement by the growth of legumes,
- increased permeability of dense sub-soils because of the deep penetrating tap roots of perennial legumes, especially alfalfa.
- a reduced rate of organic matter decomposition in the absence of tillage.

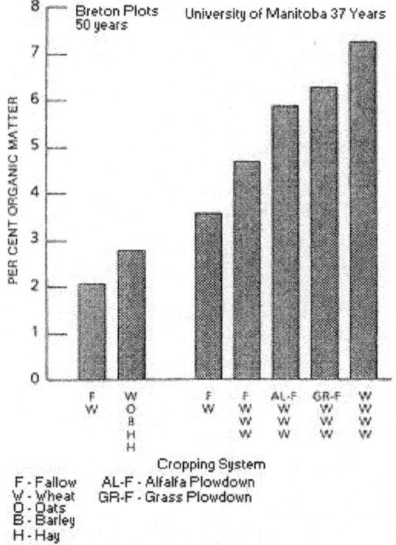

Fig. Effect of rotation on soil organic matter

In the brown and dark brown soil zones, perennial forages grown for forage production or as a plowdown crop may jeopardise subsequent cereal crops because they deplete soil moisture reserves.

Fertilization

Fertilizers will generally increase soil organic matter because the increased crop growth returns larger amounts of residues to the soil.

Data obtained from the Breton Plots is summarised in Figure. The increase in organic matter is less than what might be expected with current farming practices since all the straw had been removed from the plots. To determine the fertilizer effect, two fertilizer treatments were averaged, one involving a low rate of nitrogen and sulphur and another involving a low rate of nitrogen, phosphorus, potassium and sulphur.

One would expect that with higher rates of fertilization, higher yielding varieties, and the return of all crop residues, the effect of fertilizers on organic matter would be greater than that shown in Figure.

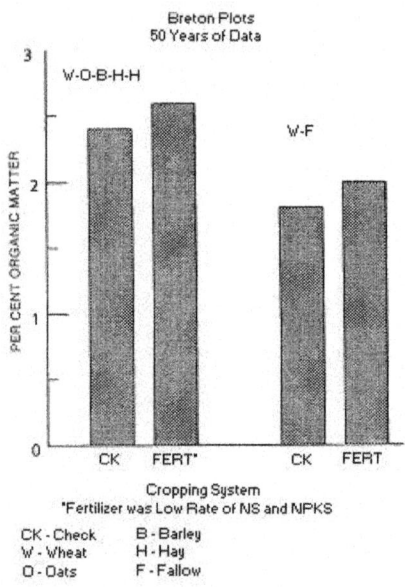

Fig. Effect of fertilizer on soil organic matter.

Plowdown

Legume plowdown has received considerable attention in recent years as an alternative to the use of nitrogen fertilizers. However, when considering this option in a cropping programme, the amount of nitrogen added by the legume, as well as the loss of one year of production, the cost of seed and the expected yield increase must be kept in mind.

Strictly as a source of nitrogen, the value of a legume plowdown is questionable. The amount of nitrogen fixed by a legume is dependent upon the type of legume, the amount of vegetative growth, the nature of the soil and environmental conditions. As a source of organic matter, legume plowdown is valuable, however, perennial forage is more effective than legume plowdown for increasing soil organic matter.

Nitrate nitrogen which accumulates following legume plowdown is subject to loss, particularly in wet, poorly drained soils. To minimise this, legumes should be plowed down in the fall rather than mid-summer to reduce nitrate accumulation and subsequent loss.

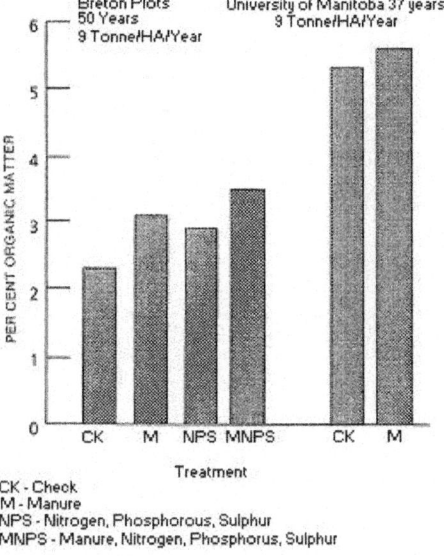

CK - Check
M - Manure
NPS - Nitrogen, Phosphorous, Sulphur
MNPS - Manure, Nitrogen, Phosphorus, Sulphur

Fig. Effect of manure on soil organic matter.

The cultivation of prairie soils has generally resulted in a decline in organic matter of 30 to 50 per cent. A product of this decline has been the release of large amounts of plant nutrients, particularly nitrogen. Crop rotations with a high frequency of summerfallow have relied on the nitrogen released from soil organic matter to supply crop requirements.

More frequent or continuous cropping, less frequent tillage, the production of high yields and the return of crop residues will help to maintain soil organic matter at a satisfactory level. Perennial forages are effective for maintaining or increasing soil organic matter.

ORGANIC MATTER DECOMPOSITION
AND THE SOIL FOOD WEB

When plant residues are returned to the soil, various organic compounds undergo decomposition. Decomposition is a biological process that includes the physical breakdown and bio-chemical transformation of complex organic molecules of dead material into simpler organic and inorganic molecules (Juma, 1998).

The continual addition of decaying plant residues to the soil surface contributes to the biological activity and the carbon cycling process in the soil. Breakdown of soil organic matter and root growth and decay also contribute to these processes. Carbon cycling is the continuous transformation of organic and inorganic carbon compounds by plants and micro- and macro-organisms between the soil, plants and the atmosphere (Figure)

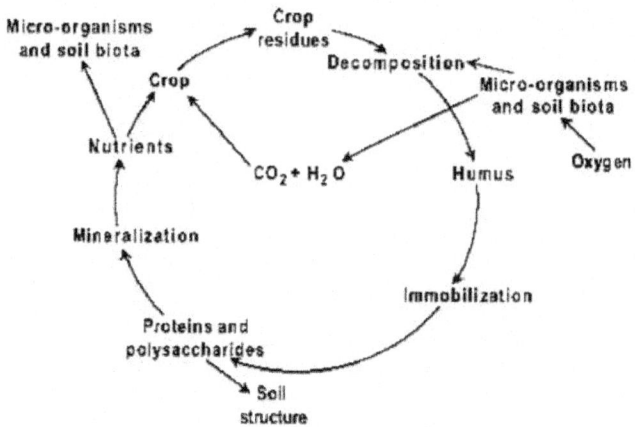

Fig. Carbon cycle.

Decomposition of organic matter is largely a biological process that occurs naturally. Its speed is determined by three major factors : soil organisms, the physical environment and the quality of the organic matter (Brussaard, 1994). In the decomposition process, different products are released : carbon dioxide (CO_2), energy, water, plant nutrients and resynthesised organic carbon compounds. Successive decomposition of dead material and modified organic matter results in the formation of a more complex organic matter called humus (Juma, 1998). This process is called humification. Humus affects soil properties. As it slowly decomposes, it colours the soil darker; increases soil aggregation and aggregate stability; increases the CEC (the ability to attract and retain nutrients); and contributes N, P and other nutrients.

Soil organisms, including micro-organisms, use soil organic matter as food. As they break down the organic matter, any excess nutrients (N, P and S) are released into the soil in forms that plants can use. This release process is called mineralisation. The waste products produced by micro-organisms are also soil organic matter. This waste material is less decomposable than the original plant and animal material, but it can be used by a large number of organisms. By breaking down carbon structures and rebuilding new ones or storing the C into their own biomass, soil biota plays the most important role in nutrient cycling processes and, thus, in the ability of a soil to provide the crop with sufficient nutrients to harvest a healthy product. The organic matter content, especially the more stable humus, increases the capacity to store water and store (sequester) C from the atmosphere.

The Soil Food Web

The soil ecosystem (Box 1) can be defined as an interdependent life-support system composed of air, water, minerals, organic matter, and macro- and micro-organisms, all of which function together and interact closely.

The organisms and their interactions enhance many soil ecosystem functions and make up the soil food web. The energy needed for all food webs is generated by primary producers : the plants, lichens, moss, photo-synthetic bacteria and algae that use sunlight to transform CO_2 from the atmosphere into carbohydrates. Most other organisms depend on the primary producers for their energy and nutrients; they are called consumers.

Some functions of a healthy soil ecosystem

- Decompose organic matter towards humus.
- Retain N and other nutrients.
- Glue soil particles together for best structure.
- Protect roots from diseases and parasites.
- Make retained nutrients available to the plant.
- Produce hormones that help plants grow.
- Retain water.

Soil life plays a major role in many natural processes that determine nutrient and water availability for agricultural productivity. The primary activities of all living organisms are growing and reproducing. By-products from growing roots and plant residues feed soil organisms. In turn, soil organisms support plant health as they decompose organic matter, cycle nutrients, enhance soil structure and control the populations of soil organisms, both beneficial and harmful (pests and pathogens) in terms of crop productivity.

The living part of soil organic matter includes a wide variety of micro-organisms such as bacteria, viruses, fungi, protozoa and algae. It also includes plant roots, insects, earthworms, and larger animals such as moles, mice and rabbits that spend part of their life in the soil. The living portion represents about 5 per cent of the total soil organic matter. Micro-organisms, earthworms and insects help break down crop residues and manures by ingesting them and mixing them with the minerals in the soil, and in the process recycling energy and plant nutrients. Sticky substances on the skin of earthworms and those produced by fungi and bacteria help bind particles together. Earthworm casts are also more strongly aggregated (bound together) than the surrounding soil as a result of the mixing of organic matter and soil mineral material, as well as the intestinal mucus of the worm. Thus, the living part of the soil is responsible for keeping air and water available, providing plant nutrients, breaking down pollutants and maintaining the soil structure.

The composition of soil organisms depends on the food source (which in turn is season dependent). Therefore, the organisms are neither uniformly distributed through the soil nor uniformly present all year. However, in some cases their biogenic structures remain. Each species and group exists where it can find appropriate food supply, space, nutrients and moisture (Plate 2). Organisms occur wherever organic matter occurs (Ingham, 2000). Therefore, soil organisms are concentrated : around roots, in litter, on humus, on the surface of soil aggregates and in spaces between aggregates. For this reason, they are most prevalent in forested areas and cropping systems that leave a lot of biomass on the surface.

Plate
Termites create their own living conditions near their
preferred food sources. Inside the colony life is highly organised.

The activity of soil organisms follows seasonal as well as daily patterns. Not all organisms are active at the same time. Most are barely active or even dormant. Availability of food is an important factor that influences the level of activity of soil organisms and thus is related to land use and management. Practices that increase numbers and activity of soil organisms include : no tillage or minimal tillage; and the maintenance of plant and annual residues that reduce disturbance of soil organisms and their habitat and provide a food supply.

Different groups of organisms can be distinguished in the soil (Brussaard and Juma, 1995). Table classifies them by size.

Decomposition Process

Fresh residues consist of recently deceased micro-organisms, insects and earthworms, old plant roots, crop residues, and recently added manures.

Crop residues contain mainly complex carbon compounds originating from cell walls (cellulose, hemicellulose, etc.). Chains of carbon, with each carbon atom linked to other carbons, form the "backbone" of organic molecules. These carbon chains, with varying amounts of attached oxygen, H, N, P and S, are the basis for both simple sugars and amino acids and more complicated molecules of long carbon chains or rings. Depending on their chemical structure, decomposition is rapid (sugars, starches and proteins), slow (cellulose, fats, waxes and resins) or very slow (lignin).

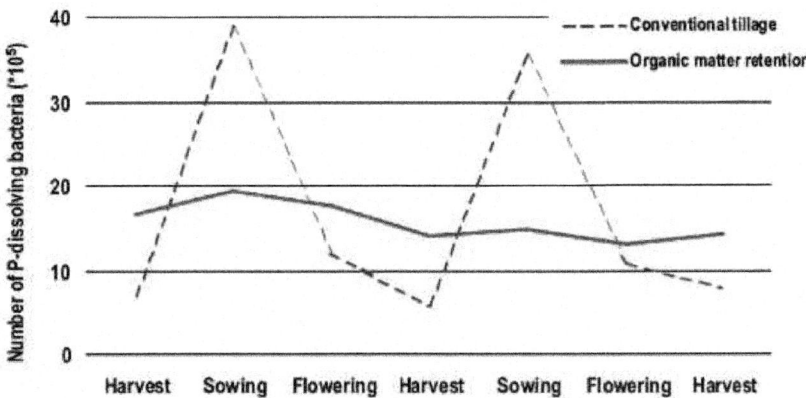

Fig. Fluctuations in microbial biomass at different stages of crop development in conventional agriculture compared with systems with residue retention and high organic matter input

Table. Classification of soil organisms.

Micro-organisms	Micro-flora	<5 μm	Bacteria Fungi
	Micro-fauna	<100 μm	Protozoa Nematodes
Macro-organisms	Meso-organisms	100 μm - 2 mm	Springtails Mites
	Macro-organisms	2 - 20 mm	Earthworms Millipedes Woodlice Snails and slugs
Plants	Algae	10 μm	
	Roots	> 10 μm	

Note : Clay particles are smaller than 2 μm.
Source : Adapted from Swift, Heal and Anderson, 1979.

Table. Essential functions performed by different members of soil organisms (biota).

Functions	Organisms involved
Maintenance of soil structure	Bio-turbating invertebrates and plant roots, mycorrhizae and some other micro-organisms
Regulation of soil hydrological processes	Most bio-turbating invertebrates and plant roots
Gas exchange and carbon sequestration (accumulation in soil)	Mostly micro-organisms and plant roots, some C protected in large compact biogenic invertebrate aggregates
Soil detoxification	Mostly micro-organisms
Nutrient cycling	Mostly micro-organisms and plant roots, some soil- and litter-feeding invertebrates

Decomposition of organic matter	Various saprophytic and litter-feeding invertebrates (detritivores), fungi, bacteria, actinomycetes and other micro-organisms
Suppression of pests, parasites and diseases	Plants, mycorrhizae and other fungi, nematodes, bacteria and various other micro-organisms, collembola, earthworms, various predators
Sources of food and medicines	Plant roots, various insects (crickets, beetle larvae, ants, termites), earthworms, vertebrates, micro-organisms and their by-products
Symbiotic and asymbiotic relationships with plants and their roots	Rhizobia, mycorrhizae, actinomycetes, diazotrophic bacteria and various other rhizosphere micro-organisms, ants
Plant growth control (positive and negative)	Direct effects : plant roots, rhizobia, mycorrhizae, actinomycetes, pathogens, phytoparasitic nematodes, rhizophagous insects, plant-growth promoting rhizosphere micro-organisms, biocontrol agents Indirect effects : most soil biota.

During the decomposition process, micro-organisms convert the carbon structures of fresh residues into transformed carbon products in the soil. There are many different types of organic molecules in soil. Some are simple molecules that have been synthesised directly from plants or other living organisms. These relatively simple chemicals, such as sugars, amino acids, and cellulose are readily consumed by many organisms. For this reason, they do not remain in the soil for a long time. Other chemicals such as resins and waxes also come directly from plants, but are more difficult for soil organisms to break down.

Humus is the result of successive steps in the decomposition of organic matter. Because of the complex structure of humic substances, humus cannot be used by many micro-organisms as an energy source and remains in the soil for a relatively long time.

Non-humic Substances : Significance and Function

Non-humic organic molecules are released directly from cells of fresh residues, such as proteins, amino acids, sugars, and starches. This part of soil organic matter is the active, or easily decomposed, fraction. This active fraction is influenced strongly by weather conditions, moisture status of the soil, growth stage of the vegetation, addition of organic residues, and cultural practices, such as tillage. It is the main food supply for various organisms in the soil.

Carbohydrates occur in the soil in three main forms : free sugars in the soil solution, cellulose and hemicellulose; complex polysaccharides; and polymeric molecules of various sizes and shapes that are attached strongly to clay colloids and humic substances (Stevenson, 1994). The simple sugars, cellulose and hemicellulose, may constitute 5-25 per cent of the organic matter in most soils, but are easily broken down by micro-organisms.

Polysaccharides (repeating units of sugar-type molecules connected in longer chains) promote better soil structure through their ability to bind inorganic soil particles into stable aggregates. Research indicates that the heavier polysaccharide molecules may be more important in promoting aggregate stability and water infiltration than the lighter molecules (Elliot and Lynch, 1984). Some sugars may stimulate seed germination and root elongation. Other soil properties affected by polysaccharides include CEC, anion retention and biological activity.

The soil lipids form a very diverse group of materials, of which fats, waxes and resins make up 2-6 per cent of soil organic matter. The significance of lipids arises from the ability of some compounds to act as growth hormones. Others may have a depressing effect on plant growth.

Soil N occurs mainly (> 90 per cent) in organic forms as amino acids, nucleic acids and amino sugars. Small amounts exist in the form of amines, vitamins, pesticides and their degradation products, etc. The rest is present as ammonium (NH_4^-) and is held by the clay minerals.

Compounds and Function of Humus

Humus or humified organic matter is the remaining part of organic matter that has been used and transformed by many different soil organisms. It is a relatively stable component formed by humic substances, including humic acids, fulvic acids, hymatomelanic acids and humins (Tan, 1994). It is probably the most widely distributed organic carbon-containing material in terrestrial and aquatic environments. Humus cannot be decomposed readily because of its intimate interactions with soil mineral phases and is chemically too complex to be used by most organisms. It has many functions (Box 2).

One of the most striking characteristics of humic substances is their ability to interact with metal ions, oxides, hydroxides, mineral and organic compounds, including toxic pollutants, to form water-soluble and water-insoluble complexes. Through the formation of these complexes, humic substances can dissolve, mobilise and transport metals and organics in soils and waters, or accumulate in certain soil horizons. This influences nutrient availability, especially those nutrients present at micro-concentrations only (Schnitzer, 1986). Accumulation of such complexes can contribute to a reduction of toxicity, *e.g.* of aluminium (Al) in acid soils (Tan and Binger, 1986), or the capture of pollutants — herbicides such as Atrazine or pesticides such as Tefluthrin — in the cavities of the humic substances (Vermeer, 1996).

Humic and fulvic substances enhance plant growth directly through physiological and nutritional effects. Some of these substances function as natural plant hormones (auxines and gibberillins) and are capable of improving seed germination, root initiation, uptake of plant nutrients and can serve as sources of N, P and S (Tan, 1994; Schnitzer, 1986). Indirectly, they may affect plant growth through modifications of physical, chemical and biological properties of the soil, for example, enhanced soil water holding capacity and CEC, and improved tilth and aeration through good soil structure (Stevenson, 1994).

About 35-55 per cent of the non-living part of organic matter is humus. It is an important buffer, reducing fluctuations in soil acidity and nutrient availability. Compared with simple organic molecules, humic substances are very complex and large, with high molecular weights. The characteristics of the well-decomposed part of the organic matter, the humus, are very different from those of simple organic molecules. While much is known about their general chemical composition, the relative significance of the various types of humic materials to plant growth is yet to be established.

Humus consists of different humic substances :

- Fulvic acids : the fraction of humus that is soluble in water under all pH conditions. Their colour is commonly light yellow to yellow-brown.

- Humic acids : the fraction of humus that is soluble in water, except for conditions more acid than pH 2. Common colours are dark brown to black.

- Humin : the fraction of humus that is not soluble in water at any pH and that cannot be extracted with a strong base, such as sodium hydroxide (NaOH). Commonly black in colour.

The term acid is used to describe humic materials because humus behaves like weak acids.

Fulvic and humic acids are complex mixtures of large molecules. Humic acids are larger than fulvic acids. Research suggests that the different substances are differentiated from each other on the basis of their water solubility.

Fulvic acids are produced in the earlier stages of humus formation. The relative amounts of humic and fulvic acids in soils vary with soil type and management practices. The humus of forest soils is characterised by a high content of fulvic acids, while the humus of agricultural and grassland areas contains more humic acids.

NATURAL FACTORS INFLUENCING THE AMOUNT OF ORGANIC MATTER

The transformation and movement of materials within soil organic matter pools is a dynamic process influenced by climate, soil type, vegetation and soil organisms. All these factors operate within a hierarchical spatial scale. Soil organisms are responsible for the decay and cycling of both macro-nutrients and micro-nutrients, and their activity affects the structure, tilth and productivity of the soil.

In natural humid and sub-humid forest ecosystems without human disturbance, the living and non-living components are in dynamic equilibrium with each other. The litter on the soil surface beneath different canopy layers and high biomass production generally result in high biological activity in the soil and on the soil surface. Mollison and Slay (1991) distinguished the following five mechanisms :

- a continuous soil cover of living plants, which together with the soil architecture facilitates the capture and infiltration of rainwater and protects the soil;

- a litter layer of decomposing leaves or residues providing a continuous energy source for macro- and micro-organisms;

- the roots of different plants distributed throughout the soil at different depths permit an effective uptake of nutrients and an active interaction with micro-organisms;

- the major period of nutrient release by micro-organisms coincides with the major period of nutrient demand by plants;

- nutrients recycled by deep-rooting plants and soil macrofauna and microfauna.

This equilibrium creates almost closed-cycle transfers of nutrients between soil and the vegetation adapted to such site conditions, resulting in almost perfect physical and hydric conditions for plant growth, *i.e.* a cool micro-climate, increased evapotranspiration, good rooting conditions with good porosity and sufficient soil moisture. This facilitates water infiltration and prevents erosion and run-off. Thus, it results in clean water in the streams emanating from the area, a relatively smooth variation in streamflow during the year, and recharge of groundwater.

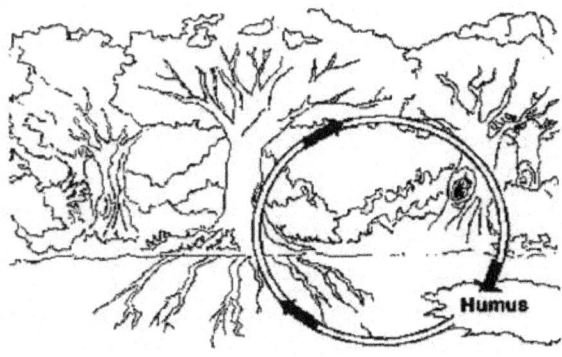

Fig. Closed cycle.

In human-managed systems, the soil biological activity is influenced by the land use system, plant types and the management practices. Chapter before outlines the influence of land management practices. The environmental and edaphic factors that control the activity of soil biota, and thus the balance between accumulation and decomposition of organic matter in the soil, are described below.

Temperature

Several field studies have shown that temperature is a key factor controlling the rate of decomposition of plant residues. Decomposition normally occurs more rapidly in the tropics than in temperate areas. Ladd and Amato (1985) reported that, despite differences in plant material and climate patterns, the decomposition of leguminous materials in southern Australian sites followed the same pattern as that of ryegrass for sites in Nigeria and the United Kingdom (Jenkinson and Ayanaba, 1977), although the time scales were different. Reaction rates doubled for each increase of 8-9 °C in the mean annual air temperature. The relatively faster rate of decomposition induced by the continuous warmth in the tropics implies that high equilibrium levels of organic matter are difficult to achieve in tropical agro-ecosystems. Hence, large annual rates of organic inputs are needed to maintain an adequate labile soil organic matter pool in cultivated soils. Soils in cooler climates commonly have more organic matter because of slower mineralisation (decomposition) rates.

Soil Moisture and Water Saturation

Soil organic matter levels commonly increase as mean annual precipitation increases. Conditions of elevated levels of soil moisture result in greater biomass production, which provides more residues, and thus more potential food for soil biota.

Soil biological activity requires air and moisture. Optimal microbial activity occurs at near "field capacity", which is equivalent to 60-per cent water-filled pore space (Linn and Doran, 1984).

On the other hand, periods of water saturation lead to poor aeration. Most soil organisms need oxygen, and thus a reduction of oxygen in the soil leads to a reduction of the mineralisation rate as these organisms become inactive or even die. Some of the transformation processes become anaerobic, which can lead to damage to plant roots caused by waste products or favourable conditions for disease-causing organisms. Continued production and slow decomposition can lead to very large organic matter contents in soils with long periods of water saturation (e.g. peat soils, and tea crops in India).

With the exception of the hyperhumid regions, the climates of vast areas of the humid, sub-humid and semi-arid tropics are characterised by distinct wet and dry seasons. In the wet-dry tropics, large amounts of nitrate often occur in the surface soil during the first part of the rainy season (Greenland, 1958; Mueller-Harvey, Juo and Wild,1989). This accelerated nitrogen mineralisation caused by a large increase in microbial activity is the result of the first few rains activating the labile soil organic matter.

Farmers who practise "slash and burn" agriculture often choose early planting in order to take advantage of this flush of inorganic N before it is lost through leaching and run-off. In these low-input systems, the amount of nitrate present in the soil during the early part of the rainy season is related closely to the organic matter content of the soil. N availability diminishes during the later part of the rainy season.

Soil Texture

Soil organic matter tends to increase as the clay content increases. This increase depends on two mechanisms. First, bonds between the surface of clay particles and organic matter retard the decomposition process. Second, soils with higher clay content increase the potential for aggregate formation. Macro-aggregates physically protect organic matter molecules from further mineralisation caused by microbial attack (Rice, 2002). For example, when earthworm casts and the large soil particles they contain are split by the joint action of several factors (climate, plant growth and other organisms), nutrients are released and made available to other components of soil micro-organisms.

Under similar climate conditions, the organic matter content in fine textured (clayey) soils is two to four times that of coarse textured (sandy) soils (Prasad and Power, 1997).

Kaolinite, the main clay mineral in many upland soils in the tropics, has a much smaller specific surface and nutrient exchange capacity than most other clay minerals. Therefore, kaolinitic soils contain considerably fewer clay-humus complexes. In addition, the unprotected labile humic substances are vulnerable to decomposition under appropriate soil moisture conditions. Thus, high levels of organic matter are difficult to maintain in cultivated kaolinitic soils in the wet-dry tropics, because climate and soil conditions favour rapid decomposition. In contrast, organic matter can persist as organo-oxide complexes in soils rich in iron and aluminium oxides. Such properties favour the formation of soil micro-aggregates, typical of many fine-textured, oxide-rich, high base-status soils in the tropics (Uehara and Gilman, 1981). These soils are known for their low bulk density, high microporosity, and high organic-matter retention under natural vegetation, but also for their high phosphate fixation capacity on the oxides when used for crop production. Current knowledge suggests that whereas organic matter contributes to the dark colour of Vertisols (Coulombe, Dixon and Wilding, 1996), it is not considered important in determining either the development, robustness or resilience of structure in these soils (McGarry, 1996). Organic matter levels tend to be low in Vertisols; even as low as 10 g/ kg (Coulombe, Dixon and Wilding, 1996).

Parent material influences organic matter accumulation not only through its effect on soil texture. Soils developed from inherently rich material, such as basalt, are more fertile than soils formed from granitic material, which contains less mineral nutrients. Moreover, the former experience more organic matter accumulation because of abundant vegetative growth.

Topography

Organic matter accumulation is often favoured at the bottom of hills. There are two reasons for this accumulation : conditions are wetter than at mid- or upper-slope positions, and organic matter is transported to the lowest point in the landscape through run-off and erosion. Similarly, soil organic matter levels are higher on north-facing slopes (in the Northern Hemisphere) compared with south-facing slopes (and the other way around in the Southern Hemisphere) because temperatures are lower.

Salinity and Acidity

Salinity, toxicity and extremes in soil pH (acid or alkaline) result in poor biomass production and, thus in reduced additions of organic matter to the soil.

For example, pH affects humus formation in two ways : decomposition, and bio-mass production. In strongly acid or highly alkaline soils, the growing conditions for micro-organisms are poor, resulting in low levels of biological oxidation of organic matter (Primavesi, 1984). Soil acidity also influences the availability of plant nutrients and thus regulates indirectly biomass production and the available food for soil biota. Fungi are less sensitive than bacteria to acid soil conditions.

Vegetation and Biomass Production

The rate of soil organic matter accumulation depends largely on the quantity and quality of organic matter input. Under tropical conditions, applications of readily degradable materials with low C : N ratios, such as green manure and leguminous cover crops, favour decomposition and a short-term increase in the labile nitrogen pool during the growing season. On the other hand, applications of plant materials with both large C : N ratios and lignin contents such as cereal straw and grasses generally favour nutrient immobilisation, organic matter ac-cumulation and humus formation, with increased potential for improved soil structure development.

Plant constituents such as lignin and other polyphenols retard decomposi-tion. In an experiment in southern Nigeria to compare management effects on soil organic matter accumulation, a three-year fallow with Guinea grass (*Panicum maximum*), which has a high lignin content, maintained a carbon level comparable to that under forest fallow. However, fallowing with leguminous species such as pigeon pea (*Cajanus cajan*) caused a significant decline in soil total C (Juo and Lal, 1977).

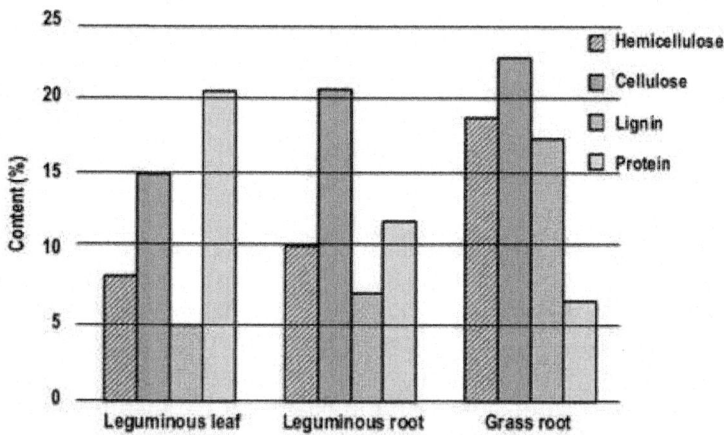

Fig. Composition of leaves and roots of leguminous and grass species

Palm and Sanchez (1990) reported that both the decomposition rate and the N-release patterns of three tropical legumes (*Inga edulis, Cajanus cajan,* and *Eryth-rina* spp.) were related to the amount of polyphenol compounds such as lignin in the leaf. *Erythrina* leaves had the lowest concentrations of polyphenols and the fastest decomposition rate of the three species studied.

Root turnover also constitutes an important addition of humus into the soil, and consequently it is important for carbon sequestration. In forests, most organic matter is added as superficial litter. However, in grassland ecosystems, up to two-thirds of organic matter is added through the decay of roots (Quideau, 2002).

Chapter 8

INFLUENCE THE AMOUNT OF ORGANIC MATTER

HUMAN INTERVENTIONS THAT INFLUENCE SOIL ORGANIC MATTER

Various types of human activity decrease soil organic matter contents and biological activity. However, increasing the organic matter content of soils or even maintaining good levels requires a sustained effort that includes returning organic materials to soils and rotations with high-residue crops and deep- or dense-rooting crops. It is especially difficult to raise the organic matter content of soils that are well aerated, such as coarse sands, and soils in warm-hot and arid regions because the added materials decompose rapidly. Soil organic matter levels can be maintained with less organic residue in finetextured soils in cold temperate and moist-wet regions with restricted aeration.

Practices that Decrease Soil Organic Matter

Any form of human intervention influences the activity of soil organisms (Curry and Good, 1992) and thus the equilibrium of the system. Management practices that alter the living and nutrient conditions of soil organisms, such as repetitive tillage or burning of vegetation, result in a degradation of their micro-environments. In turn, this results in a reduction of soil biota, both in biomass and diversity. Where there are no longer organisms to decompose soil organic matter and bind soil particles, the soil structure is damaged easily by rain, wind and sun. This can lead to rainwater run-off and soil erosion (Plate 3), removing the potential food for organisms, *i.e.* the organic matter of the topsoil. Therefore, soil biota are the most important property of the soil, and "when devoid of its biota, the uppermost layer of earth ceases to be soil".

Fig. Open cycle system.

The factors leading to reduction in soil organic matter in an open cycle system can be grouped as factors that result in :

- a decrease in biomass production;
- a decrease in organic matter supply;
- increased decomposition rates.

DECREASE IN BIOMASS PRODUCTION

Replacement of Perennial Vegetation

A consequence of clearing forest for agriculture is the disappearance of the litter layer, with a consequent reduction in the numbers and variety of soil organisms. While many temperate forest species appear to adapt well to grassland (Curry and Good, 1992), the effects of deforestation in the tropics appear to be more marked (Plate 4). Studies have shown that as soil bio-diversity declines, adapted species may take over from the indigenous species and the composition may change drastically.

Soil macro-faunal biomass and population density fell to 6 and 17 per cent, respectively, in cultivated plots, compared with primary forest in Peruvian Amazonia (Lavelle and Pashanasi, 1989). In Suriname, the number of animals per square metre has fallen to 36 per cent and the diversity of species has fallen to 28 per cent compared with primary forest (Van der Werff, 1990). The indigenous species have largely disappeared, but adapted species have been available for recolonisation. The composition of the macro-faunal community has changed drastically.

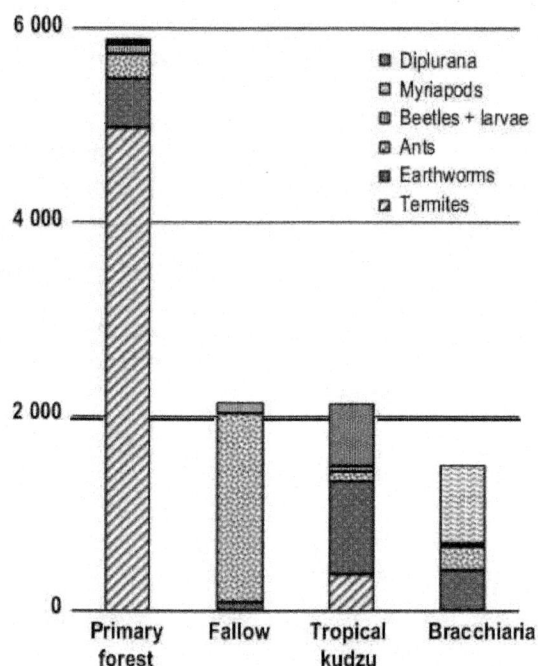

Fig. Composition of soil macro-fauna under primary forest, fallow,
kudzu and grass vegetation.

Replacement of Mixed Vegetation with Mono-culture of Crops and Pastures

The simplification of vegetation and the disappearance of the litter layer under grassland and mono-crop production systems lead to a decrease in faunal diversity. Although root systems (especially of grasses) can be extensive and explore vast areas of soil, the root exudates from one single crop will attract only a few different microbial species. This in turn will affect the predator diversity. The more opportunistic pathogen species will be able to acquire space near the crop and cause harm. Continuous cultivation and grasing also leads to compaction of soil layers, which in turn affects the circulation of air. Anaerobic conditions in the soil stimulate the growth of different micro-organisms, resulting in more pathogenic organisms.

High Harvest Index

One of the consequences of the green revolution was the replacement of indigenous varieties of species with high-yielding varieties (HYVs). These HYVs often produce more grain and less straw, compared with locally developed varieties; the harvest index of the crop (ratio of grain to total plant mass aboveground) is increased. From a production point of view, this is a logical approach. However, this is less desirable from a conservation point of view. Reduced amounts of crop residues remain after harvest for soil cover and organic matter, or for grasing of livestock (which results in manure). Moreover, where animals graze the residues, even less remains for conservation purposes.

Use of Bare Fallow

Traditionally, a fallow period is used after a period of crop production to give the land some "rest" and to regenerate its original state of productivity. Usually, this is necessary in production systems that have drawn down the nutrient supply and altered the soil biota significantly, such as in slash-and-burn systems or conventional tillage systems.

Some farmers use bare fallow to regenerate their lands. However, apart from spontaneous weed growth, this means there is no energy source for the soil biota present on the land. Instead of recovering the soil food web, the soil organic matter is degraded further and the lack of cover can result in severe erosion and run-off when the rains start after the dry season.

DECREASE IN ORGANIC MATTER SUPPLY

Burning of Natural Vegetation and Crop Residues

The burning of maise, rice and other crop residues in the field is a common practice (Plate 5). Residues are usually burned to help control insects or diseases or to make fieldwork easier in the following season. Burning destroys the litter layer and so diminishes the amount of organic matter returned to the soil. The organisms that inhabit the surface soil and litter layer are also eliminated. For future decomposition to take place, energy has to be invested first in rebuilding the microbial community before plant nutrients can be released. Similarly, fallow lands and bush are burned before cultivation. This provides a rapid supply of P to stimulate seed germination. However, the associated loss of nutrients, organic matter and soil biological activity has severe long-term consequences.

Overgrasing

There is a tendency throughout the world to overstock grasing land above its carrying capacity. Cows, draught animals and small ruminants graze on communal grasing areas and on roadsides, stream banks and other public land (Plate 6). Overgrasing destroys the most palatable and useful species in the plant mixture and reduces the density of the plant cover, thereby increasing the erosion hazard and reducing the nutritive value and the carrying capacity of the land.

Removal of Crop Residues

Many farmers remove residues from the field for use as animal feed and bedding or to make compost. Later, these residues return to contribute to soil fertility as manures or composts. However, residues are sometimes removed from the field and not returned. This removal of plant material impoverishes the soil as it is no longer possible to recycle the plant nutrients present in the residues.

INCREASED DECOMPOSITION RATES

Tillage Practices

Tillage is one of the major practices that reduces the organic matter level in the soil. Each time the soil is tilled, it is aerated. As the decomposition of organic matter and the liberation of C are aerobic processes, the oxygen stimulates or speeds up the action of soil microbes, which feed on organic matter.

This means that :

- When ploughed, the residues are incorporated in the soil together with air and come into contact with many micro-organisms, which accelerates the carbon cycle. The decomposition is faster, resulting in the formation of less stable humus and an increased liberation of CO_2 to the atmosphere, and thus a reduction in organic matter.

- The residues on the soil surface slow the carbon cycle because they are exposed to fewer micro-organisms and thus wane more slowly, resulting in the production of humus (which is more stable), and liberating less CO_2 to the atmosphere.

Table. Tillage induced flush of decomposition of organic matter.

Type of tillage	Organic matter lost in 19 days (kg/ha)
Mouldboard plough + disc harrow (2x)	4 300
Mouldboard plough	2 230
Disc harrow	1 840
Chisel plough	1 720
Direct seeding	860

In terms of short-term organic matter loss, the more a soil is tilled, the more the organic matter is broken down (Table 3). There are also longer-term losses, attributed to repeated, annual cultivation. Cropping systems that return little residue to the soil accelerate this decline. Many modern cropping systems combine frequent tillage with small amounts of residue, with resultant reductions in the organic matter content of many soils. Historically, manure application (from farm livestock) was common, and it was a dynamic way of maintaining organic matter levels despite repeated cultivation and low residue returns to the soil. Increased on-farm mechanisation has reduced livestock numbers, so this source of organic material has been reduced considerably.

Organic matter production and conservation is affected dramatically by conventional tillage, which not only decreases soil organic matter but also increases the potential for erosion by wind and water (Plate 8). The impact occurs in many ways :

- Ploughing leaves no residues on the soil surface to lessen the impact of rain.

- Ploughing reduces the quantity of food sources for earthworms and disturbs their burrows and living space, hence populations of certain species decrease drastically. Moreover, reduction of earthworm numbers reduces their impact, through burrowing, in increasing porosity and aeration (particularly continuous macropores) and lowers their ability to bury and incorporate plant residues, which facilitates rapid decomposition of organic matter.

- Tillage by repeated hoeing or discing smoothes the surface and destroys natural soil aggregates and channels that connect the surface with the sub-soil, leaving the soil susceptible to erosion. Old root channels and earthworm holes are eliminated, as are the cracks between natural aggregates. The large pores, the ones destroyed by conventional tillage practices, are necessary to conduct water into the soil during rainfall.

- The development of a plough pan or hoe pan, a layer of compacted soil resulting from smearing action at the bottom of the plough or hoe, may retard both root penetration and water infiltration.

- Ploughing or discing under dry conditions exacerbates the pulverisation of the soil, causing the soil surface to crust more easily, leading to greater water run-off and erosion. This is exacerbated by reduced soil surface roughness, which leaves few depressions for temporary storage of water during intense storms.

- Increased run-off during rainstorms may also increase the possibility of drought stress later in the season, because water that runs off the field does not infiltrate into the soil to remain available to plants.

In some circumstances, imbalances of certain soil organisms can disrupt soil structure and processes, *e.g.* certain earthworm species in rice fields or pastures.

Drainage

Decomposition of organic matter occurs more slowly in poorly aerated soils, where oxygen is limiting or absent, compared with well-aerated soils. For this reason, organic matter accumulates in wet soil environments. Soil drainage is determined strongly by topography — soils in depressions at the bottom of hills tend to remain wet for extended periods of time because they receive water (and sediments) from upslope. Soils may also have a layer in the sub-soil that inhibits drainage, again exacerbating water-logging and reduction in organic matter decomposition. In a permanently water-logged soil, one of the major structural parts of plants, lignin, does not decompose at all. The ultimate consequence of extremely wet or swampy conditions is the development of organic (peat or muck) soils, with organic matter contents of more than 30 per cent. Where soils are drained artificially for agricultural or other uses, the soil organic matter decomposes rapidly.

Fertilizer and Pesticide Use

Initially, the use of fertilizer and pesticides enhances crop development and thus production of biomass (especially important on depleted soils). However, the use of some fertilizers, especially N fertilizers, and pesticides can boost micro-organism activity and thus decomposition of organic matter. The chemicals provide the micro-organisms with easy-to-use N components. This is especially important where the C : N ratio of the soil organic matter is high and thus decomposition is slowed by a lack of N.

PRACTICES THAT INCREASE SOIL ORGANIC MATTER

Increased concern about the environmental and economic impacts of conventional crop production has stimulated interest in alternative systems. Central to such systems is the need to promote and maintain soil biological processes and minimise fossil fuel inputs in the form of fertilizers, pesticides and mechanical cultivation. All activities aimed at the increase of organic matter in the soil (Box 3) help in creating a new equilibrium in the agro-ecosystem.

For a system of natural resource management to be balanced, and thus sustainable, it must be able to withstand sharp climatic fluctuations, and to evolve steadily in response to social changes and changes in the costs and availability of inputs of land, labour and knowledge.

The more diverse and complex an agricultural system is, the more stable and sustainable it will be in the face of unpredictable vagaries of climate and market. Thus, annual crops, woody perennials and non-woody perennials may be combined in various ways with livestock or trees, or both, in what are now commonly called agrosilvipastoral systems.

Different approaches are required for different soil and climate conditions. However, the activities will be based on the same principle : increasing biomass production in order to build active organic matter. Active organic matter provides habitat and food for beneficial soil organisms that help build soil structure and porosity, provide nutrients to plants, and improve the water holding capacity of the soil.

Several cases have demonstrated that it is possible to restore organic matter levels in the soil. Activities that promote the accumulation and supply of organic matter, such as the use of cover crops and refraining from burning, and those that reduce decomposition rates, such as reduced and zero tillage, lead to an increase in the organic matter content in the soil (Sampson and Scholes, 2000).

BOX
Ways to increase organic matter contents of soils

- compost

- cover crops/green manure crops

- crop rotation

- perennial forage crops

- zero or reduced tillage

- agro-forestry

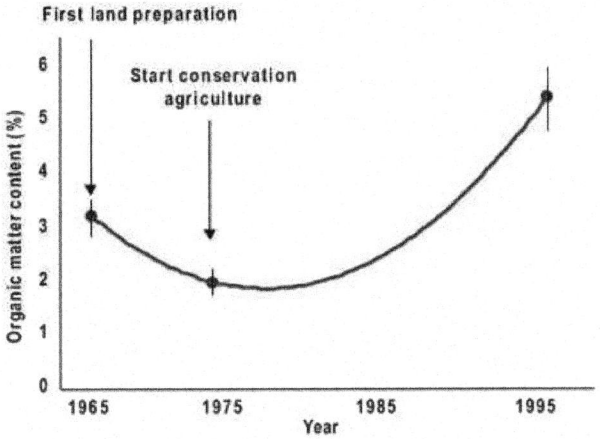

Fig. Evaluation of the organic matter content of a soil in Paraná

INCREASED BIOMASS PRODUCTION

Increased Water Availability for Plants : Water Harvesting and Irrigation

In dry conditions, water may be provided through irrigation or water harvesting. The increased water availability enhances biomass production, soil biological activity and plant residues and roots that provide organic matter.

The concept of water harvesting includes various technologies for run-off management and utilisation. It involves capture of run-off (in some cases through treating the upstream capture area), and its concentration on a runon area for use by a specific crop (annual or perennial) in order to enhance crop growth and yields, or its collection and storage for supplementary irrigation or domestic or livestock purposes. The objective of designing a water harvesting system is to obtain the best ratio of the area yielding run-off to either the area where run-off is being directed or the capacity of the storage structure (volume of water collected). In this way, the water captured for crop production during run-off periods can be stored either directly in the soil for subsequent use by plants or in small farm reservoirs or collection tanks (Plate 9). This aids stabilisation of crop production by enhancing soil moisture availability or allowing irrigation during a dry period within the rainy season or by extending crop production into the dry season. Some factors to be considered regarding these run-off farming systems and reservoirs include : site selection, watershed size and condition, rainfall distribution and run-off, and water requirements of crops. Where a minimum water depth of about 1 m can be maintained in a reservoir, fish can be raised to provide additional food (FAO, 1984).

Numerous water harvesting systems have been developed over the centuries, especially in arid areas. The principle of collecting run-off for crop production is also inherent to many other soil and water conservation technologies that apply the concept of run-off and runon areas at a micro-watershed level, such as nega-rims, trapezoidal or "eyebrow" bunds and tied ridges.

Plate 9

Balanced Fertilization

Where the supply of nutrients in the soil is ample, crops are more likely to grow well and produce large amounts of biomass. Fertilizers are needed in those cases where nutrients in the soil are lacking and cannot produce healthy crops (FAO, 2000) and sufficient biomass. Most soils in sub-Saharan Africa (SSA) are deficient in P. P is required not only for plant growth but also for N fixation. Unbalanced fertilization, for example mainly with N, may result in more weed competition, higher pest incidence and loss of quality of the product. Unbalanced fertilization eventually leads to unhealthy plants. Therefore, fertilizers should be applied in sufficient quantities and in balanced proportions. The efficiency of fertilizer use will be high where the organic matter content of the soil is also high. In very poor or depleted soils, crops use fertilizer applications inefficiently. When soil organic matter levels are restored, fertilizer can help maintain the revolving fund of nutrients in the soil by increasing crop yields and, consequently, the amount of residues returned to the soil.

Cover Crops

Growing cover crops is one of the best practices for improving organic matter levels and, hence, soil quality. The benefits of growing cover crops include :

- They prevent erosion by anchoring soil and lessening the impact of raindrops.
- They add plant material to the soil for organic matter replenishment.
- Some, *e.g.* rye, bind excess nutrients in the soil and prevent leaching.
- Some, especially leguminous species, *e.g.* hairy vetch, fix N in the soil for future use.
- Most provide habitat for beneficial insects and other organisms.
- They moderate soil temperatures and, hence, protect soil organisms.

A range of crops can be used as vegetative cover, *e.g.* grains, legumes and oil crops. All have the potential to provide great benefit to the soil. However, some crops emphasise certain benefits; a useful consideration when planning a rotation scheme. It is important to start the first years with (cover) crops that cover the surface with a large amount of residues that decompose slowly (because of the high C : N ratio). Grasses and cereals are most appropriate for this stage, also because of their intensive rooting system, which improves the soil structure rapidly.

In the following years, when soil health has begun to improve, legumes can be incorporated in the rotation. Leguminous crops enrich the soil with N and their residues decompose rapidly because of their low C : N ratio. Later, when the system is stabilised, it is possible to include cover crops with an economic function, *e.g.* livestock fodder.

The selection of cover crops should depend on the presence of high levels of lignin and phenolic acids. These give the residues a higher resistance to decomposition and thus result in soil protection for a longer period and the production of more stable.

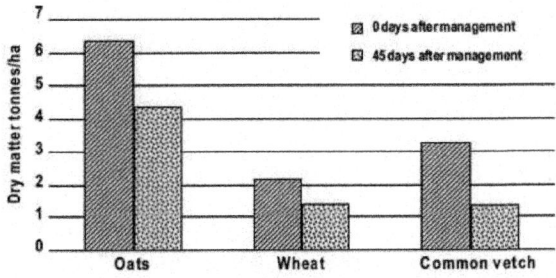

Fig. Reduction of dry matter of different cover crops.

Source : Ruedell, 1995 humus.

Another determining factor in the dynamics of residue composition is the bio-chemical composition of the residues. Depending on species, their chemical components and the time and way of managing them, there will be differences in decomposition rates. The grain species (oats and wheat) show more resistance than common vetch (legume) to decomposition. The latter has a lower C : N ratio and a lower lignin content and is thus subject to a rapid decomposition.

Agricultural production systems in which residues are left on the soil surface, such as direct seeding and the use of cover crops, stimulate the development and activity of soil fauna at many levels.

The term green manure is often used to indicate the same plant species that are used as cover crops. However, green manure refers specifically to a crop in the rotation grown for incorporation of the non-decomposed vegetative matter in the soil. While this practice is used specifically to add organic matter, this is not the most effective use of organic matter (especially in hot climates) for two reasons :

- Mechanical disturbance of the soil should be avoided as much as possible.

- When biomass is incorporated in the soil all at one time, there is a short period of high microbial activity in decomposing the material. This results in the sudden release of a large quantity of nutrients that cannot be captured by the seedlings of the following crop and is thus lost from the system.

In general, the greater the production of green manure or crop biomass, the greater is the microbial, meso-fauna and macro-fauna population of the soil — from fungi and micro-organisms to earthworms and termites. The dynamics of surface residue decomposition depend *inter alia* on the activity of micro-organisms and also on soil meso-fauna and macro-fauna. The macro-fauna consists mainly of earthworms, beetles, termites, ants, millipedes, spiders, snails and slugs. These organisms help integrate the residues into the soil and improve soil structure, porosity, water infiltration, and through-flow through the creation of burrows, ingestion and secretions.

The natural incorporation of cover-crop and weed residues from the soil surface to deeper layers in the soil by soil macro-fauna is a slow process. The activity of micro-organisms is regulated by the activity of the macro-fauna, because the latter provide them with food and air through their bio-turbation activities. In this way, nutrients are released slowly and can provide the crop with nutrients over a longer period. At the same time, the soil is covered for a long time by the residues and is protected against the impact of rain and sun.

Improved Vegetative Stands

In many places, low plant densities limit crop yields. Wide plant spacing is often practised as "a way to return power to the soil" or "to give the soil some rest", but in reality it is an indicator that the soil is impoverished. Plant spacing is usually determined by farmers in relation to soil fertility and available water or expected rainfall (unless standard recommendations are enforced by extension).

This means that plants are often spaced widely on depleted soils in arid and semi-arid regions with a view to ensuring an adequate provision of plant nutrients and water for all plants.

However, it is important to maintain the recommended plant spacing in order to optimise biomass production and rooting density and, hence, organic matter for food, moisture retention and habitat for soil organisms. Once the crop is established, reduced sunlight between closer crop rows may also reduce re-growth of weeds.

BOX
Planting pits

Planting pits achieve fast rehabilitation of severely degraded land, especially in a semi-arid climate where a short fallow period of natural grass growth (2-6 years after 2-3 years under crops) cannot be expected to maintain or restore the land's agricultural productivity (FAO, 1994).

An example of the rapid restoration of productivity of degraded land is an indigenous method in the Sahel region called "zaï" (FAO, 1994). During the dry season, farmers dig out pits 15 cm deep and 40 cm in diameter every 80 cm, tossing the earth downhill. The dry desert Harmattan wind blows various organic residues into the excavated pits. The organic materials are consumed quickly by termites, which excavate tunnels through the crusted surface, allowing the first rains to soak down deep, out of danger of direct evaporation.

Two weeks before the onset of the rains, farmers spread one or two handfuls of dry dung (1- 2.5 tonnes/ha) in the bottom of the pits and cover it with earth to prevent the rains from eroding away the organic matter.

Millet is sown into the pits at the onset of the rainy season. As the first rains wash over the surface crust (of the degraded land), the basins capture this run-off (enough to soak a pocket of soil up to 1 m in depth). The sown seeds germinate, break up the slaked surface crust and send roots down to the deeper stores of both water and nutrients (recycled by the termites).

At harvest time, stalks are cut at a height of 1 m and left *in situ* to reduce wind-speed and trap wind-borne organic matter. In the second year, the farmer either digs new basins between the first ones and dresses them with manure, or pulls up the stubble and sows again in the old basins. Stubble clumps laid between basins are in turn used as a food source by termites.

Planting pits are a way of increasing biomass production and crop yields on severely degraded land in semi-arid conditions. Rainfall is concentrated near the plants, and soil faunal activity and organic matter accumulation are concentrated in the planting pits (Box 4 and Plate 10). Planting pits have been introduced successfully in Zambia as a conservation practice for small-holder farmers, who do not have fertilizers or tractor services available to them.

Plate
Half-moons around newly planted Acacia seedlings catch and retain rainwater.
T.F. SHAXSON

Agro-forestry and Alley Cropping

Agro-forestry is a collective name for land-use systems where woody perennials (trees, shrubs, palms, etc.) are integrated in the farming system (FAO, 1989). Alley cropping is an agro-forestry system in which crops are grown between rows of planted woody shrubs or trees. These are pruned during the cropping season to provide green manure and to minimise shading of crops.

Agro-forestry covers a wide range of systems (Box 5) combining food crops, forestry and pasture species in different ways (agro-silviculture, silvipasture, agro-silvipasture and multipurpose forest production). There are two different approaches to agro-forestry. One uses agricultural crops or pasture as a transitional means of utilising the land until forest plantations are fully established. The other is to integrate trees and shrubs permanently into the crop or animal production system, to the benefit of both crop production and land resource protection. Thus, agro-forestry encompasses many traditional land-use systems such as home gardens, shifting cultivation and bush fallow systems (FAO, 1989).

BOX
Examples of agro-forestry systems worldwide

Poro (*Erythrina poeppigiana*) has been grown extensively in coffee plantations in Costa Rica for shade, soil enrichment, live mulching and live fences.

Albizzia spp. have been used in tea plantations in many Asian countries.

In Indonesia, leucaena (*Leucaena leucocephala*) has been planted as contour hedges on hillsides for erosion control, soil improvement and green mulch. It is estimated that some 20 000 ha of undulating land have been converted to these systems.

In West Africa and Rwanda, many farmers use trees, fruit trees, bushes and grasses planted with agricultural crops on their farms.

Many coconut plantations in the Caribbean are partly planted with bananas or used as pastures. Some small farms in Jamaica plant coconuts, banana and citrus together.

Alley cropping can be considered an improved bush fallow system. Small trees or shrubs are planted in cropland in rows, preferably along the contour (even where east-west orientation of the rows may minimise shading of crops). The optimal spacing between rows depends on : slope; soil type and its susceptibility to erosion; rainfall; crop species; and the soil and crop management system (FAO, 1995).

Besides adding organic matter to the system, perennial trees and shrubs recycle plant nutrients from deeper soil layers through their rooting system. Through litter and pruning, these can be used again by annual crops. Probably the most important contribution of perennials in a production system lies in the fact that throughout the whole year their roots excrete root exudates and decaying root cells, which in turn are used as an energy source by soil micro-organisms. The food web in the soil is maintained, even during dry seasons when no annual crops are grown. The result is that soil biota are in place to provide the crop with nutrients at the beginning of the next cropping season.

Direct seeding is the easiest and cheapest way of establishing hedgerows around fields or in the fields (alleys). However, emerging seedlings may not be able to compete with weeds without additional care. Therefore, starting plant growth in a nursery and transplanting may be necessary for some species. Other species may be established by cuttings. With good establishment, the plants will be better able to withstand both dry spells and browsing by livestock. Crucial to a successful establishment of the hedgerow is that the selected plants should be tall enough to outgrow the weeds at the time of the first crop harvest.

Plate

During the cropping season, hedgerow pruning is needed in order to avoid shading of the crop (Plate 11). The timing, frequency and extent of pruning depend on the species used and the season. As a general rule, the lower the hedgerows and the taller the crop, the less frequently is pruning required. Fast growing plants such as *Leucaena leucocephala* and *Gliricidia sepium* may require pruning every six weeks during the cropping season. They are often pruned to a height of about 50 cm. Care must be exercised as too frequent pruning can result in tree dieback.

The integration of trees and woody shrubs into the cropping system offers additional uses and many benefits, as mentioned by farmers using the Quezungual system in Honduras (Plate 12 and Box 6). However, farmers with short-term land tenure may not be interested in these benefits. Furthermore, the plantation of trees sometimes has an effect on the land tenure status; therefore tenants may not be allowed to establish trees on agricultural land. Agro-forestry systems can also inhibit mechanisation and may need increased labour inputs, especially for hedgerow pruning.

BOX

Farmers' perceptions of the Quezungual system : benefits and disadvantages

The Quezungual system has many benefits according to local farmers :

- improved soil moisture conservation, which permits a good development of the crop even during the dry spells of 2-4 weeks halfway through the rainy season (and during the dry period of El Niño in 1997);
- production of fuelwood and fruits from the trees and shrubs;
- agricultural production is greater than in traditionally managed plots;
- plots with the Quezungual system can be cultivated for longer periods than under the slash-andburn system;
- timber trees can be cut after about 7 years and used for construction and/ or sold;

- the mulch obtained through the pruning of the trees and shrubs protects the soil surface from the impact of rain showers, thus there is less soil erosion (even during the heavy rains produced by hurricane "Mitch" in 1998 there was little soil erosion);
- minimal labour is required to establish and maintain the Quezungual system;
- the soil becomes more fertile and the effect of fertilizers on production improves;
- the workability of the soil improves because the soil becomes softer, hence less labour is needed during sowing;
- the Quezungual system provides shade for farmers while they work the plot;
- harvested products, such as beans and maise, can be dried by hanging them over the tree trunks;
- cattle can feed on the residues after maise and sorghum harvests;
- mulch cover reduces the incidence of disease in the bean crop;
- the presence of trees and shrubs in the plot attracts animals and insects, *e.g.* birds and butterflies.

Disadvantages mentioned by the farmers are :

- in the first year, the production is the same as or slightly less than grain production obtained with the traditional system;
- in the early years of implementation, the incidence of slugs in the bean crop is greater.
- too much soil cover can impede seed germination;
- the shade of the Quezungual system can result in a higher incidence of disease during intense rainfall periods (because of greater humidity).

(FAO, 2001)

The hedgerow species have to be selected carefully in order to avoid negative impacts on crop production because of the complex relationships (competition for light, water and nutrients, allelopathy, occurrence of pest and diseases, etc.) that are inherent to agro-forestry systems. Many farmers may consider the hedgerows as not useful, especially where their positive effects are not secure or visible. Where livestock are allowed to graze freely, it can be difficult to establish hedgerows without taking special measures to protect the young plants. Fencing or control of grasing animals may require collective efforts and agreement by the local community.

The increased labour requirement, the reduced cropped area and the difficulty of mechanisation may make alley cropping uneconomic unless the hedgerow species produce direct benefits such as fruits, fuelwood or poles/timber for construction purposes (in addition to the nutrient recycling and erosion control effects).

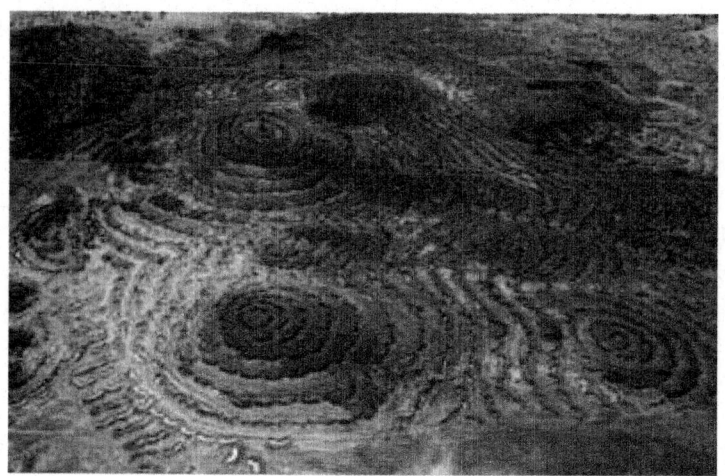

Plate

Reforestation and Afforestation

Afforestation means the establishment of a forest on land that has not grown trees recently. It can serve two principal soil and water conservation purposes : protection of erosion-prone areas, and revegetation and rehabilitation of degraded land (Plate 13). Afforestation is specifically used to provide protective cover in vulnerable, steep and mountainous areas. Afforestation helps to replenish timber resources and provide fuelwood and fodder (FAO, 1979).

The establishment of a forest cover under good management is an effective means of increasing organic matter production. However, the land must have the productive capacity to support an appropriate forest type, which differs according to climate, soil, slope and the specific purpose of the forest (timber production, livestock grasing, etc.). Therefore, the choice of species and the selection of an appropriate site are of particular importance for successful afforestation.

The procurement of adequate quantities of good quality seed of the species and provenances (adapted varieties) required is a prerequisite for any afforestation effort. However, it is often difficult to find suitable and reliable sources of such seeds.

A number of species require special pre-treatment of the seed or seedling in order to achieve satisfactory germination and uniform stand. Such treatment may consist of soaking the seed in water for varying lengths of time, alternate soaking and drying, scarifying or chipping the seed coat to render it permeable to water, plunging the seed into boiling water or even boiling it for a short time. Some tree seedlings may need a mycorrhizal treatment when planted in soils that are deprived of associated mycorrhizae species as well as rhizobium species (e.g. *Casuarina* in Senegalese sandy soils). The aim is to ensure that good numbers of plants germinate and that germination after sowing is both rapid and uniform (FAO, 1974).

Afforestation can be achieved by direct sowing or replanting young plants from a nursery. The main advantage of direct sowing is the reduced cost. However, this is usually much less reliable and is only justified where :

- seed is plentiful and cheap;
- adequate germination under field conditions can be relied on;
- the seedlings send down a deep tap root rapidly and are able to withstand adverse climatic conditions in the time after germination;
- the rate of growth is sufficiently fast to make a prolonged period of tending and weeding unnecessary.

Regeneration of Natural Vegetation

Regeneration of natural grasslands and forest areas increases biomass production and improves the plant species diversity, resulting in more diverse soil biota and other associated beneficial organisms. Natural regeneration may be more reliable where land is not very productive. In some cases, natural regeneration of a given area may lead to the infestation of plots by weeds. Increasingly, natural vegetation is being recognised for its multi-purpose benefits, for example, fuelwood, fibre, biocontrol (*e.g.* neem) and medicinal species, as well as restoration of soil fertility (*Acacia albida* and other leguminous species) and habitats for various beneficial species (pollinators and natural enemies) as well as wildlife.

INCREASED ORGANIC MATTER SUPPLY

Protection from Fire

Burning affects organic matter recycling significantly. Fire destroys almost all organic materials on the land surface except for tree trunks and large branches. In addition, the surface soil is sterilised, loses part of its organic matter, the population of soil micro-fauna and macro-fauna is reduced, and no ready-to-use organic matter is available for rapid restoration of the populations. However, this practice is widely used (*e.g.* in Africa) in order to enhance pasture regrowth for livestock (using residual P), to control pests and diseases, and even to catch small animals for food.

A specific and difficult case is the burning of sugar cane before harvesting. It has both a technical dimension (CO_2 and greenhouse gas emissions, mechanisation of harvest, sugar content, etc.) and a social dimension (manual cutting, source of survival resources for poor/landless workers). The damage depends on fire intensity, which is a function of vegetation type and climate conditions and frequency. The costs and benefits of burning and the methods to minimise harmful effects need to be identified with local populations.

Crop Residue Management

In systems where crop residues are managed well, they :

- add soil organic matter, which improves the quality of the seedbed and increases the water infiltration and retention capacity of the soil, buffers the pH and facilitates the availability of nutrients;
- sequester (store) C in the soil;
- provide nutrients for soil biological activity and plant uptake;
- capture the rainfall on the surface and thus increase infiltration and the soil moisture content;
- provide a cover to protect the soil from being eroded;
- reduce evaporation and avoid desiccation from the soil surface.

Depending on the nature of the following crop, decisions are made as to whether the residues should be distributed evenly over the field or left intact, *e.g.* where climbing cover crops (*e.g.* mucuna) use the maise stalks as a trellis.

An even distribution of residues : (i) provides homogenous temperature and humidity conditions at sowing time; (ii) facilitates even sowing, germination and emergence; (iii) minimises the development of pests and diseases; and (iv) reduces the emergence of weeds through allelopathic effects.

The most appropriate method for managing crop residues depends on the purpose of the crop residues and the experience and equipment available to the farmer. Where the aim is to maintain a mulch over the soil for as long as possible, the biomass is best managed using a knife roller, chain or sledge in order to break it down but not kill it (Plate 14). Where the decomposition process should commence immediately in order to release nutrients, the residues should be slashed or mown and some N applied because dry residues have a high C : N ratio. However, in order to avoid nitrate emission, urea should not be broadcast on the surface but injected where possible.

Plate

Utilising Forage by Grasing Rather Than by Harvesting

In many places, there is competition for the use of crop residues that can be used as fodder, for roofing, artisan handicrafts, etc. Where residues are to be used for animal feed, either the animals graze the residues directly, or they are stall- or kraal-fed.

Removal of the residues from the field can lead to a considerable loss of organic matter where animal manure is not returned to the field. By controlled grasing, the animal manure is returned in the field without a high labour input.

The experience of Guaymango, El Salvador, demonstrates that it is possible to achieve successful integration of crop and livestock components without creating competition in the allocation of crop residues (Vieira and Van Wambeke, 2002). The amount of residues produced by the system is enough to serve both as soil cover and as fodder for livestock (Choto and Saín, 1993), mainly because of the use of local sorghum varieties (instead of HYVs) that have a high straw/grain ratio (Choto, Saín and Montenegro, 1995). As farmers value crop residues as soil cover, a fodder market has developed where grasing rights, number of cattle and duration of grasing are traded.

In the northern zone of the United Republic of Tanzania, farmers have found a compromise between using the residues for grasing or soil cover, albeit one that is rather labour intensive. They separate the palatable and non-palatable parts of the crop residues. They use the non-palatable parts to cover the soil and act as food for soil organisms, while they feed the palatable parts to cattle and goats that are kept close to the homestead (Plate 15).

Integrated Pest Management

Plate

As with balanced fertilization, proper pest and disease management results in healthy crops. Healthy crops produce optimal biomass, which is necessary for organic matter production in the soil. Diversified cropping and mixed crop-livestock systems enhance biological control of pests and diseases through species interactions. Through integrated production and pest management farmers learn how to maintain a healthy environment for their crops. They learn to examine their crops regularly in order to observe ratios of pests to natural enemies (beneficial predators) and cases of damage, and on that basis to make decisions as to whether it is necessary to use natural treatments (using local products such as neem or tobacco) or chemical treatments and the required applications.

Applying Animal Manure or Other Carbon-Rich Wastes

Any application of animal manure, slurry or other carbon-rich wastes, such as coffee-berry pulp, improves the organic matter content of the soil. In some cases, it is better to allow a period of decomposition before application to the field. Any addition of carbon-rich compounds immobilises available N in the soil temporarily, as micro-organisms need both C and N for their growth and development. Animal manure is usually rich in N, so N immobilisation is minimal. Where straw makes up part of the manure, a decomposition period avoids N immobilisation in the field.

Compost

Composting is a technology for recycling organic materials in order to achieve enhanced agricultural production. Biological and chemical processes accelerate the rate of decomposition and transform organic materials into a more stable humus form for application to the soil. Composting proceeds under controlled conditions in compost heaps and pits (Müller-Sämann, 1986).

Compost heaps should have a minimum size of 1 m³ and are suitable for more humid environments where there is potential for watering the compost. Compost pits (Plate 16) should be no deeper than 70 cm and should be underlain with rough material for good aeration of the compost. Pits are suitable for drier environments where the compost may desiccate (Müller-Sämann, 1986). Dry composting relies on covering the compost with soil and creating an anaerobic environment. However, this is a slower process than the more usual moist aerobic process. The ratio of C to N in the compost pile is important for optimising microbial activity. Thus, a mixture of soft, green and brown, tougher material is used. Ash and phosphate rock are often added to accelerate the process.

Composting can complement certain crop rotations and agro-forestry systems. It can be used efficiently in planting pits and nurseries. It is very similar in composition to soil organic matter. It breaks down slowly in the soil and is very good at improving the physical condition of the soil (whereas manure and sludge may break down fairly quickly, releasing a flush of nutrients for plant growth). In many circumstances, it takes time to rejuvenate a poor soil using these practices because the amount of organic material being added is small relative to the mineral proportion of the soil.

Successful composting depends upon the sufficient availability of organic materials, water, manure and "cheap" labour. Where these inputs are guaranteed, composting can be an important method of sustainable and productive agriculture. It has ameliorative effects on soil fertility and physical, chemical and biological soil properties. Well-made compost contains all the nutrients needed by plants. It can be used to maintain and improve soil fertility as well as to regenerate degraded soil. However, materials for compost production may be in short supply and the technology demands high labour inputs for proper compost production and application. Therefore, compost application may be restricted to certain crops and limited application areas, e.g. vegetable production in home gardens.

Mulch or Permanent Soil Cover

One way to improve the condition of the soil is to mulch the area requiring amelioration. Mulches are materials placed on the soil surface to protect it against raindrop impact and erosion, and to enhance its fertility. Crop residue mulching is a system of maintaining a protective cover of vegetative residues such as straw, maise stalks, palm fronds and stubble on the soil surface.

The system is particularly valuable where a satisfactory plant cover cannot be established rapidly when erosion risk is greatest.

Mulching adds organic matter to the soil, reduces weed growth, and virtually eliminates erosion during the period when the ground is covered with mulch.

There are two principal mulching systems :

- *in situ* mulching systems — plant residues remain where they fall on the ground (Plate 17);
- cut-and-carry mulching systems — plant residues are brought from elsewhere and used as mulch .

Crop residue mulching has numerous positive effects on crop production. However, it may require a change in existing cropping practices. For example, farmers may conventionally burn crop residues instead of returning them to the soil. *In situ* mulching depends on the design of appropriate cropping systems and crop rotations, which have to be integrated with the farming system. The greater labour demands of cut-and-carry systems represent a major constraint. Mulch may be more relevant in home gardens or for valuable horticulture crops (Box 7) than in less intensive farming systems.

Mulch affects the soil life. Holland and Coleman (1987) have demonstrated that litter placement on the soil surface (as opposed to incorporation with ploughing) increased the ratio of fungi to bacteria — the reason being that fungi have a higher carbon assimilation efficiency than bacteria. In addition, it encourages bio-turbating (mixing) effects of macro-fauna that pull the materials into surface layers of the soil.

DECREASED DECOMPOSITION RATES

Reduced or Zero Tillage

Repetitive tillage degrades the soil structure and its potential to hold moisture, reduces the amount of organic matter in the soil, breaks up aggregates, and reduces the population of soil fauna such as earthworms that contribute to nutrient cycling and soil structure.

Avoiding mechanical soil disturbance implies growing crops without mechanical seedbed preparation or soil disturbance since the harvest of the previous crop. The term zero tillage is used for this practice synonymously with terms such as no-till farming, no tillage, direct drilling, and direct seeding.

Compared with conventional tillage, reduced or zero tillage has two advantages with respect to soil organic matter. Conventional tillage stimulates the heterotrophic microbiological activity through soil aeration, resulting in increased mineralisation rate. Through breakdown of soil structure, it decreases upward and downward movements of soil fauna, such as earthworms, which are largely responsible for "humus" production through the ingestion of fresh residues. Reduced or zero tillage regulates heterotrophic microbiological activity because the pore atmosphere is richer in CO_2/O_2, and facilitates the activity of the "humifiers" (Pieri *et. al.*, 2002).

BOX
Mulching in the highlands of northern Thailand

Why certain crops receive mulch and others do not

Mulching provides a particular benefit to the cultivated crop. Mulching is practised for various cash crops for specific reasons. Onion and garlic are mulched mainly to control weeds (early hand weeding would be difficult without damaging the crop). The mulch is also important to keep the soil moist and cool as these crops are usually grown during the dry season under irrigation. Mulch is also applied under flowers and strawberries, mainly to protect the fragile and valuable products from becoming soiled.

Mulching saves labour. Mulching is often seen in maise fields, before as well as after crop establishment. Maise can compete reasonably well with weeds. Therefore, some farmers plant maise without tillage in a mulch of weeds previously killed with herbicides - a system that is less labour-demanding than a tillage operation. Because maise is planted with large spacings, it generally requires less rigourous weeding. Weeding is often done by slashing, and the weed residues are left on the ground.

Plate
Mulching in the highlands of Northern Thailand.

Tillage has become the most common method to control weeds. However, mulching is a more environmentally sound practice than tillage for weed control. The loose soil that results from tilling has less structure than before; the appearance is deceptive. Subsequent traffic or heavy rain soon packs this loosened soil, not only negating the expensive cultivation that produced the loose soil but also culminating in a degraded environment for water entry, seed germination and root growth. Further cultivation is then required to re-loosen the soil; more expense with the same outcome — subsequent repacking and degraded soil structure. This is a typical "downward spiral" of conventional agriculture. Moreover, tillage when the soil is too moist or too dry leads to compaction or pulverisation of soil; but farmers may not have the option to wait for optimal conditions.

Severe, accelerated soil erosion and the high costs in terms of labour and energy associated with plough-based methods of seedbed preparation have led to the widespread adoption of no- or zero-tillage systems for cropping in temperate and tropical climates. In no-tillage systems, the crop is sown into a soil left undisturbed since the harvest of the previous crop. Crop residue mulch is maintained and anchored firmly to the ground. Weed control relies on mechanical slashing or cover crops (FAO, 1993). Contact herbicides are also used in some cases.

In reduced — or zero-tillage systems, soil fauna resume their bio-turbating activities gradually. These loosen the soil and mix the soil components (also known as biotillage). The additional benefit of the increased soil organic matter and burrowing is the creation of a stable and porous soil structure without expensive, time-consuming and potentially degrading cultivations.

In zero-tillage systems, the action of soil macro-fauna gradually incorporate cover crop and weed residues from the soil surface down into the soil. The activity of micro-organisms is also regulated by the activity of the macro-fauna, which provide them with food and air through their burrows. In this way, nutrients are released slowly and can provide the following crop with nutrients.

Several authors have demonstrated that some crop rotations and zero tillage favour *Bradyrhizobia* populations, nodulation and thus N fixation and yield. Figure indicates a 200-300-per cent increase in population size of root nodule bacteria in a zero-tillage system compared with conventional tillage. The presence of soybean in the crop rotation resulted in a fivefold to tenfold increase in population size of the same bacteria compared with cropping systems without soybean.

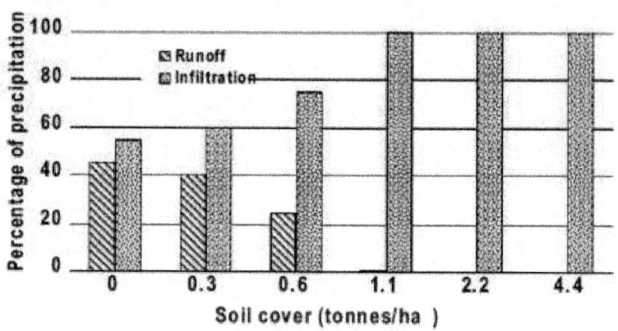

Fig. Population size of root nodule bacteria with different crop rotations

Note : S = soybean; W = wheat; M = maise.
Source : Voss and Sidirias, 1985.

Strictly speaking, the term zero tillage applies to methods involving no soil disturbance whatsoever, a condition that may be difficult to achieve. Broadcasting of seed is one way of applying zero tillage. The seed is broadcast over the previous crop residues and, where necessary, the residues are shaken to ensure that the seed falls on the soil surface.

In direct drilling, seeds such as maise, sorghum, soybean, wheat and barley are sown directly into shallow furrows cut into the previous crop residues (Plate 21). Weeds are controlled mechanically with a knife, which knocks down the plants and breaks their stems, or chemically with herbicides.

Traditional practices such as the burning of crop residues may inhibit the introduction of no-tillage systems. In many situations, a conflict exists between leaving crop residues on the surface or feeding them to livestock in the dry season when there is a shortage of fodder.

Mechanical soil disturbance also includes soil compaction through wheel impact of machinery, especially important in large-scale mechanised agriculture, *e.g.* plantations (sugar cane) or biannual crops (cotton). In a zero-tillage farming system, consideration must be given to reducing both the random placement of tyres/wheels in fields as well as the potential for compaction from animal hooves. Pietola, Horn and Yli-Halla (2003) reported the destructive effect of cattle trampling on the soil structure. Proffitt, Bendotti and McGarry (1995) demonstrated the almost total loss of soil porosity in the soil surface as a result of trampling by sheep. There is a belief that draught animals cause less land degradation than tractors. However, there are reports of soil compaction on small-holder farming enterprises in both Malawi (Douglas *et. al.*, 1999) and Bangladesh (Brammer, 2000). The hooves of draught animals and the shearing effect of ploughs or hand hoes, which are used repeatedly at a constant depth, can cause severe compacted layers. Grasing animals should be removed from zero-till fields in moist-wet soil conditions as the compaction risk is greatest at these times.

Plate 1
Maise seedling directly drilled in residues of wheat.

Controlled traffic, where the wheels of all in-field equipment follow permanent, defined tracks, ensures that compaction is restricted to specific known areas (Plate 22). Alternatively, flotation tyres (low ground-pressure tyres) should be fitted to all large tractors, harvesters, in-field grain bins, etc. in order to reduce their compacting potential.

Recent research has demonstrated the devastating effects of compaction from wheel impact on the occurrence and survival of eartworms (Pangnakorn *et. al.*, 2003). Earthworm incidence was greater under controlled traffic than under wheeled traffic. Figure shows the immediate effects of wheeling and tillage on the earthworm population. It appears that wheeling has the most detrimental effect on earthworm survival and that where wheeling is followed by tillage the survival rate is much greater. This may indicate that earthworms are able to survive an initial compaction in the field as long as it is relieved immediately. Where it is not, earthworms are inmobilised and unable to find air and nutrients.

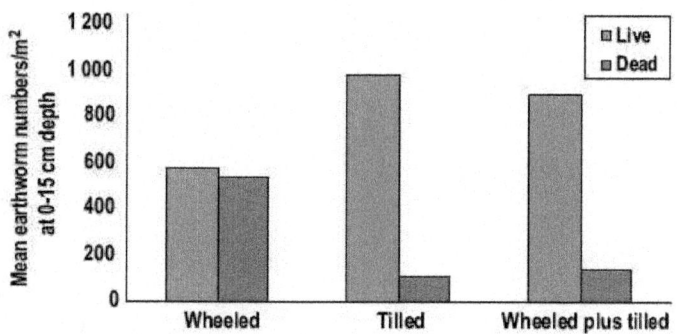

Fig. Live and dead earthworm numbers per square metre at 0-15 cm of soil depth sampled immediately after treatment.

Chapter 9

CREATING DROUGHT-RESISTANT SOIL

EFFECT OF SOIL ORGANIC MATTER ON SOIL PROPERTIES

Organic matter affects both the chemical and physical properties of the soil and its overall health. Properties influenced by organic matter include : soil structure; moisture holding capacity; diversity and activity of soil organisms, both those that are beneficial and harmful to crop production; and nutrient availability. It also influences the effects of chemical amendments, fertilizers, pesticides and herbicides. This chapter focuses on those properties related to soil moisture and water quality.

Inefficient Use of Rainwater

Drylands may have low crop yields not only because rainfall is irregular or insufficient, but also because significant proportions of rainfall, up to 40 per cent, may disappear as run-off. This poor utilisation of rainfall is partly the result of natural phenomena (relief, slope, rainfall intensity), but also of inadequate land management practices (*i.e.* burning of crop residues, excessive tillage, eliminating hedges, etc.) that reduce organic matter levels, destroy soil structure, eliminate beneficial soil fauna and do not favour water infiltration. However, water "lost" as run-off for one farmer is not lost for other water users downstream as it is used for recharging groundwater and river flows.

Where rainfall lands on the soil surface, a fraction infiltrates into the soil to replenish the soil water or flows through to recharge the groundwater. Another fraction may run off as overland flow and the remaining fraction evaporates back into the atmosphere directly from unprotected soil surfaces and from plant leaves.

The above-mentioned processes do not occur at the same moment, but some are instantaneous (run-off), taking place during a rainfall event, while others are continuous (evaporation and transpiration).

To minimise the impact of drought, soil needs to capture the rainwater that falls on it, store as much of that water as possible for future plant use, and allow for plant roots to penetrate and proliferate. Problems with or constraints on one or several of these conditions cause soil moisture to be one of the main limiting factors for crop growth.

The capacity of soil to retain and release water depends on a broad range of factors such as soil texture, soil depth, soil architecture (physical structure including pores), organic matter content and biological activity. However, appropriate soil management can improve this capacity.

Practices that increase soil moisture content can be categorised in three groups : (i) those that increase water infiltration; (ii) those that manage soil evaporation; and (iii) those that increase soil moisture storage capacities. All three are related to soil organic matter.

In order to create a drought-resistant soil, it is necessary to understand the most important factors influencing soil moisture.

Increased Soil Moisture

Organic matter influences the physical conditions of a soil in several ways. Plant residues that cover the soil surface protect the soil from sealing and crusting by raindrop impact, thereby enhancing rainwater infiltration and reducing run-off. Surface infiltration depends on a number of factors including aggregation and stability, pore continuity and stability, the existence of cracks, and the soil surface condition. Increased organic matter contributes indirectly to soil porosity (via increased soil faunal activity). Fresh organic matter stimulates the activity of macro-fauna such as earthworms, which create burrows lined with the glue-like secretion from their bodies and are intermittently filled with worm cast material.

The proportion of rainwater that infiltrates into the soil depends on the amount of soil cover provided. The figure shows that on bare soils (cover = 0 tonnes/ha) run-off and thus soil erosion is greater than when the soil is protected with mulch. Crop residues left on the soil surface lead to improved soil aggregation and porosity, and an increase in the number of macropores, and thus to greater infiltration rates.

Increased levels of organic matter and associated soil fauna lead to greater pore space with the immediate result that water infiltrates more readily and can be held in the soil (Roth, 1985). The improved pore space is a consequence of the bio-turbating activities of earthworms and other macro-organisms and channels left in the soil by decayed plant roots.

On a site in southern Brazil, rainwater infiltration increased from 20 mm/h under conventional tillage to 45 mm/h under no tillage (Calegari, Darolt and Ferro, 1998). Over a long period, improved organic matter promoted good soil structure and macroporosity. Water infiltrates easily, similar to forest soils.

The consequence of increased water infiltration combined with a higher organic matter content is increased soil storage of water. Organic matter contributes to the stability of soil aggregates and pores through the bonding or adhesion properties of organic materials, such as bacterial waste products, organic gels, fungal hyphae and worm secretions and casts. Moreover, organic matter intimately mixed with mineral soil materials has a considerable influence in increasing moisture holding capacity. Especially in the top-soil, where the organic matter content is greater, more water can be stored.

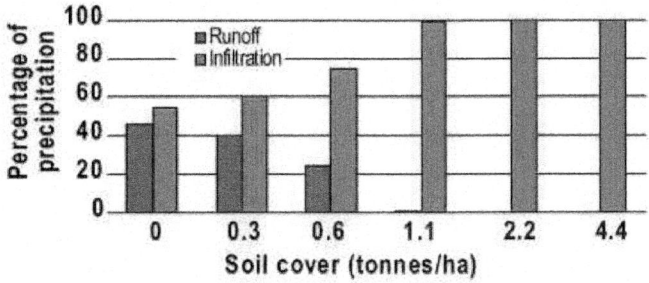

Fig. Effect of amount of soil cover on rainwater run-off and infiltration.

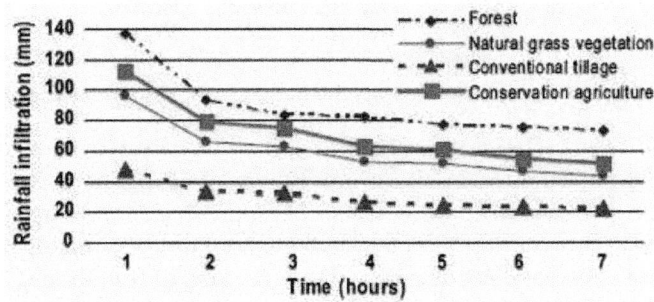

Fig. Water infiltration under different types of management.

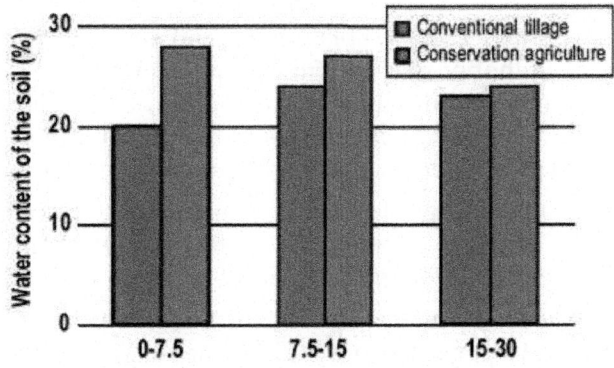

Fig. Quantity of water stored in the soil under conventional tillage and conservation agriculture.

The quality of the crop residues, in particular their chemical composition, determines the effect on soil structure and aggregation. Blair *et. al.* (2003) report a rapid breakdown of medic (*Medicago truncatula*) and rice (*Oryza sativa*) straw residues resulting in a rapid increase in soil aggregate stability through the release of many soil-binding components. As these compounds undergo further breakdown, they will be lost from the system resulting in a decline in soil aggregate stability over time. The slow release of soil-binding agents from flemingia (*Flemingia macrophylla*) residues resulted in a slower but more sustained increase in the stability of soil aggregates. This indicates that continual release of soil-binding compounds from plant residues is necessary for continual increases in soil aggregate stability to occur.

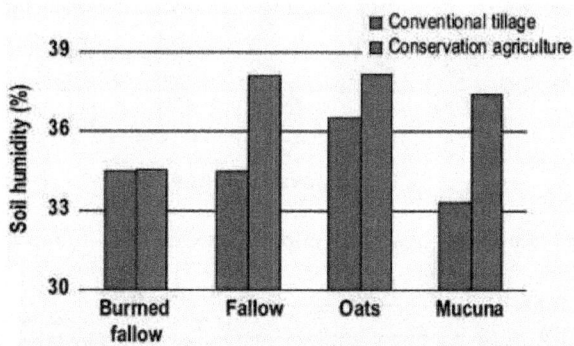

Fig. Effect of different soil covers on in-soil storage of water.

Elliot and Lynch (1984) showed that soil aggregation is caused primarily by polysaccharide production in situations where residues have a low N content. There is a strong relationship between soil carbon content and aggregate size. An increase in soil carbon content led to a 134-per cent increase in aggregates of more than 2 mm and a 38-per cent decrease in aggregates of less than 0.25 mm (Castro Filho, Muzilli and Podanoschi, 1998). The active fraction of soil C (Whitbread, Lefroy and Blair, 1998) is the primary factor controlling aggregate breakdown (Bell *et. al.*, 1999).

In addition, although they do not live long and new ones replace them annually, the hyphae of actinomycetes and fungi play an important role in connecting soil particles (Castro Filho, Muzilli and Podanoschi, 1998). Gupta and Germida (1988) showed a reduction in soil macro-aggregates correlated strongly with a decline in fungal hyphae after six years of continuous cultivation.

The in-soil storage of water depends not only on the type of land preparation but also on the type of cover or previous vegetation on the soil. Figure indicates the effect of burning vegetation on the amount of water stored in the soil.

Conserving fallow vegetation as a cover on the soil surface, and thus reducing evaporation, results in 4 per cent more water in the soil. This is roughly equivalent to 8 mm of additional rainfall. This amount of extra water can make the difference between wilting and survival of a crop during temporary dry periods.

A study conducted in 1999 in Guatemala, Honduras and Nicaragua to evaluate the resilience of agro-ecosystems showed that 3-15 per cent more water was stored in the soil under more ecologically sound practices (Table).

Unger (1978) showed that high wheat-residue levels resulted in increased storage of fallow precipitation, which subsequently produced higher sorghum grain yields. High residue levels of 8-12 tonnes/ha resulted in about 80-90 mm more stored soil water at planting and about 2.0 tonnes/ha more of sorghum grain yield compared to no residue management.

Table. Average soil depth at which moisture starts,
and difference in moisture stored.

Soil as a source of CO_2	Soil as a sink of CO_2
Soil properties : coarse textured soil, excessive drainage, high susceptibility to erosion	Soil properties : clayey soil, poorly drained ecosystems, depositional sites, including footslopes
Land use : seasonal crops, simple ecosystem, shallow roots and low root-shoot ratio	Land use : perrenial crops, diverse ecosystem, deep roots and high root-shoot ratio
Soil management : intensive tillage based on plough, negative nutrient balance, residue removal and/or burning, continuous cropping, loss of soil and water by run-off and erosion	Soil management : no tillage, positive nutrient balance, mulch farming cover crops in rotation, cycle, soil and water conservation

The addition of organic matter to the soil usually increases the water holding capacity of the soil. This is because the addition of organic matter increases the number of micropores and macropores in the soil either by "gluing" soil particles together or by creating favourable living conditions for soil organisms. Certain types of soil organic matter can hold up to 20 times their weight in water (Reicosky, 2005). Hudson (1994) showed that for each 1-per cent increase in soil organic matter, the available water holding capacity in the soil increased by 3.7 per cent. Soil water is held by adhesive and cohesive forces within the soil and an increase in pore space will lead to an increase in water holding capacity of the soil. As a consequence, less irrigation water is needed to irrigate the same crop (Table).

Table. Economy of irrigation water through soil cover, the Brazilian Cerrados.

Country	Agro ecologically sound practices (cm)	Conventional practices	Difference (per cent)
Honduras	9.98	10.28	2.9
Guatemala	2.44	2.99	15.0
Nicaragua	15.81	17.80	11.2

Reduced Soil Erosion and Improved Water Quality

The less the soil is covered with vegetation, mulches, crop residues, etc., the more the soil is exposed to the impact of raindrops. When a raindrop hits bare soil, the energy of the velocity detaches individual soil particles from soil clods. These particles can clog surface pores and form many thin, rather impermeable layers of sediment at the surface, referred to as surface crusts. They can range from a few millimetres to 1 cm or more; and they are usually made up of sandy or silty particles. These surface crusts hinder the passage of rainwater into the profile, with the consequence that run-off increases. This breaking down of soil aggregates by raindrops into smaller particles depends on the stability of the aggregates, which largely depends on the organic matter content.

Increased soil cover can result in reduced soil erosion rates close to the regeneration rate of the soil or even lower, as reported by Debarba and Amado (1997) for an oats and vetch/maise cropping system.

Soil erosion fills surface water reservoirs with sediment, reducing their water storage capacity. Sedimentation also reduces the buffering and filtering capacity of wetlands and the flood-control capacity of floodplains. Sediment in surface water increases wear and tear in hydroelectric installations and pumps, resulting in greater maintenance costs and more frequent replacement of turbines. Sediments can also reach the sea (Plate 23), harming fish, shellfish and coral. Eroded soil contains fertilizers, pesticides and herbicides; all sources of potentially harmful off-site impacts.

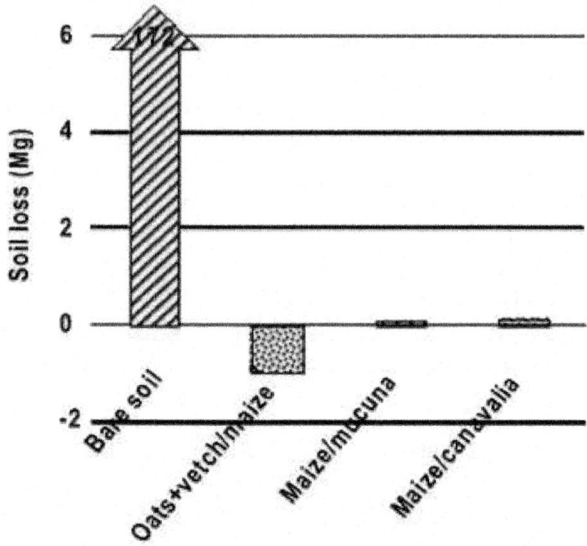

Fig. Soil loss due to water erosion for different maise cropping systems.

Note : Corrected with soil regeneration = 1.7 tonnes/ha/year.
Source : Debarba and Amado, 1997.

When the soil is protected with mulch, more water infiltrates into the soil rather than running off the surface. This causes streams to be fed more by subsurface flow rather than by surface run-off. The consequence is that the surface water is cleaner and resembles groundwater more closely compared with areas where erosion and run-off predominate. Greater infiltration should reduce flooding by increased water storage in soil and slow release to streams. Increased infiltration also improves groundwater recharge, thus increasing well supplies.

Bassi (2000) reported significant reductions in water turbidity and sediment concentration over a period of ten years (1988-1997) in different catchment areas in southern Brazil. The reductions varied between 50 and 80 per cent depending on locally predominant soil types. These reductions were caused by increases in the incidence of planting perennial crops (banana and pasture) on hillsides, thereby decreasing erosion potential. Total sediment loss decreased by 16 per cent and the cost of fertilizers declined by 21 per cent; an indication of the previous loss of fertilizers with the eroded soil. Guimarães, Buaski and Masquieto (2005) illustrate the same effect for one specific catchment. The catchment area of Rio do Campo, Paraná, provides 80 per cent of the water supply for Campo Mourão, a city with an urban population of 357 000. In the period 1982-1999, a drastic reduction in water turbidity was measured.

Sediment and dissolved organic matter in surface water have to be removed from drinking-water supplies. Reduced erosion, and hence fewer soil particles in suspension, lead to lower costs for water treatment. Data from Chapecó, Brazil, indicate that the quantity of aluminium sulphate used for flocculating suspended solids fell by 46 per cent in five years. Where water is chlorinated to kill disease organisms, the chlorine reacts with dissolved organic matter to form trihalomethane (THM) compounds such as chloroform. THMs are suspected of causing cancers (Fawcett, 1997). Reductions in run-off and erosion should lead to reduced formation of THMs during the chlorination process.

Erosion may also have long-lasting secondary consequences through effects on plant growth and litter input (Gregorich et. al., 1998). If erosion suppresses productivity, thereby limiting replenishment of organic matter, the amount of organic matter may spiral downwards in the long-term.

Soil cover protects the soil against the impact of raindrops, prevents the loss of water from the soil through evaporation, and also protects the soil from the heating effect of the sun. Soil temperature influences the absorption of water and nutrients by plants, seed germination and root development, as well as soil microbial activity and crusting and hardening of the soil.

Roots absorb more water at higher soil temperatures up to a maximum of 35 °C. Higher temperatures restrict water absorption. Soil temperatures that are too high are a major constraint on crop production in many parts of the tropics. Maximum temperatures exceeding 40 °C at 5 cm depth and 50 °C at 1 cm depth are commonly observed in tilled soil during the growing season, sometimes with extremes of up to 70 °C. Such high temperatures have an adverse effect not only on seedling establishment and crop growth but also on the growth and development of the micro-organism population. The ideal rootzone temperature for germination and seedling growth ranges from 25 to 35 °C. Experiments have shown that temperatures exceeding 35 °C reduce the development of maise seedlings drastically and that temperatures exceeding 40 °C can reduce germination of soybean seed to almost nil.

Mulching with crop residues or cover crops regulates soil temperature. The soil cover reflects a large part of solar energy back into the atmosphere, and thus reduces the temperature of the soil surface. This results in a lower maximum soil temperature in mulched compared with unmulched soil and in reduced fluctuations.

Plate
Run-off and soil loss immediately after a rainstorm,
Naisi catchment. Zomba Mountain, Malawi.

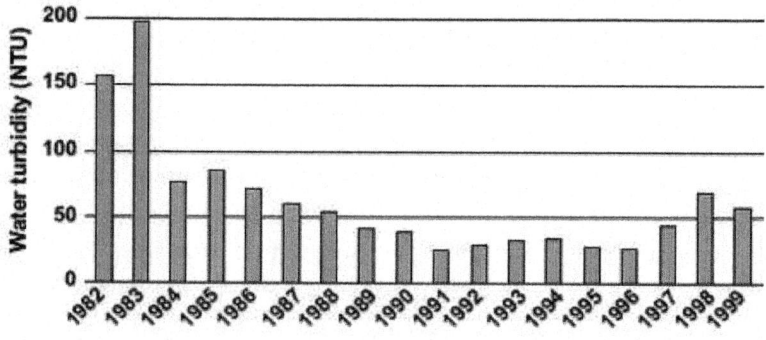

Fig. Development of water turbidity rates in the catchment
area of Rio do Campo, Paraná.

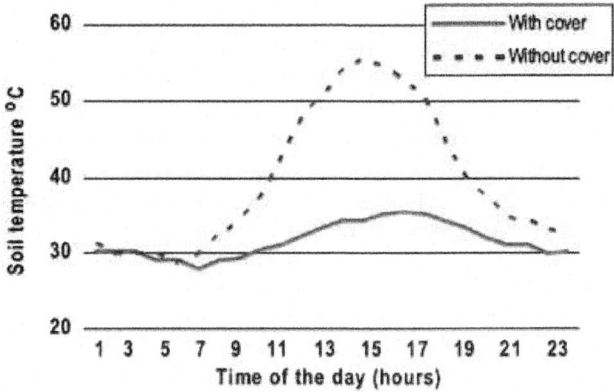

Fig. Temperature fluctuations at a soil depth of 3 cm in a cotton crop with and without a soil cover of mucuna.

KEY FACTORS IN SUSTAINED FOOD PRODUCTION

Increased Plant Productivity

Plant productivity is linked closely to organic matter (Bauer and Black, 1994). Consequently, landscapes with variable organic matter usually show variations in productivity. Plants growing in well-aerated soils are less stressed by drought or excess water. In soils with less compaction, plant roots can penetrate and flourish more readily. High organic matter increases productivity and, in turn, high productivity increases organic matter.

Increased Fertilizer Efficiency

The two major soil fertility constraints of the West African savannah and in the sub-humid and semi-arid regions of SSA are low inherent nutrient reserve and rapid acidification under continuous cultivation as a consequence of low buffering or cation exchange capacity (Jones and Wild, 1975). Generally, these constraints are tackled by applying chemical fertilizers and lime. However, the application of inorganic fertilizers on depleted soils often fails to provide the expected benefits. This is basically because of low organic matter and low biological activity in the soil.

The chemical and nutritional benefits of organic matter are related to the cycling of plant nutrients and the ability of the soil to supply nutrients for plant growth. Organic matter retains plant nutrients and prevents them leaching to deeper soil layers. Micro-organisms are responsible for the mineralisation and immobilisation of N, P and S through the decomposition of organic matter (Duxbury, Smith and Doran, 1989). Thus, they contribute to the gradual and continuous liberation of plant nutrients. Available nutrients that are not taken up by the plants are retained by soil organisms. In organic-matter depleted soils, these nutrients would be lost from the system through leaching and run-off.

Phosphate fixation and unavailability is a major soil fertility constraint in acid soils containing large amounts of free iron and aluminium oxides. In comparing the P-sorption capacity of surface and subsurface soil samples, Uehara and Gilman (1981) provided indirect evidence that soil organic matter can reduce the P-sorption capacity of such soils. This implies that for high P-fixing soils, i.e. oxide-rich soils derived from volcanic and ferro-magnesian rocks, management systems that are capable of accumulating and maintaining greater amounts of calcium-saturated soil organic matter in the surface horizon would increase P availability from both organic and fertilizer sources.

Weak acids, such as the organic acids in humus, do not relinquish their hydrogen (H) easily. H is part of the humus carboxyl (-COOH) under acidic conditions. When a soil is limed and the acidity decreases, there is a greater tendency for the H+ to be removed from humic acids and to react with hydroxyl (OH-) to form water. The carboxyl groups on the humus develop negative charge as the positively charged H is removed. When the pH of a soil is increased, the release of H from carboxyl groups helps to buffer the increase in pH and at the same time creates the CEC (negative charge). With an increase in organic matter, the soil recovers its natural buffer capacity; this means an increase in pH in acid soils.

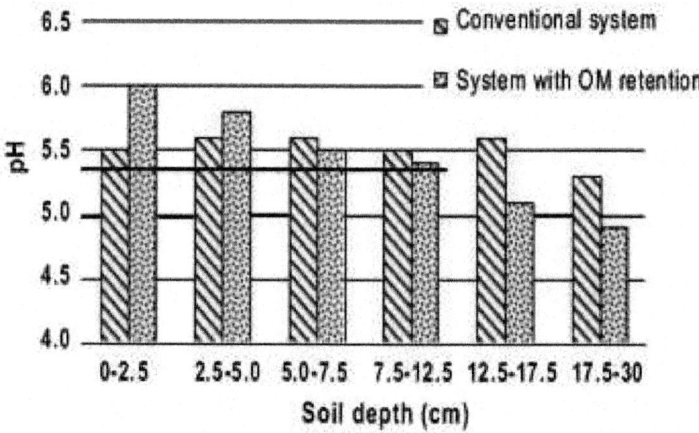

Fig. Effect of organic matter on soil pH.

Note : Original pH level was 5.3.

Table. Incidence of lime in the soil profile under different soil covers over the same period after surface application.

Cover	Soil depth (cm)
Bare soil	0-7
Black oats	0-20
Oil radish	0-22

CEC is linked closely to the organic matter content of the soil. It increases gradually with time where organic residues are retained, first in the top-soil and later also at greater depth. Crovetto (1997) reported an increase in CEC of 136 per cent (from 11 to 26 meq/100 g of soil) as a consequence of humus increase in the top-soil after 20 years of residue retention.

To overcome acidity, lime is usually incorporated in the soil. However, organic matter on the soil surface favours the transport of calcium carbonate (lime) to deeper soil layers after surface application (Table).

The crop residues release organic acids that cause the lime to penetrate deeper into the profile much more rapidly than when applied on bare soil. Thus, it is no longer necessary to mix lime intensively into the soil, which is appropriate for farming systems based on reduced or zero tillage.

Reduced Water-logging

Examined the water storage capacity of soils under improved organic management. However, in case of water-logging, organic matter plays also an important role. The bio-turbating activity of the macro-fauna leaves various so-called conducting macropores in the soil, which are responsible for the drainage of water to deeper soil layers.

Chan *et. al.* (2003) found a significant reduction in water-logging after three years under no tillage compared with conventional tillage. The reduction was related to higher density of conducting macropores ($140/m^2$ and $5/m^2$ for no tillage and conventional tillage, respectively), which was associated with higher population density of earthworms ($240/m^2$ and $36/m^2$ for no tillage and conventional tillage, respectively). Plate 24 provide a demonstrationof this effect.

Increased Yields

Agronomic practices that influence nutrient cycling, especially mineralisation and immobilisation, result in an immediate productivity gain or loss, which is reflected in the economics of the agricultural system.

Plate
Under no-tillage conditions, the internal pore system of the soil is not distroyed through land preparation activities and able to drain rainwater from the surface to deeper layers (left). On bare soil the impact of raindrops results in sealing of the surface pores and thus poor drainage.

Crop yields in systems with high soil organic matter content are less variable than those in soil that are low in organic matter. This is because of the stabilising effects of favourable conditions of soil properties and micro-climate. Improvements in crop growth and vigour stem from direct and indirect effects. Direct effects stem from improvements in nutrient and water content, as described above. Indirect effects stem from a favourable rooting environment and possible weed suppression and a reduction in pests and diseases.

BOX
Does improved organic matter management pay?

There are several ways to calculate the rentability of a farm. In general, key parameters such as expenses, income, yield, cost of production and gross margin are used to analyse how well farmers have managed to reach an income. However, in areas where water is a limiting factor, it may be more useful to analyse the conversion of water into yield or even money. The following example illustrates different ways to compare outcomes of farms and their cropping systems. It is based on the cropping results of 16 farms with a cropped surface of 60 000 ha in Australia. One-third of this surface is cropped in a conventional way (CT), using intensive tillage practices, fire and several passages with herbicides for weed control. Two-thirds of the area is cropped with the aim to retain as much ground cover as possible, and thus improve the organic matter content of the soil, using specialised planters and herbicides to control weeds, also known as conservation farming (MT). An analysis of the cropping system data for three years yielded the following gross margins for three crops in rotation (wheat, chickpea and sorghum) :

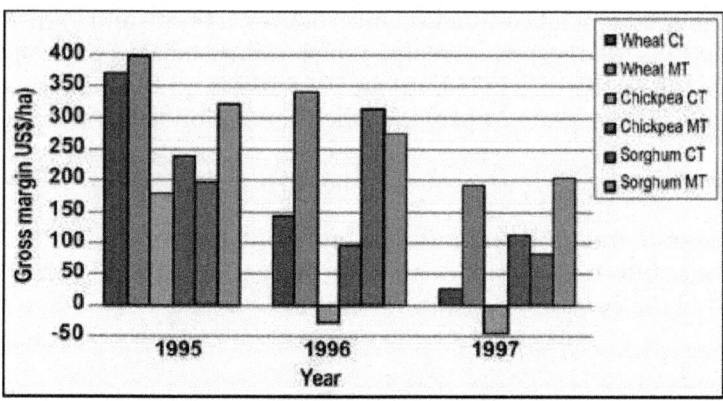

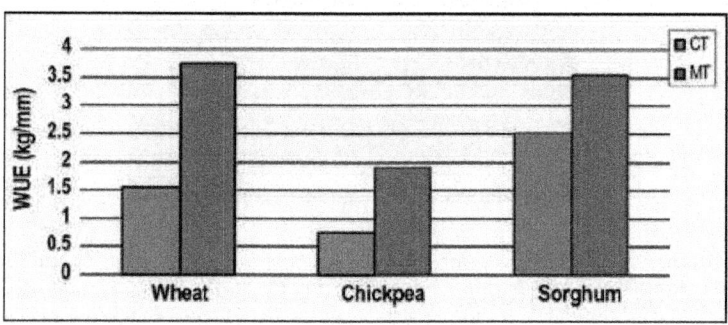

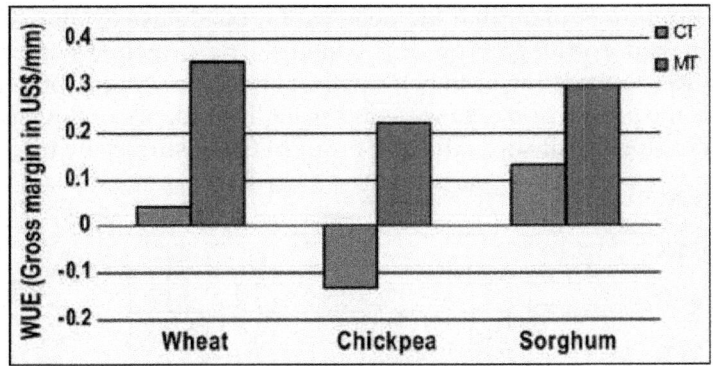

However, when using water-use efficiency (WUE) as means of comparing outputs across farming systems, the results are even more drastic. In this case, rather than the common method of determining WUE by breaking the season into fallow and in-crop components, a total water-use efficiency factor was used. Both grain yield and gross margin are divided by total water to obtain an insight into how well farmers managed the conversion of water into yield and money in the year 1997.

Immobilisation of N may occur in systems with crop residue management, especially where C : N ratios of the residues are high (tough, woody materials). This can cause a decrease in maise yield. Figure shows the effect of tillage and the preservation of crop residues on maise yield. The preservation of wheat and horse-radish residues on the soil surface led to immobilisation of N, which was overcome through the application of N fertilizer. Based on these data, maise with oats, lupine and vetch as a winter cover crop (without fertilization) can produce a yield that is comparable with or higher than those obtained with conventional tillage and a fertilizer treatment of 90 kg/ha. In this case, the yield increase was highly correlated with the P content of the leaves and the P availability in the soil. This occurred because of the higher moisture content in the soil under the mulch layer, which led to a higher P uptake by plant roots.

Grain crops can also have residual effects on each other through the decomposition of chemical compounds in the residues.

Reduced Herbicide and Pesticide Use

Some people are concerned that intensified systems with reduced or zero tillage will increase herbicide use and in turn lead to increased contamination of water by herbicides. According to Fawcett (1997), total herbicide use in the United States of America declined during the period of adoption of no-tillage systems. He concludes that herbicides are important, but that farmers using conventional tillage methods use similar amounts of herbicides to no-tillage farmers. In Honduras, a strong decline in the use of herbicides has been observed. Farmers who no longer burn their fields prior to preparation spend less money on herbicides. Farmers who have adopted the Quezungual system spend less on herbicides and make savings, both in terms of land preparation and total costs.

It is becoming evident that the need for herbicide use diminishes over time in well managed no-tillage cropping systems. The principal reason is that the system reduces the existing seed bank in the soil by the synergy of two activities : reduction of the production of new seeds through avoidance of flowering and fruit setting; and reduction of seeds that are brought to the surface by tillage practices.

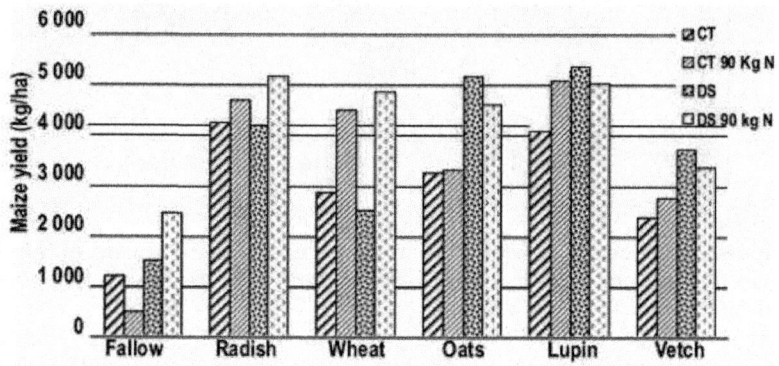

Fig. Maise yield under conventional tillage and direct sowing, with and without 90 kg of N fertilizer.

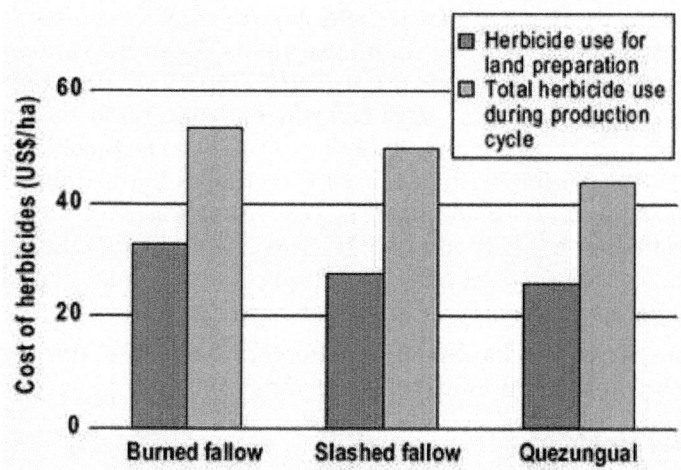

Fig. Herbicide costs in different production systems in Lempira Sur, Honduras.

With direct seeding, the reservoir of seeds differs from conventional tillage because :

- the weed seeds remain on the soil surface, where they are susceptible to attacks from insects, birds and soil organisms and to atmospheric influences;

- the soil remains covered by residues, which prevents light reaching the seeds and thus reduces germination;

- weed seeds already at certain depths are not brought to the surface again, where they could germinate;
- perennial weeds are no longer redistributed by equipment.

The result is that the soil weed seed store diminishes in time and, consequently, the weed problem also diminishes, as does the need to use herbicides. The concentration of soil organic matter in the top-soil layer plays an important role in the absorption of herbicides. When the concentration of organic matter in the top-soil decreases, contamination of the environment by herbicides is likely to increase. Enhanced levels of organic matter cause enhanced adsorption of pesticides, followed by gradual degradation. Herbicides, like other pesticides, can be used by micro-organisms as a substrate to feed on (Haney, Senseman and Hons, 2002). Herbicides are broken down in soil and water by micro-organisms into natural acids, NH_4^-, amino acids, carbohydrates, phosphate and CO_2 (Schuette, 1998). As microbes cause more rapid degradation of pesticides, enhancing microbial activity may reduce leaching of pesticides.

Many herbicides, including glyphosate and paraquat, which are the most common herbicides used in reduced-tillage systems, are bound tightly to clay and organic matter by electrostatic forces and hydrogen bonding. Once they are bound to soil organic matter, the herbicides become inactive and no longer affect plants. Moreover, they can form insoluble complexes with metals in the soils. This also contributes to their rare stability in the soil and low potential for leaching into groundwaters (Ahrens, 1994a, 1994b). Some studies have shown that there is no reason to believe that glyphosate may cause any unexpected damage to the environment (Torstenson, 1985). However, other studies illustrate negative effects on soil life or its functions.

Farming systems that increase soil organic matter content (*e.g.* no tillage) reduce the probability of environmental contamination by herbicides.

Increased Biodiversity

Conventional agriculture tends to reduce aboveground and belowground diversity. Thus, it brings about significant changes in the vegetation structure, cover and landscape. The change in vegetal cover during the conversion of forest and pastures to cropping affects plants, animals and micro-organisms. Through increasing specialisation of certain plant species (food and fibre crops, pasture and fodder crops, and tree crops) and livestock species, some functions may be affected severely, *e.g.* nutrient cycling and biological control. Some non-harvested or associated species profit from the change and become pests. However, many organisms either disappear completely or their numbers are reduced drastically, *e.g.* pollinators and beneficial predators, unless efforts are made to retain a suitable habitat (Box 9).

Associated species can be managed to a certain extent. Through appropriate crop rotations, crop-livestock interactions and the conservation of soil cover, a habitat can be created for a number of species that feed on pests. This will in turn attract more insects, birds and other animals. Thus, rotations and associations of crops and cover crops as well as hedgerows and field borders promote biodiversity and ecological functions. Because of the complexity and richness of soil biodiversity, the effects of crop and pasture management are less well understood. However, the effects on certain functional groups and, hence, specific soil functions are being recognised increasingly as vital for agricultural productivity and system sustainability.

BOX
Effect of different tillage practices on scarab beetle-grub holes and their volumes

Before the transformation of native grasslands and forests into agricultural areas, a large number of species of scarab beetles and their larvae (white grubs) inhabited the soils in southern Brazil. With the transformation of these areas, some of these beetles disappeared, while others became so well adapted and, lacking biological control agents, became important soil-dwelling pests. However, there are other species that can be considered as facultative pests. The larvae prefer feeding on surface litter or surface deposited animal excrements, but may become pests when not enough surface litter is present for them to feed on, like the genera *Diloboderus* and *Bothynus*. The large beetles create large, permanent galleries (holes) in the soil, down to a depth of more than 1 m, in which they spend most of their lives. The holes may serve as preferential pathways for water infiltration and root growth, and the chambers become niches of increased soil fertility. Both chambers and galleries provide temporary and permanent refuge for many other soil-dwelling invertebrates and micro-fauna.

Research revealed that beetle grub holes were more abundant under no tillage (NT) (8.8-9.6 holes/ m^2) than conventional tillage (CT) (0.7-1.3 holes/ m^2). The largest and deepest holes were also found in NT (up to 33.5 mm in diameter and 117 cm deep). Consequently, the total volume of pores opened in NT (450-503 cm^3) was up to almost 10 times greater than in CT (53-107 cm^3).

Soil has the ability to restore its life-support processes provided that the disturbance is not too drastic and that sufficient time is allowed for such recovery. Organic matter and biodiversity of soil organisms are the driving factors in this restoration. Decreases in numbers and types of soil organisms and available substrate (organic matter) lead to a decrease in resilience, which in turn can result in a downward spiral of degradation.

Resilience

Resilience can be defined as the ability of a system to recover after disturbance (Elliot and Lynch, 1994). Soil resilience depends on a balance between restorative and degrading processes. Factors affecting resilience can be grouped in two categories : endogenous and exogenous. Endogenous factors are related to inherent soil properties (rooting depth, texture, structure, topography and drainage) and micro-climate and meso-climate. Exogenous factors include land use and farming system, technological innovations and input management (Lal, 1994). Hence, appropriate agricultural practices can influence these factors in order to enhance soil resilience.

Organic matter and soil organisms play important roles in conserving and improving soil properties that are related to soil resilience. In addition to creating more pores through biological activity, organic matter plays an important role in the formation and stabilisation of soil aggregates through bonds between the organic matter and the mineral soil particles. Soil aggregation can take place through two binding agents :

- waste products of bacteria - polysaccharides;
- fungal and bacterial hyphae.

The preservation of aggregate stability is important in order to reduce surface sealing and increase water infiltration rates (Whitbread, Lefroy and Blair, 1998). With increased stability, surface run-off is reduced (Roth, 1985).

CONSERVATION AGRICULTURE IN ORGANIC MATTER

Principles of Conservation Agriculture

Conservation agriculture makes use of soil biological activity and cropping systems to reduce the excessive disturbance of the soil and to maintain the crop residues on the soil surface in order to minimise damage to the environment and provide organic matter and nutrients. It is based on four principles :

- minimal mechanical soil disturbance, mainly through direct seeding;
- permanent soil cover, organic matter supply through the preservation of crop residues and cover crops;
- crop rotation for biocontrol and efficient use of the soil profile;
- minimal soil compaction.

Although the principles are not new (except for that of minimal disturbance to the soil), it is the fact that they are applied together in conservation agriculture that generates positive outcomes. All the practices (minimal tillage, soil cover and crop rotation) are combined for synergy and added value. In the past, farmers may have tried but abandoned the use of cover crops or zero tillage because of weed problems or yield declines. There is also a need for improved weed control and rotations for biocontrol of pests and diseases and nutrient uptake. Integration of the conservation agriculture principles provides a win-win situation for both people and the environment, which has catalysed successful expansion of the area under conservation agriculture worldwide.

Conservation agriculture aims to :

- provide and maintain optimal conditions in the rootzone (maximum possible depth for crop roots) in order to enable them to grow and function effectively and without hindrance in capturing plant nutrients and water;

- ensure that water enters the soil so that : (i) plants have sufficient water to express their potential growth; and (ii) excess water passes through soil to groundwater and streamflow, not over the surface as run-off where it can cause erosion. There is greater potential for increased cropping efficiency as more water is held in the soil profile than under conventional systems;

- increase beneficial biological activity in the soil in order to : (i) maintain and rebuild soil architecture for enhanced water entry and distribution within the soil profile; (ii) compete with potential soil pathogens; (iii) contribute to decomposition of organic materials to soil organic matter and various grades of humus; and (iv) contribute to the capture, retention and gradual release of plant nutrients;

- avoid physical or chemical damage to roots and soil organisms that would disrupt their effective functioning.

Organic Matter Deposition

The reduction of soil disturbance through zero-tillage, the use of cover crops and the preservation of crop residues on the soil surface result in increased activity of the soil and in the accumulation of organic matter, mainly in the top-soil.

An argument often heard in the discussion on conservation agriculture is that it is only feasible in the humid and sub-humid tropics and that the generation of sufficient biomass in semi-arid regions is the limiting factor to start implementing conservation agriculture.

However, recent research has shown that even in semi-arid areas of Morocco the application of the principles of conservation agriculture bears its fruits. Mrabet (2000) reports higher yields through better water use and improved soil quality; the latter caused by an increase in soil organic C and N and a slight pH decline in the seedzone (Bessam and Mrabet, 2003; Mrabet *et. al.*, 2001a, 2001b).

Increased Carbon Sequestration

World soils are important reservoirs of active C and play a major role in the global carbon cycle. As such, soil can be either a source or sink for atmospheric CO_2 depending on land use and the management of soil and vegetation (Lal, 2005). The conversion of native ecosystems (*e.g.* forests, grasslands and wetlands) to agricultural uses, and the continuous harvesting of plant materials, has led to significant losses of plant biomass and C (Davidson and Ackerman, 1993), thereby increasing the CO_2 level in the atmosphere.

In particular, the practice of burning agricultural fields before cultivation has a disastrous effect on soil organic carbon content. Figure shows the reduction in soil organic carbon in agricultural fields after 100 years of burning crop residues and weeds compared with an area that was not burned or ploughed during the same period. The top-soil layer (0-5 cm) represented the greatest carbon loss

(36 per cent) compared with the area that was not burned. Soil N stock in the same layer was reduced by 16 per cent. The carbon stock was reduced not only through burning, but because of the whole land-use management, especially a drastic reduction in diversity of species as mono-cropping was practised.

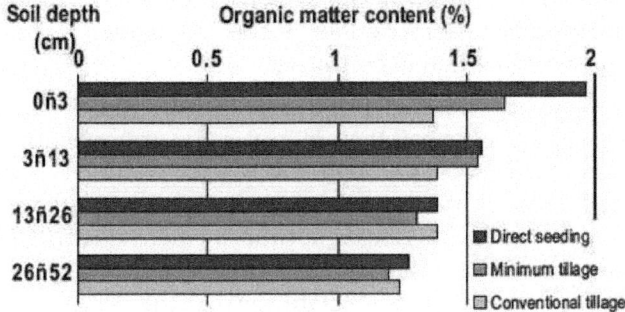

Fig. Organic matter content of a soil under different tillage management.

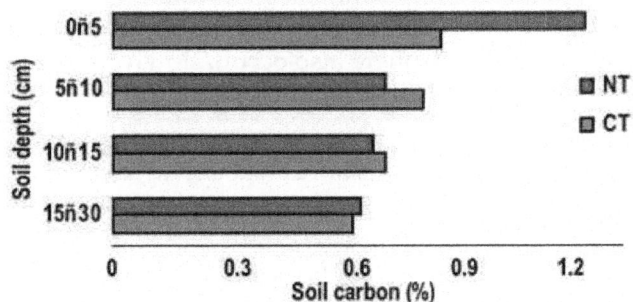

Fig. Soil carbon concentration at various soil depths affected by management system.

Note : Conventional (CT) and conservation (NT) agrculture, after two complete cropping cycles (4 years).

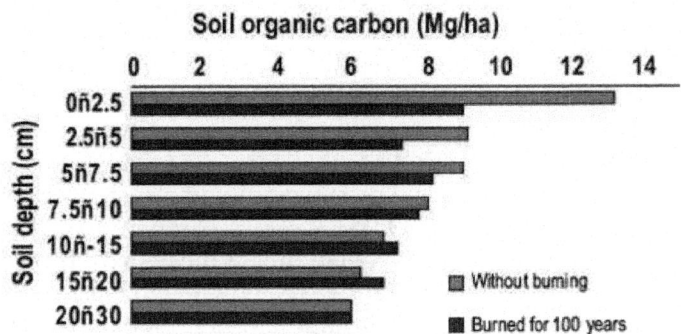

Fig. Effect of crop residue burning once every two years on soil carbon stock.

Table lists general practices that determine whether soil will be a sink or a source of atmospheric CO_2.

Soil can play a part in mitigating CO_2 levels (Paustian, 2002). This removal process is achieved naturally, and quite effectively, through photo-synthesis. Living plants take CO_2 from the air in the presence of sunlight and water, convert it into seeds, leaves, stems and roots. Part of the CO_2 is retained or "sequestered", or stored as C in the soil when decomposed.

In particular, systems based on high crop-residue addition and no tillage tend to accumulate more C in the soil than is lost to the atmosphere. Carbon sequestration in managed soils occurs when there is a net removal of atmospheric CO_2 because C inputs (crop residues, litter, etc.) exceed C outputs (harvested materials, soil respiration, C emissions from fuel and the manufacture of fertilizers, etc.) (Izaurralde and Cerri, 2002). Management practices that increase soil C comply with a number of principles of sustainable agriculture : reduced tillage, erosion control, diversified cropping system, balanced fertilization, etc.

In the early years of no-tillage systems, the organic matter content of the soil is increased through the decomposition of roots and the contribution of vegetative residues on the surface. This organic material decomposes slowly, and thus the liberation of C to the atmosphere also occurs slowly. In the total balance, net fixation or sequestration of C takes place; the soil is a net sink of C.

Table. Land use and land management determining whether soil will be a sink or source of atmospheric CO_2.

Soil as a source of CO_2	Soil as a sink of CO_2
Soil properties : coarse textured soil, excessive drainage, high susceptibility to erosion.	Soil properties : clayey soil, poorly drained ecosystems, depositional sites, including footslopes.
Land use : seasonal crops, simple ecosystem, shallow roots and low root-shoot ratio.	Land use : perrenial crops, diverse ecosystem, deep roots and high root-shoot ratio.
Soil management : intensive tillage based on plough, negative nutrient balance, residue removal and/or burning, continuous cropping, loss of soil and water by run-off and erosion.	Soil management : no tillage, positive nutrient balance, mulch farming cover crops in rotation, cycle, soil and water conservation.

Figure illustrates the fact that some cropping systems can act as a sink for CO_2. In this example, the carbon stock in soils under natural vegetation is used as a reference (steady state : DC = 0). In eight years, the fallow/maise system liberated 4.3 tonnes of CO_2 per hectare. The maise/mucuna system showed a positive balance of almost 20 tonnes of CO_2 per hectare compared with fallow/maise. Compared with soils under natural vegetation, this means a capture of atmospheric CO_2 of more than 15 tonnes/ha in eight years. Lovato (2001) found an increase of 2 tonnes/ha/year over 13 years in a rotation system of oats — common vetch/ maiz — cowpea. These figures confirm the potential of conservation agriculture for carbon sequestration. However, the "simple" change from soil tillage to zero tillage is not enough. According to Lovato *et. al.* (2004), a minimum addition of 4.2 tonnes/ha/year of carbon in vegetative residues in cropping systems and 4.5 tonnes/ha/year in mixed systems of pastures and crops (Nicoloso, Lovato and Lanzanova, 2005) is necessary for maintaining soil organic matter at stable levels. This means that below these values CO_2 emission will or can take place and, thus, that for conservation agriculture systems to become successful in promoting carbon sequestration, it is necessary to include crops and pastures in the rotation that add large quantities of biomass.

Even more C can be stored by adding leguminous cover crops to the rotation cycle. This is shown in Figure, based on two long-term experiments in Rio Grande do Sul, Brazil. Besides addition of C to the soil, legumes add a substantial quantity of N to the soil, which results in increased biomass production of succeeding crops.

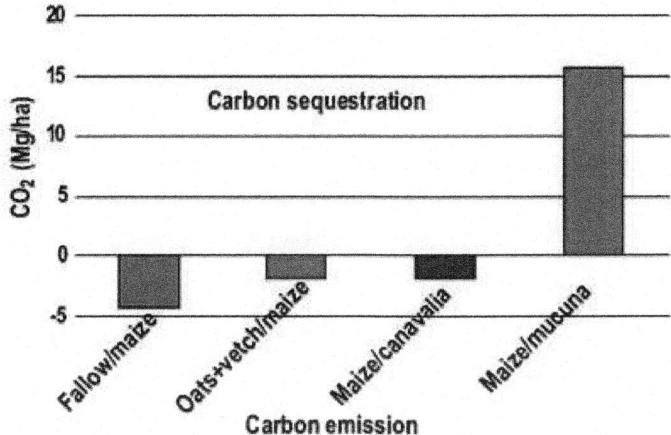

Fig. Estimation of emission and sequestration of CO_2 (total over 8 years) under different maise production systems with cover crops under direct sowing compared with natural vegetation in southern Brazil.

Assuming an average carbon accumulation of 0.5 tonne/ha/year, an area like southern Brazil (Rio Grande do Sul, Santa Catarina and Paraná) under conservation agriculture would have the potential to sequester 5 million tonnes of C annually, which corresponds to 18 million tonnes of atmospheric CO_2. To compare, Brazil as emitted an estimated 84 million tonnes of CO_2 in 2000 (Carbon Dioxide Information Analysis Center, 2003).

Recent studies have shown that soil temperature is one of the main climate factors that influence CO_2 emission. High soil temperatures accelerate soil respiration and thus increase CO_2 emission (Brito *et. al.*, 2005). This has implications for the landscape and land use in a certain area. On convex slopes and hilltops, emission is greater than in foothills, where temperatures are normally lower. The foregoing indicates that not only for soil and water conservation is it important to protect the soil with vegetation (reduction in soil temperature), but that it is advisable to cover the soil also with a view to reducing greenhouse gas emissions.

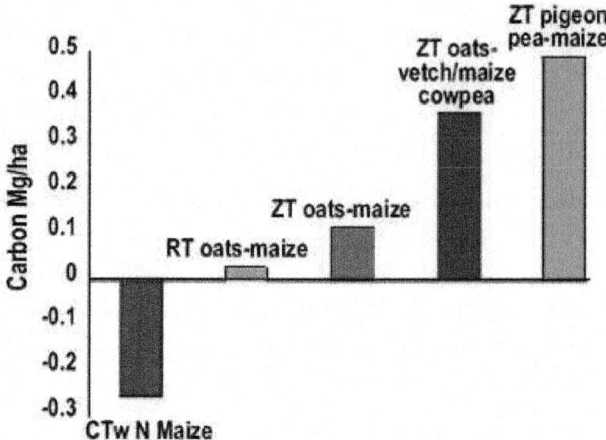

Fig. Potential for carbon sequestration in long-term experiment in southern Brazil.

Notes : CT = conventional tillage, RT = reduced tillage, ZT = zero tillage,
CTwN = conventional tillage without N fertilizer.

Chapter 10

ORGANIC FORM OF SOIL

Nitrogen

Over 90 per cent of the nitrogen N in the surface layer of most soils occurs in organic forms, with most of the remainder being present as NH_4^- whichis held within the lattice structures of clayminerals. The surface layer of most cultivated soils contains between 0.06 and 0.3 per cent N. Peat soils have high N contents to 3.5 per cent .

Plant remains and other debris contribute nitrogen N in the form of :

Amino acids

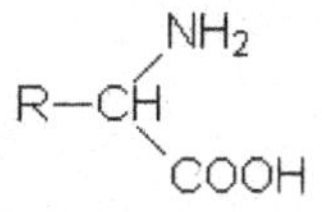

$$R-\underset{\diagdown COOH}{\overset{\diagup NH_2}{CH}}$$

The general formula of
amino acids.
(Stevenson 1982)

Amino acids exist in soil in several different forms, like :

1. As free amino acids
   o in the soil solution
   o in soil micropores
2. As amino acids, peptides or proteins bound to clay minerals :
   o on external surfaces
   o on internal surfaces

3. As amino acids, peptides or proteins bound to humic colloids :

    o   H-bonding and van der Waals' forces

    o   in covalent linkage as quinoid-amino acid complexes

4. As mucoproteins

5. As a muramic acid.

Amino acids, being readily decomposed by micro-organisms, have only an-ephemeral existence in soil. Thus the amounts present in the soil solution at any one time represent a balance between synthesis and destruction by micro-organisms.

The free amino acids content of the soil is strongly influenced by weathercon-ditions, moisture status of the soil, type of plant and stage of growth, additions of organic residues, and cultural conditions.

Amino Sugars

Amino sugars occur as structal components of a broad group of substances, the mucopolysaccharides and they have been found in combination with muco-peptides and mucoproteins. Some of the amino sugar material in soil may exist in the form of an alkali-insoluble polysaccharide referred to as chitin.

Generally the amino sugars in soil are of microbial origin. From 5 to 10 per cent of the N in the surface layer of most soils can be accounted for in N-containing carbohydrates or amino sugars.

D - Glucosamine
(Stevenson 1982)

Nucleic Acids

Nucleic acids, which occur in the cells of all living organisms, consist of individual mononucleotide units (base-sugar-phosphate) joined by a phosphoricacid ester linkage through the sugar.Two types : ribonucleic acid (RNA) and deoxyribonucleic acid (DNA).They have pentose sugar (ribose or deoxyribose),the purine : adenine, guanine and the pyrimidine : cytosine, thymine. RNA contains also the uracil.

The N in purine and pyrimidine bases is usually considered to account forless than 1 per cent of the total soil N.

Small amounts of N are extrcted from soil in the form of **glycerophosphatides, amines, vitamins, pesticide** and **pesticide degradation products.**

Nitrogen Transformation

A key feature of the internal cycle is the biological turnover of N between-mineral and organic forms through the opposing processes of mineralisation and immobilisation. The latter leads to incorporation of N into microbial tissues. Whereas much of this newly immobilised N is recycled through mineralisation, some is converted to stable humus forms.

The overall reaction leading to incorporation of inorganic forms of N into-stable humus forms is depicted on the picture.

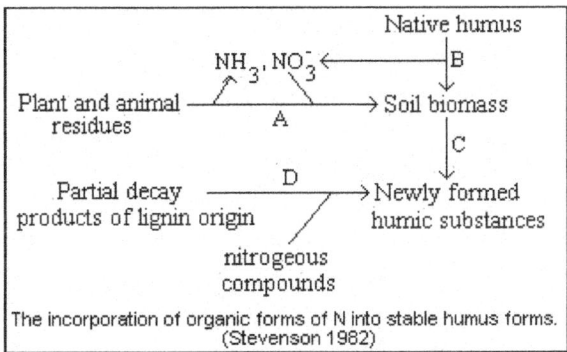

The incorporation of organic forms of N into stable humus forms.
(Stevenson 1982)

Thus the decay of plant and animal residues by micro-organisms results in the formation of mineral forms of N (NH_4^+ and NO_3^-) and assimilation of part of the C into microbial tissue (reaction A). Part of the native humus undergoes a similar fate (reaction B). Subsequent turnover through mineralisation-immobilisation leads to incorporation of N into stable humus forms (reaction C). Stabilisation of N may also occur through the reaction of partial decay products of lignin with nitrogenous constituents (raection D). Except under unusual circumstances, both mineralisation and immobilisational-ways function in soil, but in opposite direction.

Chemical Reaction of Ammonia and Nitrite with Organic Matter

The fate of mineral forms of N in soil is determined to some extent by non-biological reactions involving NH_4^+, NH_3 and NO_2^- as depicted in fig.

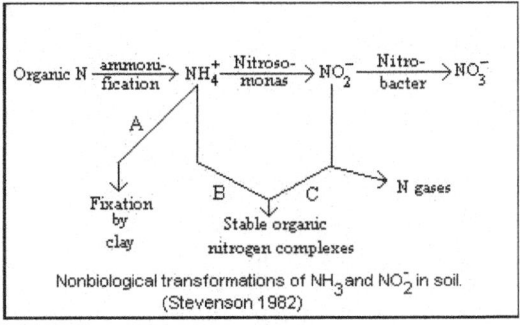

Nonbiological transformations of NH_3 and NO_2^- in soil.
(Stevenson 1982)

In addition to NH_4^+ fixation by clay minerals (reaction A), NH_3 and NO_2^- react chemically with organic matter to form stable organic N complexes (reaction B and C). The chemical interaction of NO_2^- with organic matter may lead to the generation of N gases. Although both types of reactions can proceed over a wide pH range, fixation of NH_3^+ is favoured by a high pH (>7.0). In contrast, NO_2^- -organic matter interactions occur most readily under highly acidic conditions (pH of 5.0 to 5.5 or below).

Stability of Soil Organic Nitrogen

1. Proteinaceous constituents are stabilised through their reaction with other organic constituents, such as lignins, tanins, quinons.
2. Biologically resistant complexes are formed in soil by chemical reactions involving NH_3^+ or NO_2^- with lignins or humic substances.
3. Adsorption of organic N compounds by clay minerals (pariculary montmorillinitic types) protects the molecule from decomposition.
4. Complexes formed between organic N compounds and polyvalent cations, such as Fe, are biologically stable.
5. Some of the organic N occurs in small pores or voids and is physicallyinaccessible to micro-organisms.

C/N ratio

For surface soils, and for the top layer of lake and marine sediments, the ratio generally falls within well-defined limits, usually from about 10 to 12. In most soils, the C/N ratio decreases with increasing depth, often attaining values less than 5.0. Native humus would be expected to have a lower C/N ratio than most undecayedplant residues for following reasons. The decay of organic residues by soil-organisms leads to incorporation of part of the C into microbial tissue with the remainder being liberated as CO_2. As a general rule, about one-third of the applied C in fresh residues will remain in the soil after the first few months of decomposition. The decay process is accompanied by conversion of organic form of N to NH_3 and NO_3^- and soil micro-organisms utilise partof this N for synthesis of new cells. The gradual transformation of plantraw material into stable organic matter (humus) leads to the establishment of reasonably consistent relationship between C and N. Other factors which may be involved in narrowing of the C/N ratio include chemical fixation of NH_3 or amines by ligninlike substances.

The C/N ratio of virgin soils formed under grass vegetation is normally lower than for soils formed under forest vegetation, and for the latter, the C/N ratio of the humus layers is usually higher than for the mineral soil proper. Also the C/N ratio of a well-decomposed muck soil is lower than for a fibrous peat.

As a general rule it can be said that conditions which encourage decomposition of organic matter result in narrowing of the C/N ratio. The ratio nearly always narrows sharply with depth in the profile; for certain subsurface soils C/N ratios lower than 5 are not uncommon.

ORGANIC PHOSPHORUS

Phosphorus rank in importance with N and K as major plant nutrients.

Phosphorus compounds in soil can be placed into the following three classes :

1. organic compound of the soil humus,

2. inorganic compounds in which the P is combined with Ca, Mg, Fe, Al and with clay minerals,

3. organic and inorganic P compounds associated with the cells of living matter. Micro-organisms are involved in transformations of phosphorus between organic and mineral forms.

From 15 to 80 per cent of the phosphorus in soils occurs in organic forms, the exact amount being dependent upon the nature of the soil and its composition. The higher percentages are typical of peats and uncultivated forest soils.

From the standpoint of plant nutrion, phosphorus is adsorbed by plants largely as the negatively charged primary and secondary orthophosphate ions ($H_2PO_4^-$ and HPO_4^{2-}) which are present in the soil solution. Small quantities of soluble organic P compounds are also present in water extracts of soil.

The Phosphorus Cycle

In a broad sense, the phosphorus cycle in soil involves the uptake of phosphorus by plants and its return t. the soil in plant and animal residues.

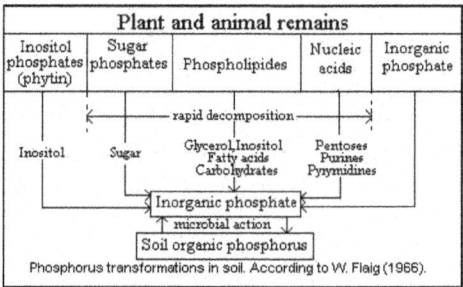

Phosphorus transformations in soil. According to W. Flaig (1966).

As can be seen from picture three general types of compounds make up the bulk of the organic phosphorus in plants, namely : phytin, phospholipids, and nucleic acids.

Approximate recoveries of organic phosphorus in these forms are as follows :

Inositol phosphates	2-50 per cent
Phospholipids	1-5 per cent
Nucleic acids	0.2-0.5 per cent
Phosphoproteins	trace
Metabolic phosphates	trace

When crop residues are returned to the soil, net immobilisation of P will occur when the C/organic- P ratio is 300 or more; net mineralisation will result when the ratio is 200 or less.

FUNCTION OF ORGANIC MATTER IN SOIL

Organic matter contributes to plant growth through its effect on the physical, chemical, and biological properties of the soil. It has a :

- *nutritional* function in that it serves as a source of N, P for plant growth
- *biological* function in that it profoundly affects the activities of micro-flora and micro-faunal organisms
- *physical* and *physico-chemical* function in that it promotes good soil structure, thereby improving tilth, aeration and retention of moisture and increasing buffering and exchange capacity of soils.

Humus also plays an indirect role in soil through its effect on the uptake of micro-nutrients by plants and the performance of herbicides and other agricultural chemicals. It should be emphasised that the importance of any given factor will vary from one soil to another and will depend upon such environmental conditions as climate and crpping history.

Availability of Nutrients for Planth Growth

Organic matter has both a direct and indirect effect on the availability of nutrients for plant growth. In addition to serving as a source of N, P, S through its mineralisation by soil micro-organisms, organic matter influences the supply of nutrients from other sources (for example, organic matter is required as an energy source for N-fixing bacteria).

A factor that needs to be taken into consideration in evaluating humus as a source of nutrient is the cropping history. When soils are first placed under cultivation, the humus content generally declines over a period of 10 to 30 years until a new equilibrum level is attained. At equilibrium, any nutrients liberated by microbial activity must be compensated for by incorporation of equal amounts into newly formed humus.

Effect on Soil Physical Condition, Soil Erosion and Soil Buffering and Exchange Capacity

Humus has a profound effect on the structure of many soils. The deterioration of structure that accompanies intensive tillage is usually less severe in soils adequately supplied with humus. When humus is lost, soils tend to become hard, compact and cloddy.

Aeration, water-holding capacity and permeability are all favourably affected by humus.

The frequent addition of easily decomposable organic residues leads to the synthesis of complex organic compounds that bind soil particles into structural units called aggregates. These aggregates help to maintain a loose, open, granular condition. Water is the better able to infiltrate and percolate downward through the soil.The roots of plants need a continual supply of O_2 in order to respire and grow. Large pores permit better exchange of gases between soil and atmosphere.

Humus usually increases the ability of the soil to resist erosion. First, it enables the soil to hold more water. Even more important is its effect in promoting soil granulation and thus maintaining large pores through which water can enter and percolate downward.

From 20 to 70 per cent of the exchange capacity of many soils is caused by colloidal humic substances. Total acidities of isolated fractions of humus range from 300 to 1400 meq/100g.As far as buffer action is concerned, humus exhibits buffering over a wide pH range.

Effect on Soil Biological Condition

Organic matter srves as a source of energy for both macro- and micro-faunal organisms.

Numbers of bacteria, actinomycetes and fungi in the soil are related in a general way to humus content. Earthworms and other faunal organisms are strongly affected by the quantity of plant residue material returned to the soil.

Organic substances in soil can have a direct physiological effect on plant growth. Some compounds, such as certain phenolic acids, have phytotoxic properties; others, such as the auxins, enhance plant growth.

It is widely known that many of the factors influencing the incidense of pathogenic organisms in soil are directly or indirectly influenced by organic matter. For example, a plentiful supply of organic matter may favour the growth of saprophytic organisms relative to parasitic ones and thereby reduce populations of the latter. Biologically active compounds in soil, such as antibiotics and certain phenolic acids, may enhance the ability of certain plants to resist attack by pathogens.

TYPES OF HUMUS IN SOILS

Humus occurs in soils in many types, differentates in regard to morphology and fractional composition.

A type of humus is it a morphological form of naturals accumulation of humic substances in profile or on the surface of soil, conditioned by general direction of soil-forming process and humification of organic matter.

A types of humus in terrestrial enviroment are following :

1. mor

2. moder

3. mull

Mor is a type of humus, which occur largely in coniferous forest soils and the moorlands soils.

This humus arise under conditions of low-biological activity in soil. The mineralisation of organic matter proceed slowly and create layers, which maintain a structure of vegatable material.Acidophilic fungi and low active invertebrates participates in transformations of plant residues.

Under these circumstances forms a litter of large thickness. C/N ratio of mor humus is always more than 20, or even 30-40, whereas pH is acid.

Moder is a transitional form of humus between mull and moder, characteristic for sod-podzolic soils, loesses and mountain grassland soils.

The organic horizons with moder humus consist of low-thicknessed litter (2-3 cm), which gradually, without bounds, pass on to humus-accumulative horizons.

Moder is a type of medium humified humus. Acidophilic fungi and arthropodan participates in transformations of plant residues. C/N ratio equal 15-25. Produced mineral-organic complexes are labile and weakly bounded with mineral portion of soil.

Mull is a type of humus characteristic for chestnut soils, phaeozems, rendzinas and others soils.

This type of humus arise under grass vegetation.

Mull is a well humified organic matter, which is produce in very biologically active habitat. This type of humus is characterised by neutral pH, C/N ratio nearing to 10 and ability to creation stable mineral-organic complexes. Mull is a type of humus which occurs in soils under cultivation.

According to Kononova, the types of humus are divide as follows :

First type of humus is characteristic for podzolic soils, grey brown soils and lateric soils under forest communities. In this humus predominate humic acids, thus humic acid/fulvic acid ratio is below 1. Humic acid indicate small extent of aromatic rings condensation and they are approximate to fulvic acids. Considerable hydrophilic properties of humic acids favour to creation of chelates with polyvalent cations and ability to displacement deep into profile of soil. Considerable mobility of this humus favour process of podsolisation.

Second type of humus is characteristic for phaeozems, rendzinas, black earths and brown soils. Humic acid/fulvic acid ratio is upper than 1, Extent of aromatic rings condensation is high in humic acids, which cause their hydrophobic properties and inability to creation of chelates. Humic acids are strongly connected with mineral portion of soil in this type of humus.

Third type of humus is characteristic for semi-desert soils. In this humus predominate fulvic acids fraction, whereas arise of humic acids is limited. Beyond this, humic acids are largely bounded with mineral portion of soil.

HUMUS CONTENT OF SOIL

Humus content in soils fluctuating in broad range.

On humus content have influence the following factors :

*   amount and quality of humus, which get at soil in given bio-ecological zone
*   tempo of humification process of organic matter
*   tempo of mineralisation of humus, which is contain in soil
*   chemical, physico-chemical and physical soil properties
*   amount and quality of mineral compounds contained in soil.

Humus content in accumulation horizons of the main soil units in Poland.

Division and order	Type,genera and kind	Humus content per cent
Calcisols :	Calcarious Rendzinas	3.4 2.1 - 6.3
	Jurasic Rendzinas	4.4 1.5 - 7.0
Phaeozems	Haplic Phaeozems	2.8 1.8 - 4.0
Cambisols :	Brown soils formed from sands	1.5 0.9 - 2.2
	Brown soils formed from light and medium loams	1.8 1.1 - 3.0
	Brown soils formed from heavy loam	2.5 1.6 - 3.7
	Brown soils formed from silt formations	1.7 1.3 - 1.9
	Brown soils formed from loess and loess-like materials	1.9 1.4 - 2.6
Luvisols :	Grey brown soils formed from silt formations	1.9 1.4 - 2.4
	Grey brown soils formed from loess and loess-like materials	1.8 1.0 - 2.5
	Grey brown soils formed from light loam	1.6 1.0 - 2.6
Podzols	Podzolic soil formed from sands	1.5 1.1 - 2.0
Gleysols :	Boggy soils formed from silts	1.6 1.2 - 2.1
Gleysols :	Black earth formed from sands	2.8 1.2 - 4.1
	Black earth formed from light and medium loams	2.6 1.2 - 5.7
	Black earth formed from heavy loams and clays	4.9 2.5 - 5.6
Fluvisols :	Alluvial soils formed from sands	2.9 1.5 - 5.2
	Alluvial soils formed from silts	3.5 1.7 - 5.8
	Alluvial soils formed from clays	4.2 2.4 - 6.8

Chapter 11

SOIL CARBON AND NITROGEN STOCKS OF DIFFERENT HAWAIIAN SUGARCANE CULTIVARS

Rebecca Tirado-Corbalá[1,2]*, Ray G. Anderson[1,3], Dong Wang[1] and James E. Ayars[1]

[1] USDA, Agricultural Research Service, San Joaquin Valley Agricultural Sciences Center, Water Management Research Unit, 9611 S. Riverbend Ave., Parlier, CA 93648-9757, USA; E-Mails: ray.anderson@ars.usda.gov (R.G.A.); dong.wang@ars.usda.gov (D.W.); james.ayars@ars.usda.gov (J.E.A.)

[2] Department of Agro-environmental Sciences, University of Puerto Rico, Mayagüez, Puerto Rico, USA

[3] USDA-Agricultural Research Service, U.S. Salinity Laboratory Contaminant Fate and Transport Unit, 450 W. Big Springs Rd., Riverside, CA 92507-4617, USA

* Author to whom correspondence should be addressed; E-Mail: rebeccatiradocorbala@gmail.com or rebecca.tirado@upr.edu; Tel.: +1-787-370-9179.
  Academic Editor: Yantai Gan

ABSTRACT

Sugarcane has been widely used as a biofuel crop due to its high biological productivity, ease of conversion to ethanol, and its relatively high potential for greenhouse gas reduction and lower environmental impacts relative to other derived biofuels from traditional agronomic crops. In this investigation, we studied four sugarcane cultivars (H-65-7052, H-78-3567, H-86-3792 and H-87-4319) grown on a Hawaiian commercial sugarcane plantation to determine their ability to store and accumulate soil carbon (C) and nitrogen (N) across a 24-month growth cycle on contrasting soil types. The main study objective establish baseline parameters for biofuel production life cycle analyses; sub-objectives included (1) determining which of four main sugarcane cultivars sequestered the most soil C and (2) assessing how soil C sequestration varies among two common Hawaiian soil series (Pulehu-sandy clay loam and Molokai-clay). Soil samples were collected at 20 cm increments to depths of up to 120 cm using hand augers at the three main growth stages (tillering, grand growth, and maturity) from two experimental plots at to observe total carbon (TC), total nitrogen (TN), dissolved organic carbon (DOC) and nitrates (NO^{-3}) using laboratory flash combustion for TC and TN and solution filtering and analysis for DOC and NO^{-3}. Aboveground plant biomass was collected and subsampled to determine lignin and C and N content. This study determined that there was an increase of TC with the advancement of growing stages in the studied four sugarcane cultivars at both soil types (increase in TC of 15–35 kg·m^2). Nitrogen accumulation was more variable, and NO^{-3} (<5 ppm) were insignificant. The C and N accumulation varies in the whole profile based on the ability of the sugarcane cultivar's roots to explore and grow in the different soil types. For the purpose of storing C in the soil, cultivar H-65-7052 (TC accumulation of ~30 kg m^{-2}) and H-86-3792 (25 kg m^{-2}) rather H-78-3567 (15 kg m^{-2}) and H-87-4319 (20 kg m^{-2}) appeared to produce more accumulated carbon in both soil types.

Keywords

Hawaii sugarcane; cultivars; soil carbon; soil nitrogen; carbon sequestration; biofuel.

1. INTRODUCTION

Sugarcane is widely used as a biofuel crop due to its high biomass, ease of conversion to ethanol, and the higher potential for greenhouse gas reductions and lower environmental impacts relative to other biofuels derived from traditional agronomic crops [1,2]. There is an increasing interest in converting sugarcane to biofuel using advanced cellulosic approaches, particularly in the Pacific Basin [3,4]. The Hawaiian Islands have been identified as a potential location for growing biofuels due to the very high potential productivity of Hawaiian sugarcane and the availability of land following large scale closures of sugarcane plantations [5,6]. Several notable climatic factors are in favor for Hawaiian sugarcane productivity and efficiency, including high solar irradiance (>20 MJ·m^{-2}·day^{-1}), mild maximum daily temperatures (<30 °C), and low vapor pressure deficit (<1.5 kPa) [7,8]. However, management of Hawaiian sugarcane production is challenging due to high variability in soil fertility, farmland slopes, and other elevation/slope/aspect, and climatic aspects [7,9]. In order to achieve high yields, Hawaiian sugarcane production systems have been improved with the use of numerous practices that are distinctive from other major sugarcane growing regions. One of the most distinctive practices is a ~24 month cropping system with a greater rate of biomass accumulation in the first 15 months of growth and sucrose accumulation thereafter [5]. Other important agricultural practices include: tilling (sub-soiling) the soil to 60 cm depth before the seed canes are planted, using local (Hawaiian), high yielding, disease-resistant cultivars, and using natural predators to control insects and pests. Also, the improved farming practices include the addition of water and fertilizers through drip irrigation systems, incorporation of sand and gypsum to improve the physical properties of the soil before planting, and weed removal in the first six months of sugarcane growth (chemically with glyphosate or hexazinone, or mechanically). The practices of inducing ripening at 12 months by depleting nitrogen in the soil and crop, withholding the irrigation and applying glyphosate for desiccation have been used to enhance sucrose accumulation [9]. Putting those practices together has resulted in high sugar yields of up to 35 t·ha^{-1} [10].

Hawaiian sugarcane cultivars have been studied for increased yield [5,11], improved pest resistance [12,13], and salinity tolerance [14] in relation to sugar production. However, parameters relevant to biofuel production, such as total carbon and nitrogen accumulation, nitrogen fertilizer recovery, and soil organic and inorganic carbon sequestration [15] are less understood, particularly since the rise of drip irrigation in the 1980s [16]. These parameters are critical for biofuel production since they affect fossil fuel inputs via nitrogen fertilizers [17], potential emissions of greenhouse gases such as nitrous oxide that counteract the greenhouse gas benefits of reduced carbon emissions [18], water quality and ecosystem services [19], and energy and economic feasibility [20]. Along with organic carbon sequestration, the long cultivation history (>100 years), continuous monoculture [21], and extensive soil amendments may have created conditions for inorganic carbon sequestration or emissions [22] in Hawaiian sugarcane soils that warrant examination. Furthermore, soil organic matter (SOM) is involved in the maintenance of soil quality, sustainability of natural and agricultural systems and the natural balance of greenhouse gases [23]. For example, pre-harvest straw burning reduces SOM [24,25], thereby affecting the chemical, physical, and biological features of soil. Although SOM represents only a small parcel of the total mass of mineral soils, it is essential for many chemical, physical and biological processes of terrestrial ecosystems [26,27].

Baseline data needed to conduct life cycle analyses [17], required to verify greenhouse gas reductions under current biofuel mandates is lacking [28,29]. In this study, we determined the effect of soil type and cultivar on the carbon and nitrogen accumulation and storage across the 24 month sugarcane growth. We hypothesize that Molokai silty clay soils will have better C sequestration potential due to better protection of organic matter in this tropical environment and that cultivars differ in crop yield, carbon sequestration and the contribution to the carbon emissions.

2. MATERIALS AND METHODS

2.1. Study Area

The study was conducted on a commercial sugarcane plantation (CSP) on the island of Maui, Hawaii (20°54' N and 156°26' W). We selected two, ~1 ha experimental plots with contrasting soil types: Pulehu series-cobbly silt loam (fine-loamy, mixed semiactive, isohyperthermic Cumulic Haplustolls) and Molokai series-silty clay loam (Very-fine, kaolinitic, isohyperthermic Typic Eutrotorrox) [30]. These soils were selected because both are common soil series encountered in Maui's agricultural lands. Both plots were planted with four different commercial sugarcane cultivars, H65-7052, H78-3567, H86-3792 and H87-4319. The Pulehu series plot was planted on 19 July 2011 and harvested on 9 June 2013. A total of 375 kg of N ha^{-1} in the form of Urea (46% N) was applied in 7 applications in the first 300 days after planting (DAP). The Molokai series plot was planted on 23 June 2011 and harvested on 7 May 2013. A total of 345 kg of N ha^{-1}, in the form of Urea, was applied in 10 applications in the first 300 DAP. The CSP uses drip irrigation to supplement rain and tried to maximize limited surface and ground water resources [10]. The drip irrigation system consists of drip laterals spaced at 2.74 m intervals with a row of sugarcane planted on both sides of each drip tape at 46 cm distance away from the tape. The system is pressure compensated to 82.7 kPa (12 pounds per square inch) at risers at the head of the tape lines. The discharge rate is 1.58 L/hour/meter of tape (12.7 US gallons/ hour/ 100 feet of tape). In total 2500 mm of water was drip-applied during the two-year growth cycle.

2.2. Baseline Soil Properties of the Experimental Plots

At the beginning of the experiment, 12 samples were collected randomly from each soil depth (0–20, 20–40, 40–80 and 80–120 cm) from the both Pulehu and Molokai soils, oven dried at 65 °C for 48 h, ground and sieved through a 2-mm screen. Soil pH was measured in 1:1 solid/DI water suspension [31]. Electrical conductivity (EC) of the soil samples was determined from a 1:1 soil: DI water suspension. Exchangeable macronutrients such as Calcium (Ca^{2+}), Magnesium (Mg^{2+}), Potassium (K^+) and Sodium (Na^+) were measured using 1-molar ammonium acetate (NH4OAc) as the extractant (pH 7) [32] and determined using Inductively Coupled Plasma-Optical Emission Spectrometry analysis (Varian, Palo Alto, CA, USA. (Mention of trade names or commercial products in this publication is solely for the purpose of providing specific information and does not imply recommendation or endorsement by the U.S. Department of Agriculture). Soil texture was determined by the hydrometer method [33]. Soil bulk density was determined by using a 5.7 cm diameter bulk density soil sampler (0200 Soil Core Sampler from Soil Moisture Equipment Corp., Ventura, CA, USA).

2.3. Soil Variables

In each experimental plot, three samples from each soil depth (0–20, 20–40, 40–80 and 80–120 cm) and sugarcane cultivar were collected at tillering, grand growth, and maturity sugarcane growing stages, oven dried at 65 °C for 48 h, ground and sieved through a 2-mm screen. Total carbon (TC) and total nitrogen (TN) contents were determined by dry combustion with a Flash 2000 N and C Soil Analyzer (Thermo Scientific®, Pittsburgh, PA, USA). Organic carbon (OC) was determined after eliminating all the inorganic carbon (IC) in the samples in the form of carbon dioxide (CO_2), after the samples were acidified with

1:1 HCL: DI water in silver container. The soluble organic compounds were then dried and combusted using the Flash 2000 N and C Soil Analyzer (Thermo Scientific®, Pittsburgh, PA, USA). IC was determined by subtracting OC from the TC. Soil carbon and nitrogen stocks (kg·m^{-2}) were calculated by multiplying the carbon concentration by the thickness of the soil layer (m) and the soil bulk density (kg·m^{-3}) for each layer.

Dissolved organic carbon (DOC) was determined after saturating the soil with DI water (1:1 soil:DI water) for 24 h, shaken for one hour on a reciprocal shaker, and filtered through a Whatman no. 42 filter. Carbon recovered in the water extract was determined using a Fusion Total Organic Carbon Analyzer ™ (Teledyne Tekmar, Mason, OH). Nitrate (NO3-N) content (1:1 soil: DI water) was determined by using Nitrate-Nitrite Astoria Pacific 2 analyzer (Portland, OR, USA).

2.4. Plant Measurements

In each experimental plot, plant samples from three locations per sugarcane cultivar were collected to measure aboveground and belowground biomass less than two weeks prior to harvest. Aboveground biomass was determined by using two m^2 rectangular frames. The long dimension (2 m) of the frame was installed along the row. With the aboveground measurements, we only considered the biomass from the cane stalks growing inside the frame (*i.e.*, green tops, dried leaves and trash on the ground) because sugarcane tends to lodge. Before cutting the cane, the plant height and dewlap of three representative sugarcane plants were measured. Aboveground dry biomass was determined after oven drying the samples at 65 °C for 5 days. Aboveground dry biomass samples were shipped to our laboratory and ground to 2 mm size particles using a grinder (Thomas Scientific 174931 grinder, Swedesboro, NJ, USA). Lignin content analysis of the different sugarcane cultivars was conducted at an independent laboratory (the Soil, Water, and Forage Testing Laboratory at Texas A and M Agri Life Extension Service College Station, TX, USA).

Belowground (root) biomass was measured at the time when aboveground biomass was sampled. Root biomass for each soil depth interval (0–20, 20–40, 40–80 and 80–120 cm) was determined after collecting soil samples at intervals of 0 m (next to the cane row), 0.75 m and 1.5 m from the sugarcane row using a 7 cm diameter mud auger (Signature Series 350.19, AMS Inc., American Falls, ID, USA). Probe method (mud auger) was selected to minimize disturbance to the soil. All soil samples were stored in plastic containers and were frozen in preparation for root sieving. After thawing, soils were hand sieved by using a 1.4 mm mesh sieve (# 14) to ensure the collection of the majority of the roots. Collected roots from each sample were oven dried overnight at 65 °C to determine root dry weight.

2.5. Statistical Analysis

The treatments were arranged in randomized complete design that included four sugarcane cultivars and at two soil types. For each experimental plot (Molokai and Pulehu) and sugarcane cultivar, soil samples were collected at matching sugarcane growing stages (tillering, grand growth and maturity). The soil results from the growing stages for each cultivar were compared for: TC, OC, IC, DOC, TN and NO_3-N. Plant samples from all the sugarcane cultivars, collected two weeks prior harvest, were compared for aboveground biomass and lignin content. The data were analyzed using the Mixed model of JMP Version 10 (SAS Institute, Cary, NC, USA). Two-way analysis of variance (ANOVA) followed by means separation using Tukey's Honestly Significant Difference (HSD) test at $Pr < 0.05$ was utilized to examine the significant differences among cultivars and between soil types in soil chemical properties and other variables.

3. RESULTS

3.1. Soil Chemical Properties

Our results showed that both soils, Pulehu and Molokai, are moderately alkaline (i.e., pH = 7.5–8.2), and there were no acidity problems present in the whole soil profile at the experimental fields (Table 1). Pulehu soils have a higher pH than Molokai soils. At deeper soil depths, we observed an increase in pH (i.e., 8.1 to 8.2) for Pulehu soils and a decrease in pH (i.e., ~8.0 to 7.5) for Molokai soils. However, for electrical conductivity (EC), an opposite pattern of pH is observed in both soils (Table 1). Molokai soils have more than twice the EC (i.e., 0.8 to 1.5 dS·m⁻¹) and exchangeable sodium (Na⁺) concentration (i.e., 1.7 to 6.8 cmolc kg⁻¹) than Pulehu soils for the whole soil profile, while Pulehu soils have higher exchangeable calcium (Ca²⁺) (i.e., 16 to 22 cmolc·kg⁻¹), cation exchange capacity (CEC) (i.e., 22 to 25 cmolc·kg⁻¹), and bulk density than Molokai soils (Tables 1 and 2). This high pH and low EC and Na⁺ in Pulehu soils can affect the availability of favorable nutrients such as Ca²⁺ and potassium (K⁺), (necessary for sugar cell structure) and soil C. Pulehu soils have approximately 18 times more exchangeable Ca²⁺ than magnesium (Mg²⁺) and, 1.5–2 more exchangeable Ca²⁺ than Mg²⁺ is observed in Molokai soils (Table 1). Higher total C and N were observed in samples collected from all soil depths from Pulehu soils compared with Molokai soils (Table 2). However, lower values of BD were obtained in Molokai soils (dominated sandy clay loam texture) compared with Pulehu soils (dominated clay texture) (Table 2).

Table 1. Soil Chemical Properties of the Experimental Fields.

Soil	Soil Depth	pH [1]	EC [a,1]	[Ca²⁺	Mg²⁺	K⁺	Na⁺	CEC][a,2]
	cm	1:1	dS·m⁻¹	----------------cmolc·kg⁻¹-------------				
Pulehu	0–20	8.10 [b]	0.47	19.4	1.30	0.40	0.60	21.8
	20–40	8.19	0.32	21.1	0.90	0.30	0.60	22.9
	40–80	8.23	0.30	22.4	1.80	0.30	0.80	25.3
	80–120	8.23	0.29	16.0	4.40	0.60	2.30	23.4
Molokai	0–20	7.97	0.81	9.16	4.70	0.65	1.69	16.2
	20–40	7.97	0.94	7.69	3.98	0.39	1.60	13.7
	40–80	7.55	1.13	4.11	2.41	0.13	4.12	10.8
	80–120	7.51	1.50	2.69	2.37	0.18	6.84	12.1

[a] EC = Electrical conductivity and CEC = Cation exchange capacity; [b] Mean values from 12 samples;
[1] 1:1 Soil: DI water suspension; [2] 1M Ammonium acetate extraction.

Table 2. Soil Total Carbon and Nitrogen and Physical Properties
of the Experimental Fields.

Soil	Soil Depth	TC	TN	BD [a]	Sand [c]	Silt [c]	Clay [c]	Texture [d]
	cm	---- kg·m⁻²----		---g·cm⁻³---		---------%---------		
Pulehu	0–20	4.26[b]	0.90	1.37	53.8	15.8	30.4	SCL
	20–40	3.98	0.78	1.39	58.9	11.6	29.4	SCL
	40–80	5.40	1.04	1.34	43.5	25.2	31.3	C
	80–120	5.25	0.78	1.32	41.6	22.0	36.4	L
Molokai	0–20	2.91	0.31	1.26	27.0	44.2	28.0	C
	20–40	2.86	0.26	1.23	22.0	50.2	27.8	C
	40–80	4.16	0.36	1.26	28.7	46.4	24.8	L
	80–120	3.38	0.21	1.31	36.5	36.6	26.8	L

[a]BD = Bulk density; [b]Mean values from 12 samples; [c]Calculated by using Hydrometer method; [d]USDA
Soil Classification: SCL-sandy clay loam, C-clay, L-loam.

3.2. Soil Carbon

In general, total carbon (TC) increased with the growth stages in the four sugarcane cultivars on both soil types (Figures 1 and 2). Cultivar H-65-7052 showed consistent increases in TC (Pr < 0.05) with growing stages for the whole soil profile and at both soil types (Figures 1 and 2). In Pulehu soils, around 45%–60% of the TC of H-65-7052 was in the organic carbon (OC) form and 40%–55% in the inorganic carbon (IC) form. In depth intervals 0–20 and 40–80 cm of Pulehu soils, there was a noticeable increase in OC with the sugarcane growing stages (Figure 1). However, at 20–40 cm depth interval in the Pulehu soils, TC response varied and a higher OC was found during the grand growth stage (Figure 1). At 80–120 cm depth in the Pulehu soils, TC was four times higher than at 0–20 cm depth interval at tillering (Figure 1).

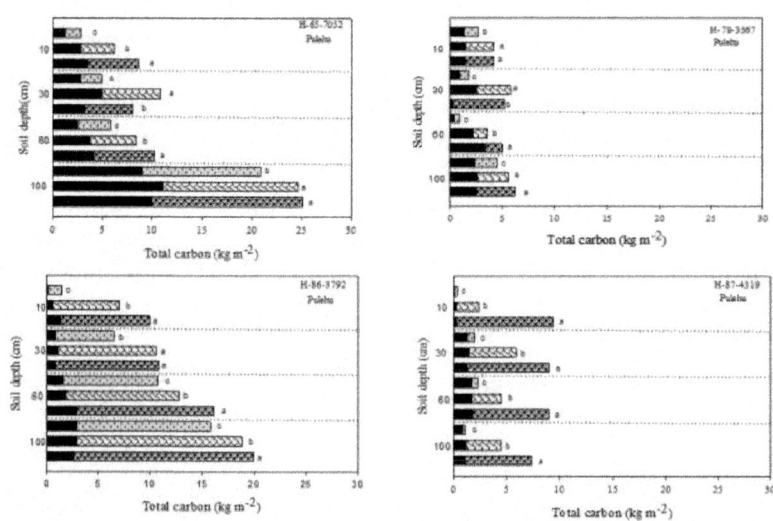

Figure 1. Total organic (OC) and inorganic (IC) carbon for the four sugarcane cultivars grown in Pulehu soils. Black (solid) portion of the bars represent the IC and color portion represent OC for each soil depth and sugarcane growing stages (*i.e.*, blue (open arrows)-tillering, yellow (irregular shape)-grand growth, green (light circles)-maturity) when the soil samples were collected. Means followed by the same letter or no letters for the bars in each soil depth are not significantly different in the Tukey's test at Pr < 0.05.

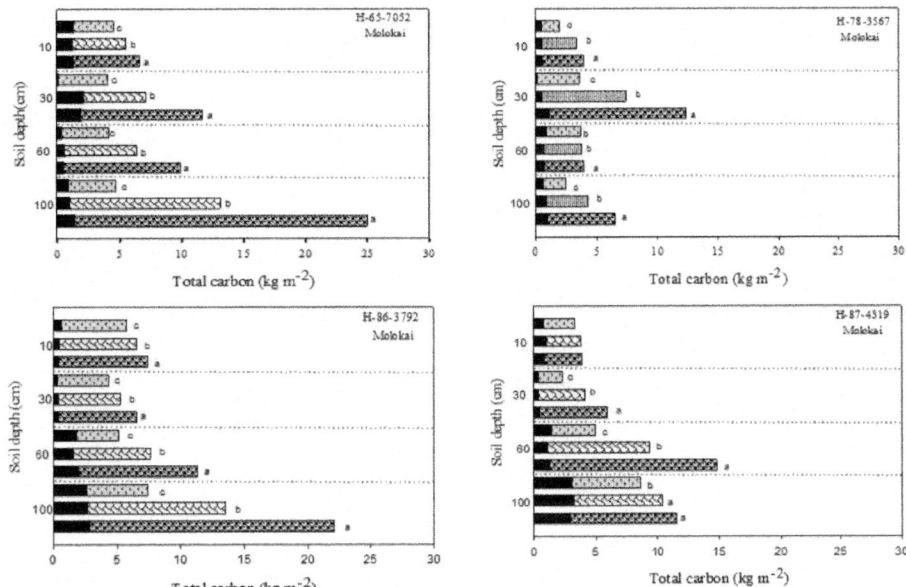

Figure 2. Total organic (OC) and inorganic (IC) carbon for the four sugarcane cultivars growing in Molokai soils. Black (solid) portion of the bars represent the IC and color portion represent OC for each soil depth and sugarcane growing stages (*i.e.*, blue (open arrows)-tillering, yellow (irregular shape)-grand growth, green (light circles)-maturity) when the soil samples were collected. Means followed by the same letter or no letters for the bars in each soil depth are not significantly different Tukey's test at Pr < 0.05.

In Molokai soils, cultivar H-65-7052 had more than 70% of the TC in the OC and less than 30% in IC form (Pr < 0.05). Also, a 1.5 fold increase was observed in OC from tillering to maturity (Figure 2). In the first 40 cm of soil, TC increases with respect to soil depth and sugarcane growing stages (Figure 2). However, a decrease in TC was observed at 40–80 cm soil depth while the TC increases almost twice at 80–120 cm soil depth from grand growth to maturity (Figure 2).

With respect to dissolved organic carbon (DOC), significantly higher concentrations (Pr < 0.05) were found on samples collected from soil depth intervals 20–40 and 80–120 cm at grand growth and maturity stages and 0–20 cm at maturity stage with respect of tillering in Pulehu soils (Figure 3). About twice the DOC concentrations (Pr < 0.05) were found on samples collected in the first 20 cm of soil at grand growth and maturity for cultivar H-65-7052 growing in Molokai soils compared with the samples collected on tillering (Figure 4). However at deeper depth, no specific pattern was observed for H-65-7052 growing in Molokai soils (Figure 4).

In the first 20 cm of Pulehu soils, cultivar H78-3567 showed no statistical difference between the samples collected in grand growth and maturity compared with tillering (Figure 1). However, an increase of 1.6 times more OC content was observed over two years growth (Pr = 0.04). At depths deeper than 20 cm, there was noticeably higher accumulation of TC (4.4, 6, and 1.4 times more TC at maturity than tillering,

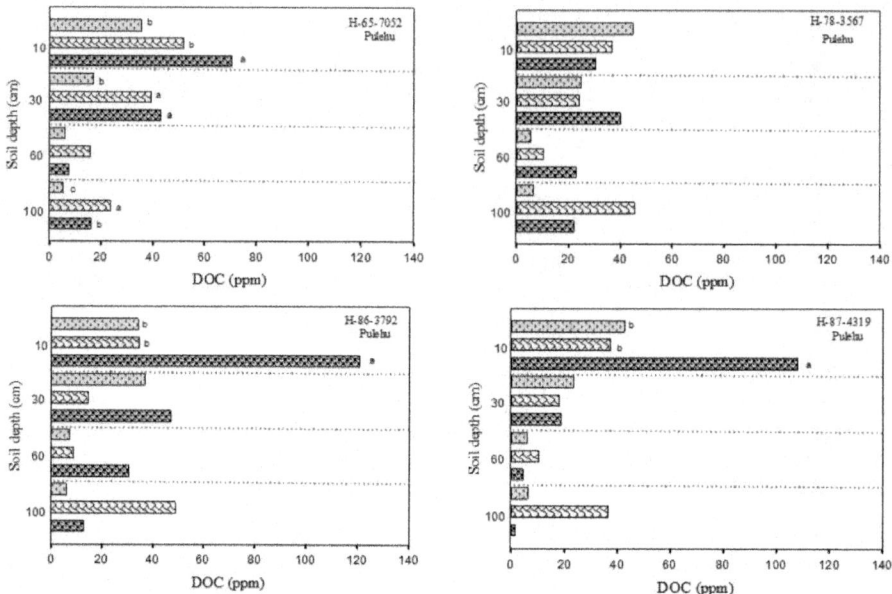

Figure 3. Dissolved organic carbon (DOC) in Pulehu soils. Color represents sugarcane growing stages (*i.e.*, blue-tillering, yellow-grand growth, green-maturity) when the soil samples were collected. Means followed by the same letter or no letters for the bars in each soil depth are not significantly different Tukey's test at Pr < 0.05.

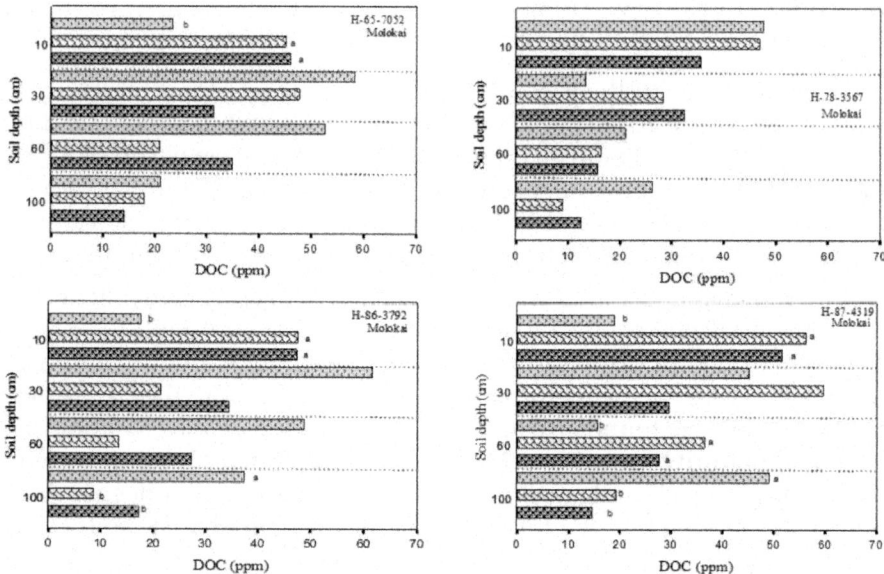

Figure 4. Dissolved organic carbon (DOC) in Molokai soils. Color represents sugarcane growing stages (*i.e.*, blue-tillering, yellow-grand growth, green-maturity) when the soil samples were collected. Means followed by the same letter or no letters for the bars in each soil depth are not significantly different Tukey's test at Pr < 0.05.

for 20–40 cm, 40–80 cm, and 80–120 cm respectively) (Figure 1). Carbon as OC represents more than 50% of the TC in all the samples collected at the different sugarcane growing stages. Significantly higher (Pr < 0.05) IC was observed in samples collected at maturity compared with grand growth and tillering growth stages at 20–80 cm depth (Figure 1).

Cultivar H-78-3567 growing in Molokai soils, showed a 1.2-fold increase in OC collected at grand growth and maturity compared with tillering (Figure 2). The 20-40 cm soil depth interval has the highest observed accumulation of C (threefold increase in OC) from tillering to maturity (Figure 2). However at 40-80 cm soil depth interval, C content remains constant during the whole sugarcane growth (Pr = 0.55). At deeper depths (> 80 cm), observed C increases with sugarcane growth stages (Pr = 0.02) (Figure 2). Around 71%–95% of the TC in Molokai soils comes from the OC form, while 5%–19% is in the IC form (Figure 2). For DOC, there was no statistical difference between sugarcane growth stages in all soil depths and soil types (Figure 3 and 4). Although, cultivar H86-3792 has a significant increase (Pr ≤ 0.02) for TC and OC between the growths stages for the whole Pulehu soil profile (Figure 1). More than 80% of the TC in Pulehu soil is OC (Figure 1). However, a higher increase in OC was found in the first 20 cm of soil compared with samples from deeper depths. At deeper depths (21-120 cm), OC increased by 50% or 1.5 times, while inorganic C increased at 40-80 cm soil depth interval (Pr = 0.05). Almost twice the amount of IC was encountered at maturity compared with grand growth and tillering (Figure 1), whereas DOC was 3.5 times higher at maturity compared with the samples collected from 0-20 cm depth interval at the other two stages (Pr = 0.03) (Figure 3).

Cultivar H86-3792 growing in Molokai soil exhibits the highest amount of TC in the first 20 cm of soil during tillering and has the least change in TC with depth compared to other cultivars (Figure 2). In the first 40 cm of soil (Pr = 0.04), more than 88% of the carbon is present as OC. At deeper depths (> 40 cm), 65%-87% of TC is in OC. At maturity, TC decreases with respect to soil depth in the first 80 cm of soil. However, at deeper depths, there is almost double amount of TC (Pr = 0.02) (Figure 2). While for DOC, in the first 20 cm of soil, DOC increases (Pr = 0.05) with respect of sugarcane growing stages. While at 80–120 cm soil depth, lower DOC was found at grand growth (Pr = 0.02) (Figure 4).

Cultivar H-87-4319 has consistent increases for TC and OC between the growing stages for the whole Pulehu soil profile (Figure 1) with a wider range of the portion of TC as OC (30%–90%). Cultivar H-87-4319 samples collected on Pulehu soils exhibit lower amounts of C compared with the other cultivars growing in both soils. No statistical difference was found for IC for the whole profile (Figure 1). Whereas, H87-4319 has similar response as H-86-3792 for DOC at the first 20 cm of soil (Pr = 0.04). DOC was 2.5 times higher at maturity compared with the samples collected during other two stages (Figure 3).

Cultivar H-87-4319 growing in Molokai soils is the third sugarcane cultivar with higher TC (Figure 2). Irregular response was found in the first 20 cm of soil with respect to sugarcane growing stages for H-87-4319 (Pr = 0.25), but consistent increases in TC with growing stage (P $\leq$ 0.05) at deeper depths (>20 cm) were observed (Figure 2). An increase in TC was detected with respect to growth stages and more evenly distributed. In the first 80 cm of soil, more than 70% of TC comes from the OC form, while at deeper depths; 64%–74% comes from the OC form (Figure 2). Higher DOC is observed on samples collected from 0–20 and 40–80 cm depth intervals in grand growth and maturity (P $\leq$ 0.05) (Figure 4). However at deeper depths (>80 cm), DOC decreases with respect to sampling date. Also, there was a clear reduction of DOC with respect of soil depth in grand growth and maturity compared with tillering stage (Figure 4).

3.3. Soil Total Nitrogen and Nitrates

In our study, cultivar H-65-7052 growing in Pulehu soils, higher total nitrogen (TN) (Pr $\leq$ 0.04) was observed in samples collected at 20–80 cm depth at maturity (Figure 5). Cultivar H-65-7052 growing in Molokai soils has higher TN (Pr = 0.04) in samples collected in the first 40 cm of soil at grand growth (Figure 6). At deeper depths (>41 cm), higher TN (Pr $\leq$ 0.04) was measured during tillering (Figure 6). In general, a decrease is observed in TN with respect to depth at grand growth and maturity *versus* tillering (Figure 6).

Cultivar H-78-3567 growing in Pulehu soils showed significant variation between growth stages at all soil depths (Pr < 0.05); however, there is not a consistent trend between sugarcane growth stages for each soil depth interval (Figure 5). A decrease in TN is observed at tillering with respect of soil depth (Figure 5). In samples collected at grand growth almost the same amount of N is observed in the whole profile. Whereas, on samples collected at maturity, the TN content is observed to increase with respect to soil depth or at deeper depths (Figure 5). However, cultivar H-78-3567 growing in Molokai soils exhibits higher TN (Pr < 0.05) on samples collected at maturity in the first 40 cm of soil and 80–120 cm soil depths (Figure 6). Although, no statistical difference (Pr = 0.78) was observed for sugarcane H-78-3567 growing in Molokai soils at 40–80 cm soil depth interval. While cultivar H-86-3792 growing in both soils (Figure 5 and 6) and cultivar H87-4319 growing in Pulehu soils exhibits an increase (Pr < 0.05) in TN with respect of sugarcane growth stages (Figure 5). Cultivar H87-4319 growing in Molokai soils had increases (Pr $\leq$ 0.04) in TN at depths deeper than 20 cm (Figure 6).

Soil nitrate concentration does not exceed 5 ppm in all sampling dates. Maximum NO_3-N concentrations (around 5 ppm), was encountered in the first 20 cm of the Molokai soil samples collected at grand growth in cultivar H-65-7052 and H-87-4319. This low concentration of nitrate indicates that most nitrates were taken up by the sugarcane or denitrified.

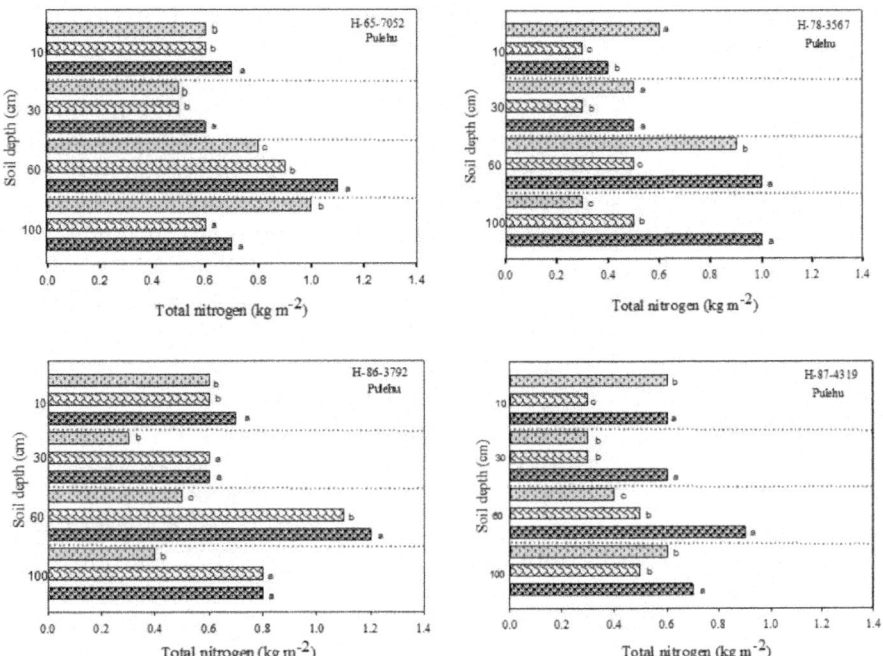

Figure 5. Total Nitrogen in Pulehu soils. Color represents sugarcane growing stages (*i.e.*, blue-tillering, yellow-grand growth, green-maturity) when the soil samples were collected. Means followed by the same letter or no letters for the bars in each soil depth are not significantly different Tukey's test at Pr < 0.05.

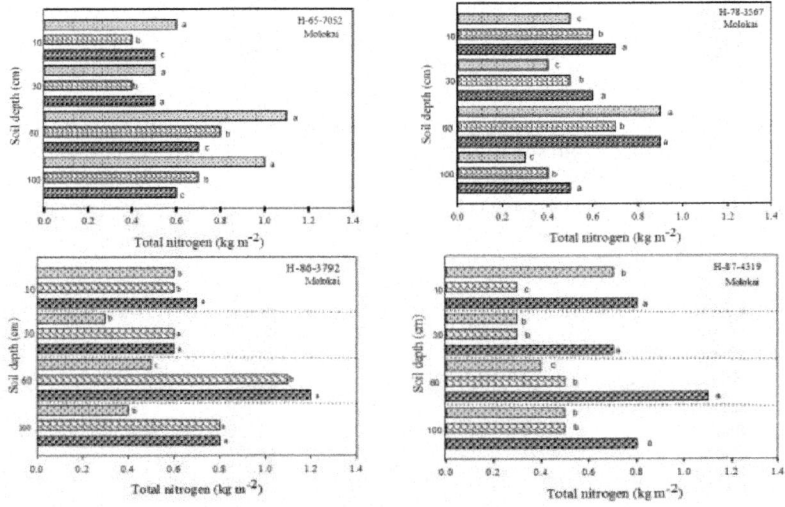

Figure 6. Total Nitrogen in Molokai soils. Color represents sugarcane growing stages (*i.e.*, blue-tillering, yellow-grand growth, green-maturity) when the soil samples were collected. Means followed by the same letter or no letters for the bars in each soil depth are not significantly different Tukey's test at Pr < 0.05.

3.4. Aboveground Biomass, Lignin Content, Carbon and Nitrogen in Plant Samples

No water and/or N stress conditions were observed throughout the crop growth cycle (*i.e.*, tillering, grand growth and maturity) in both experimental fields (*i.e.*, Pulehu and Molokai soils). In general, no statistical difference were found for aboveground dry biomass (~8–11 kg m^{-2}) (Pr = 0.19, Figure 7) and lignin content (~ 9%–11%) (Pr = 0.31, Figure 8) between sugarcane cultivars growing in Pulehu soils. Even though there was no statistical difference for aboveground biomass between sugarcane cultivars growing in Pulehu soils, it was observed that cultivar H-65-7052 (~10.7 kg m^{-2}) has the highest and cultivar H-87-4319 (~8.7 kg m^{-2}) has the lowest aboveground dry biomass (Figure 7). Similar response to aboveground dry biomass was observed for plant canopy and dewlap of cultivars growing in Pulehu soils. Greater plant canopy height and dewlap length were observed for cultivar H-65-7052 (6.8 ± 0.05 m/ 5.8 ± 0.39 m) compared with cultivar H-78-3567 (5.6 ± 0.4 m/ 4.3 ± 0.29 m), cultivar H-86-3792 (5.1 ± 0.05 m/ 4.0 ± 0.5 m), and cultivar H-87-4319 (4.8 ± 0.2 m / 3.36 ± 0.25 m), respectively. Plant moisture content was around 64%–68% on samples collected two weeks prior harvest in Pulehu soils (Data not shown).

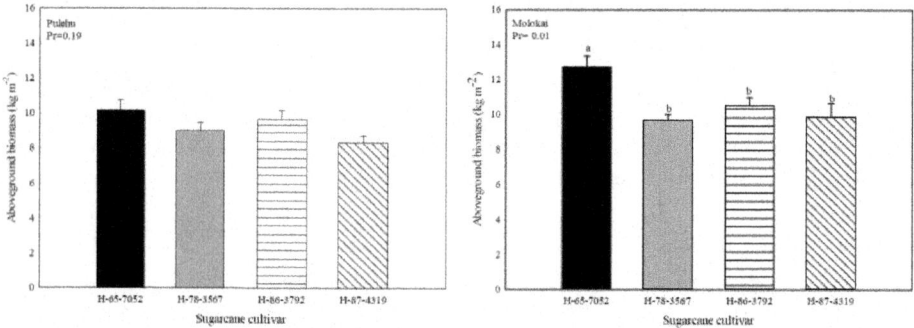

Figure 7. Aboveground biomass from sugarcane cultivars plant samples collected less than two weeks prior to harvest date. Means from sugarcane cultivars bars followed by the same letter or no letters are not significantly different Tukey's test at Pr < 0.05.

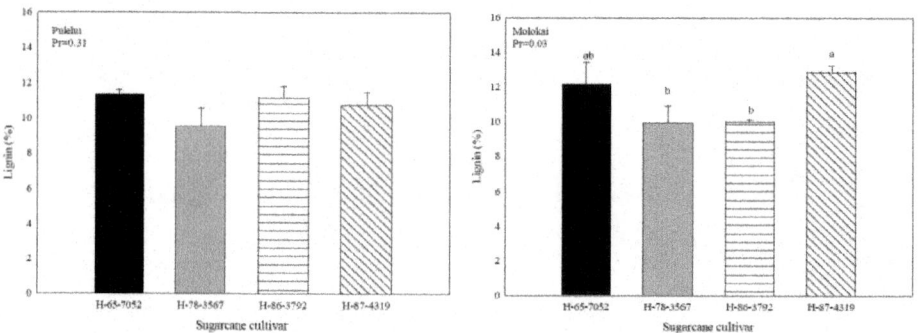

Figure 8. Lignin content (%) from sugarcane cultivars plant samples collected less than two weeks prior to harvest date. Means from sugarcane cultivars bars followed by the same letter or no letters are not significantly different Tukey's test at Pr < 0.05.

In Molokai soils, higher aboveground dry biomass (Pr = 0.01) was found for cultivar H-65-7052 (~13.8 kg·m^{-2}) compared with the other three sugarcane cultivars (~ 10.5 kg·m^{-2}) (Figure 7). However, higher lignin content (Pr = 0.03) was found in cultivar H-87-4319 and H-65-7052 (~13%) compared with cultivar H-78-3567 and H-86-3792 (~10.5%) (Figure 8). A more uniform response was observed for all sugarcane cultivars height and dewlap length: H-65-7052 (4.7 ± 0.05 m/ 4.0 ± 0.20 m), H-78-3567 (4.9 ± 0.02 m/ 3.5 ± 0.15 m), H-86-3792 (4.4 ± 0.07 m/ 3.0 ± 0.18 m) and H-87-4319 (4.2 ± 0.05 m/ 3.1 ± 0.13 m). Plant samples collected two weeks prior harvest from the experimental field with the Molokai soil had 35%–38% moisture content.

With respect to plant total carbon (TC) and nitrogen (TN) percentages between cultivars growing in both soils, we observed lower total N (i.e., 0.256- Pulehu and 0.181%-Molokai soils) and variable C (i.e., 50.9- Pulehu and 49.2%- Molokai soils) percentages for cultivar H-65-7052 growing in both soils compared to the other cultivars in each soil type (Table 3). Cultivar H-78-3567 has the 2nd highest amount of total N (0.334%) and C (50.6%) when it grows in Pulehu soils compared with Molokai soils (0.277% and 46.8%, N and C, respectively) (Table 3). Total plant nitrogen and lignin contents are important variables in determining N mineralization kinetics in the soil. When a high C/N ratio is present, intense N immobilization is expected. The N present in sugarcane residues follows a slow decay rate once deposited in the soil which varies from 3% to 30% in one year. Based on our Total C and N results in plants presented in Table 3, we can observe cultivar H-86-3792 has the lower C/N ratio followed by cultivar H-87-4319, H-78-3567, and H-65-7052 on Pulehu soils. Cultivar H-65-7052 has the highest C/N ratio when grows in both soils.

Table 3. Total Nitrogen and Carbon and C/N Ratio of Sugarcane Cultivars Plant Samples Collected Less Than Two Weeks Prior to Harvest Date.

Sugarcane Cultivar	Total Nitrogen		Total Carbon		C/N Ratio	
	Pulehu	Molokai	Pulehu	Molokai	Pulehu	Molokai
	---------%--------		---------%--------			
H-65-7052	0.256 c [1] BC	0.181 c D	50.9 a A [2]	49.4 a B	199 a B	273 a A
H-78-3567	0.334 b A	0.277 a B	50.6 b AB	46.8 b C	151 b C	169 c C
H-86-3792	0.367 a A	0.214 b CD	50.4 b AB	50.1 a Ab	137 c C	234 ab A
H-87-4319	0.328 b A	0.232 ab CD	49.2 c B	49.2 a B	150 b C	212 b B

[1] Lower case letter represents difference between sugarcane cultivars within the field (i.e., field with Pulehu or Molokai soils) for each analyzed parameter at Pr < 0.005; [2] Upper case letter represents difference sugarcane cultivars between fields for each analyzed parameter at Pr < 0.0001.

3.5. Soil Carbon Accumulation

In general it was observed that cultivar H-65-7052 and H-86-3792 accumulated more TC (~50 kg·m^{-2}) than H-78-3567 and H-87-4319 (~25 and 35 kg·m^{-2}, respectively) in the two year sugarcane cycle in both soil types. However, soil total carbon increases (from tillering to maturity) 17% and 36% for cultivar H-65-

7052 growing in Pulehu and Molokai soils, respectively (Figure 9). TC increases to around 22% on cultivar H-86-3792 and ~15% on cultivar H-78-3567 growing in both soils. Therefore, cultivar H-87-4319 has the ability to store more carbon in Pulehu soils (~29%) than Molokai soils (~17%); even though at the end the accumulation is ~35 kg·m^{-2} in both soils.

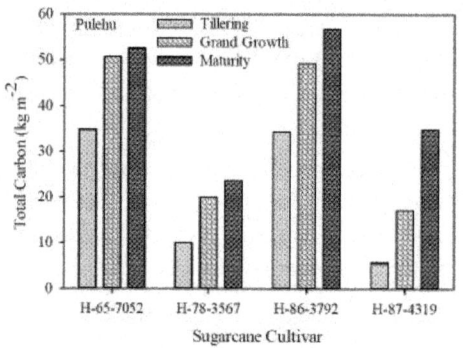 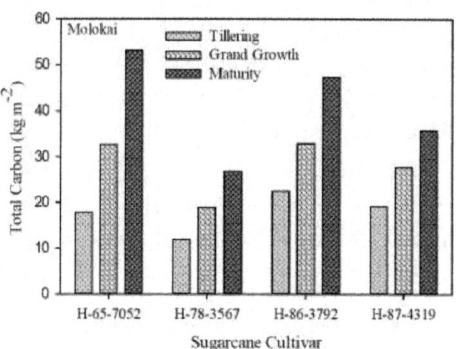

Figure 9. Total Carbon (kg·m^{-2}) in the first 120 cm of soil accumulation in the different sugarcane cultivars over the sugarcane growing stages.

Soil organic carbon accumulation from tillering to maturity changes with the inclusion of the root carbon fraction (Table 4). Opposite responses (Pr < 0.0001) in soil carbon accumulation were observed for cultivars H-65-7052 and H-87-4319 growing in different soil types (Table 4). The predominant cultivar H-65-7052 has the highest and second lowest accumulation (Pr < 0.0001) of soil organic carbon compared with the other cultivars when grown in Molokai soils (~34 kg·m^{-2}) and Pulehu soils (~13 kg·m^{-2}), respectively (Table 4). Cultivar H-65-7052 can store 2.6 fold increases in carbon on Molokai soils (~34 kg·m^{-2}) compared with Pulehu soils (~13 kg·m^{-2}). While cultivar H-87-4319 can store 1.7 fold increases in carbon on Pulehu soils (~29 kg·m^{-2}) compared with Molokai soils (~17 kg·m^{-2}). Cultivar H-86-3792 stores similar amounts of soil organic carbon (~18–20 kg·m^{-2}) when grown in both soils. While, cultivar H-78-3567 is the lowest organic carbon accumulator (~8-10 kg·m^{-2}) when grown in both soils (Table 4).

Table 4. Soil Organic Carbon Accumulation In Soil from Tillering to Maturity*.

Sugarcane Cultivar	Pulehu Soils	Molokai Soils
	------------- kg·m^{-2}------------	
H-65-7052	12.65 c[1] E	33.62 a A[2]
H-78-3567	8.23 d G	10.42 c F
H-86-3792	20.34 b C	18.44 b D
H-87-4319	29.01 a B	17.18 b D

[1] Lower case letter represents difference between sugarcane cultivars within the field (*i.e.,* field with Pulehu or Molokai soils) at Pr < 0.0001; [2] Upper case letter represents difference sugarcane cultivars between fields at Pr < 0.0001; * Total Carbon in maturity stage includes soil and roots carbon.

Figure 10 shows root carbon content from root samples collected two weeks prior to sugarcane harvest after two years of growth. We found significantly different root carbon density (Pr < 0.05), but no consistent trends, between cultivars for all soil depths and soil types (Figure 10). Our results have demonstrated different root C content are based on cultivar's root distribution growing in contrasting soil types (*i.e.*, Pulehu *vs.* Molokai soils). Moreover, our results confirmed that a cultivar's root distribution is affected by soil texture. Cultivar H-65-7052 exhibits similar response in both soils on samples collected in 0–20 (~1.4 kg·m^{-2}) and 80–120 (~0.15 kg·m^{-2}) cm soil depth intervals, respectively. However, for 20–80 cm soil depths, higher root C was observed on Pulehu soils (~1.2 kg·m^{-2}) compared with Molokai soils (~1.0 kg·m^{-2}). Total root C for cultivar H-65-7052 was 2.5 kg·m^{-2} in Molokai soils and ~ 2.7 kg·m^{-2} in Pulehu soils. While for the rest of cultivars, there is no consistent or similar response in both soils. Cultivar H-78-3567 exhibits higher root C (sum 0–120 cm) when grown in Pulehu soils (~4.7 kg·m^{-2}) compared with Molokai soils (~2.1 kg·m^{-2}). Cultivar H86-3792 and H-87-4319 exhibits higher root C (sum 0–120 cm) when grown in Molokai soils (~4.4, 3.6 kg·m^{-2}) compared with Pulehu soils (~3.1, 2.2 kg·m^{-2}), respectively.

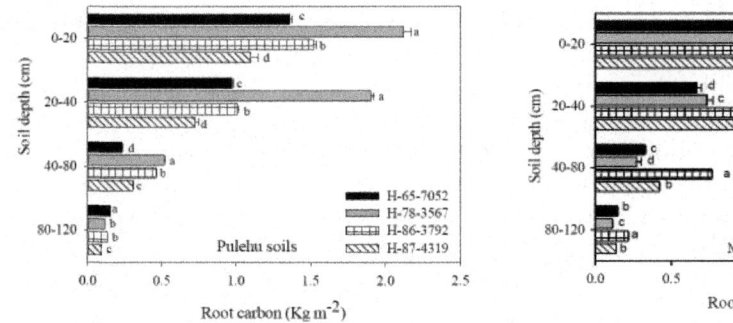 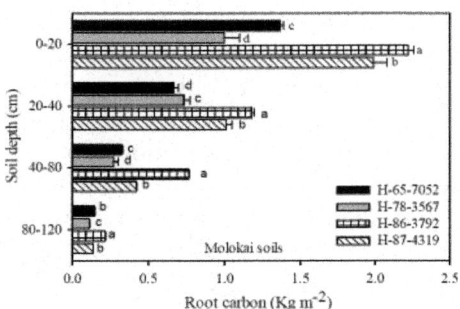

Figure 10. Root carbon (kg·m^{-2}) from sugarcane cultivars collected less than two weeks prior harvest date. Means from sugarcane cultivars bars followed by the same letter or no letters are not significantly different Tukey's test at Pr < 0.05.

4. DISCUSSION

4.1. Impacts of Hawaiian Sugarcane Practices Compared to Other Sugarcane Growing Regions

Nutrient efficiencies are crucial agricultural concerns in many regions of the world [34]; especially in tropical settings where the soils tend to be highly weathered. It is well known that the amounts of soil nutrients, such as carbon and nitrogen, are affected by climate factors, management practices (*i.e.*, tilling, liming, fertilization, burning, and irrigation type), vegetation and soil type [35,36]. Rozeff [37] reported that EC greater than 3 dS·m^{-1} may represent a problem tissue sugar concentration [38] and biomass in sugarcane [39,40]. Watcharapirak [41] estimated C storage in various growing stages of Thai sugarcane. The author found that C storage in plant and soil increased with sugarcane growth stages by storing C in the soil, plant, and ground cover. In sugarcane, C was mostly stored in stems (1700-5150 kg ·ha^{-1}) at all the growing stages, but the C accumulation rates varied with soil properties and farming management (*i.e.*, fertilization, amendment and irrigation). More carbon is stored in well managed fields [41]. The present study showed a correlation between biomass stored in plants and the quantity of C stored in primary stems [41]. Also, they found that C accumulation rate in the soil increased from tillering to stalk elongation stage (grand growth) and then decreased with plant maturity. In our study, higher organic C content at Molokai series (Oxisol) can be attributed to clay protection of the organic matter as reported by Dominy *et al.* [42]. In Brazil, similar results were found in very clayey [43] and medium-textured soil [44], but not in sand-textured soil [45]. Additionally, a recent review by Cerri *et al.* [46] shows that non-burned areas accumulate more carbon in top soil than burned areas, and this accumulation depends on soil texture with a 3-fold higher accumulation in clayey than in sandy soil. Robertson and Thorburn [47] showed that soil organic C and total N at 10 and 25 cm soil depth were up to 21% greater than in burned soils. In our study, an increase of C based on sugarcane growing stages was up to two years of growth; however, this pool is affected again by the following burning of cane. According to Robertson and Thorburn [47] and Souza *et al.* [44], crop residues that could be returned to soil are lost in fires, preventing accumulation of nutrients and organic matter from litter and thereby compromising C sequestration and microbial activity in the soils. Also, Thorburn *et al.* [48] found changes in soil C concentrations are highly site specific and not in proportion to the residues were retained (*i.e.*, soil C decrease (up to 2.5 cm)) by 0.9 g ·kg^{-1} and 0.5 g ·kg^{-1} at sites where residues had been retained for one and 17 years, respectively, but increased by 2.0 g ·kg^{-1} at a site with residues retained for six years.

Plant burning produces C in charcoal form, which is non-reactive biologically and chemically when compared to other organic materials of soil, thus representing an inert carbon fraction in C-cycling models [49]. A 57-year study on a Cambissolo (Inceptsol) soil in northern Rio de Janeiro state, Brazil, managed with pre-harvest sugarcane burning crops exhibited TOC levels of 13.3 g ·kg^{-1} at a depth of 0–20 cm and 11.80 g ·kg^{-1} at 20–40 cm [50]. Compared to these values, the TOC levels detected in the present study were higher in the 10- and 20-year crop areas but lower in the 1- and 5-year crop areas. Sugarcane crops managed with straw burning for 50 years in the northern São Paulo state contained between 15.4 and 19.2 g ·kg^{-1} of soil TOC at the 0–40 cm depth in clay textured Latossolo Vermelho Distroférrico (Oxisol) and 6.0–8.4 g ·kg^{-1} in an area with loamy-sand textured Argissolo Vermelho-Amarelo Distroférrico (Ultisol) [51]. The TOC values found in the present study were higher in the area with loamy-sand textured soils. Pre-harvest sugarcane burning in crops planted in Latossolo Vermelho-Amarelo Distrófico (Oxisol) produces lower TOC than crude sugarcane with or without straw incorporation, with no differences between the two last techniques [44]. In Cambissolo (Inceptsol) areas, long-term sugarcane cropping without straw burning produces 78% higher TOC in the 0–20 cm soil layer and 48% higher carbon in the 20–40 cm layer than in areas managed with pre-harvest straw burning [50].

In sugarcane managed with pre-harvest straw burning, Correia and Alleoni [25] found higher TOC (22.7 g kg^{-1} at the 0–5 cm depth and 20.8 g kg^{-1} at the 5–10 cm depth), probably because the area of the present study exhibited higher sand content. Comparing sugarcane crops managed with straw burning for 55 years with non-burned crop areas, Canellas *et al.* [50] found a 40% decrease in TOC stocks in top soil and 35% decrease in the subsurface layer. Canellas *et al.* [50] argued that sugarcane straw burning promoted SOM oxidation and exposed the soil surface to erosion, thereby decreasing TOC stocks over time. We showed that carbon stocks in the area with 1-year sugarcane crop were not different from those in pasture areas at the depths of 10–20 cm and 20–30 cm. Reductions in carbon stocks were also found in pasture and Cerrado areas. These results indicate that the longer the period of time with stalk burning management, the greater the losses of C stocks. The values of bulk density (Bd) (Table 2) were higher in areas planted with sugarcane than those without sugarcane, leading to increased Bd content and decreased soil aggregation and water infiltration because of soil compacting. Given the increase in soil resistance, root penetration becomes more difficult compromising root system development and ultimately diminishing crop yield in areas with high Bd. In fact, productivity in 2009 and 2010 was 113 and 110 Mg biomass ha^{-1} in areas under 1-year old crop; 111 and 106 Mg ha^{-1} for five-year old crop; 85 and 108 Mg ha^{-1} for 10-year old crop and 96 and 85 Mg ha^{-1} for 20-year old crop [47]. The increase in soil Bd, such observed in the present study, is commonly observed in areas converted from natural vegetation into cropland [52]. In such cases, soil aggregates are broken by soil tilling and agricultural machines, causing organic matter loss.

4.2. Relative Nitrogen Uptake Efficiency and Potential for Nitrate Leaching in Hawaiian Sugarcane

Large amounts of N in the form of Urea (345 to 375 kg of N ha^{-1}) have been applied to our experimental fields at the CSP plantation in Hawaii, to minimize crop stress and maximize biomass growth and sugar production. Historically, recommendations of N application rates have been 150–200 kg ha^{-1} in the US [53] and ~ 220 kg ha^{-1} of N globally [54]. Kwong and Deville [55] found the active period of N uptake by sugarcane occurs in the first six months after N application. However, less than 50% of the annually applied N fertilizer is taken up by the sugarcane crops [56,57]. Historically, low efficiencies of N uptake in sugarcane crops were observed in South Africa (9%–31%) [58] and in Taiwan (10%–25%) [57]; however, Australian sugarcane cultivars are able to recover between 6%–54%. Chang and Wang [57] found differences in N uptake efficiency between cultivars. A non-responsive cultivar recovered only 24% of the N *versus* a responsive cultivar which is able to recovered 45% of the N. A further sample was observed in Hawaii [58], where one cultivar (H-49-3533) showed a linear response in N uptake over a wide range of fertilizer levels while the other cultivar (H-50-7209) showed a typical non-linear response.

Nitrogen (N) pollution is considered one of the major threats to ecosystem integrity and biochemical cycles on sugarcane plantations [59]. When sugarcane is burnt either pre or post-harvest, 70%–95% of the dry matter and N are lost from the system [60], with nitrate (NO_3-N) leaching being one of the main pathways [61]. The magnitude of leaching varies with soil type, cropping system, weather conditions, and fertilizer regimes [62,63]. Nitrate leaching is associated with percolation of water and fertilizer application [36]. However, less information in regard to N leaching is available for tropical and sub-tropical, undeveloped agricultural regions. Nitrate produced through nitrification processes in the upper layers can subsequently move downward and accumulate in deeper layers [64,65]. However, in our study, no statistical difference was found for dissolved nitrogen in the form of nitrates (NO_3-N) in the whole soil profile (0–120 cm) (Data not shown). The NO_3-N concentration (around 5ppm) found in the present study was similar to those encountered in grassland temperate regions [36] and nitrate concentration increases in the soil profile (up to 120 cm) with the increase of N fertilizer (150–350 kg $\cdot$ha^{-1}) and irrigation. Maximum nitrate level 18 ppm accrued in 350 kg $\cdot$ha^{-1} of N at deeper depth (30–60 cm- clay loam) and irrigation treatment. With passing time (*i.e.*, sampling dates with respect of growing stages) the nitrate concentration increased at deeper depths (clay loam to clay) and decreased in the upper layers (sandy clay loam 0–30 cm).

4.3. Root Dynamics and Soil C Sequestration

Root systems play an important role in the organic matter and nutrient dynamics of the sugarcane growth [66]. In harvest systems without burning, most of the organic matter is returned to the soil with trash harvesting and reincorporation into the soil, whereas in burned systems, the main return of C is through root turnover [67]. Grahan and Haynes [68] found total root biomass was similar under burning and trashing management practices based on a redistribution of roots towards the first 10 cm of soil in inter-row space versus in row as root proliferated beneath the trash mulch. Also, they found soil C content decreased in response to increasing plant row distance (up to 20 cm) and change of management practice (*i.e.*, burning versus trashing up to 10 cm). Annual C inputs from fine roots frequently equal or exceed those from leaves and can occur to great depths and transfer C deep into subsoil horizons [69,70]. Root systems C exertion ranges from 8% to 26% for sugarcane at 124 DAP, and this response varies depending on sugarcane cultivar, root and air temperature [71]. Rostron [72] established a value of 17% of root system C exertion for South African sugarcane cultivar NCo376 growing under irrigation at 224 DAP. Our study found a similar value of 33% C exertion by the root systems for the four sugarcane cultivars two years after planted. Also, our results demonstrated different root C content are based on cultivar's root distribution growing in different soil texture. Van Anterwerpen [66] found an effect of soil texture on well watered sugarcane cultivar NCo376 root distribution per depth interval. They found highest root biomass in the first 45 cm of clay soil while in sandy soil it was at the 45–75 cm depth. At deeper soil layers (*i.e.*, 75 to 120 cm) there was no difference in root biomass between soil types. Smith *et al.* [73] found that the maximum rooting depth of sugarcane varied with genotype.

5. CONCLUSIONS

This study evaluated the efficacy of Hawaiian sugarcane cultivars in the accumulation and storage of carbon and nitrogen across the two years of sugarcane growth cycles on two contrasting soils. At both soils, total carbon was increased with the advancement of growing stages in all the four sugarcane cultivars. The carbon (*i.e.*, total, organic and dissolved organic carbon) and total nitrogen accumulation varied in the whole profile (up to 120 cm) depending on the ability of the sugarcane roots to explore and grow in the different soils. Nitrate concentration did not exceed 5 ppm in all sampling dates for the four Hawaiian sugarcane cultivars growing in both soil types; the low concentration of nitrate indicates that most of the applied nitrogen was taken up by the sugarcane plant or little being leached. Based on the results we recommend that the selectively use of sugarcane cultivars with improved traits (such as the cultivars H-65-7052 and H-86-3792 evaluated in the study) can help improve soil carbon and nitrogen cycles, provided that improved farming practices are employed.

ACKNOWLEDGMENT

The authors would like to acknowledge the Office of Naval Research and USDA-Agricultural Research Service, National Program 211: Water Availability and Watershed Management (project numbers: 2034-13000-011-00 and 2036-61000-015-00) for providing funding, Mae Nakahata, Agronomist for Hawaiian Commercial and Sugar Company helped with all aspects of the field experiments. We would also like to acknowledge Adel Youkahana, Jim Gartung, Jason Drogowski, Michael Ross, Ronald Cadiz, Justin Lau, Huihui Zhang, Neil Abrayni and Meghan Pawlowski for field assistance and to Matthew Gonzales, Don Tucker, Phyllis Ukatu and Julianne Anaya for laboratory assistance at the USDA-ARS San Joaquin Valley Agricultural Sciences Center in Parlier, CA. USDA is an equal opportunity employer and provider.

Author Contributions

Dong Wang and Rebecca Tirado- Corbalá designed the experiment with feedback from James Ayars. Rebecca Tirado-Corbalá conducted most field sampling and analysis in the experimental work with sampling assistance and feedback from Ray Anderson. Rebecca Tirado-Corbalá and Ray Anderson wrote the manuscript, and all the authors revised the manuscript.

Conflicts of Interest

The authors declare no conflict of interest.

REFERENCES

1. Kingston, G.; Annink, M.C.; Allen, D. Acquisition of nitrogen by ratoon crops of sugarcane as influence by waterlogging and split applications. *Proc. Conf. Aust. Soc. Sugar Cane Technol.* **2008**, *26*, 238-246.

2. Furlan, F.F.; Tonon-Filho, R.; Pinto, F.H.P.B.; Costa, C.B.B.; Cruz, A.J.G.; Giordano, R.L.C.; Giordano, R.C. Bioelectricity *versus* bioethanol from sugarcane bagasse: It is worth being flexible? *Biotechnol. Biofuels* **2013**, *6*, 142. doi:10.1186/1754-6834-6-142.

3. Chandel, A.K.; da Silva, S.S.; Carvalho, W.; Singh, O.V. Sugarcane bagasse and leaves: Foreseeable biomass of biofuel and bio-products. *J. Chem. Technol. Biotechnol.* **2012**, *87*, 11-20.

4. Steiner, J.L. Biofuels: No Single Answer, Many Possibilities. *Agric. Res.* **2012**, *60*, 2.

5. Evensen, C.I.; Muchow, R.C; El-Swaify, S.A.; Osgood, R.V. Yield accumulation in irrigated sugarcane: I. Effect of crop age and cultivar. *Agronomy J.* **1997**, *89*, 638-646.

6. Heinz, D.J.; Osgood, R.V. A History of the Experiment Station: Hawaiian Sugar Planters' Association. *Hawaii. Planters' Record.* **2009**, *61*, 1-108.

7. Anderson, R.G.; Tirado-Corbalá, R.; Wang, D.; Ayars, J.E. Long-Rotation sugarcane in Hawaii sustains high carbon accumulation and radiation use efficiency in 2nd year of growth. *Agric. Ecosyst. Environ.* **2015**, *199*, 216-224.

8. Giambelluca, T.W.; Chen, Q.; Frazier, A.G.; Price, J.P.; Chen, Y.-L.; Chu, P.-S.; Eischeid, J.K.; Delparte, D.M. Online Rainfall Atlas of Hawai'i. *Bull. Am. Meteorol. Soc.* **2013**, *94*, 313-316.

9. Crop Profile for Sugarcane in Hawaii. 2000. Available online: www.ipmcenters.org/ crop-profiles/docs/his sugarcane.html (accessed on 28 August 2011).

10. Moore, R.C; Fitschen, J.C. The drip irrigation revolution in the Hawaiian sugarcane industry. In Visions of the Future: Proceedings of the 3rd National Irrigation Symposium, Held in Conjunction with the 11th Annual International Irrigation Exposition, 1990, Phoenix Civic Plaza, Phoenix, AZ, USA. (accessed on 6 October 2011).

11. Koehler, P.H.; Moore, P.H.; Jones, C.A.; dela Cruz, A.; Maretzki, A. Response of Drip-Irrigated Sugarcane to Drought Stress. *Agronomy J.* **1982**, *74*, 906.

12. Comstock, J.C; Ferreira, S.A.; Tew, T.L. Hawaii's Approach to Control of Sugarcane Smut. *Plant Dis.* **1983**, *67*, 452.

13. Sugihara, R.T.; Tobin, M.E.; Koehler, A.E. Zinc Phosphide Baits and Prebaiting for Controlling Rats in Hawaiian Sugarcane. *J. Wildland Manag.* **1995**, *59*, 882-889.

14. Plaut, Z.; Meinzer, F.C.; Federman, E. Leaf development, transpiration ad ion uptake and distribution in sugarcane cultivars grown under salinity. *Plant Soil* **2000**, *218*, 59–69.

15. Lal, R. Sequestration of atmospheric CO2 into global carbon pool. *Energy Environ. Sci.* **2008**, *1*, 86–100.

16. Yamauchi, H.; Bui, W. *Drip irrigation and the survival of the Hawaiian sugarcane, Research Extension Series*; 113. Hitahr: College of Tropical Agriculture and Human Resources, Honolulu, Hawai'i, USA, 1990.

17. Davis, S.C.; Anderson-Teixeira, K.J.; DeLucia, E.H. Life-Cycle analysis and the ecology of biofuels. *Trends Plant Sci.* **2009**, *14*, 140–146.

18. Crutzen, P.J.; Mosier, A.R.; Smith, K.A.; Winiwarter, W. N2O release from agro-biofuel production negates global warming reduction by replacing fossil fuels. *Atmos. Chem. Phys.* **2008**, *8*, 389–395.

19. Bengtsson, J.; Lundkvist, H.; Saetre, P.; Sohlenius, B.; Soldreck, B. Effects of organic matter removal on the soil food web: forestry practices meet ecological theory. *Appl. Soil Ecol.* **1998**, *9*, 137–143.

20. Ulgiati, S. A. Comprehensive Energy and Economic Assessment of Biofuels: When "Green" is not enough. *Crit. Rev. Plant Sci.* **2001**, *20*, 71–106.

21. Mikhailova, E.A.; Post, C.J. Effect on land use on soil inorganic carbon stocks in the Russian Chernozen. *J. Environ. Qual.* **2006**, *35*, 1384–1388.

22. Wu, H.B.; Guo, Z.T.; Gao, Q.; Peng, C.H. Distribution of soil inorganic carbon storage and its changes due to agricultural land use activity in China. *Agric. Ecosyst. Environ.* **2009**, *129*, 413–421.

23. De Figueiredo, E.B.; la Scala, N., Jr. Greenhouse gas balance due to the conversion of sugarcane areas from burned to green harvest in Brazil. *Agric. Ecosyst. Environ.* **2011**, *141*, 77–85.

24. Galdos, M.V.; Cerri, C.C.; Cerri, C.E.P. Soil carbon stocks under burned and unburned sugarcane in Brazil. *Geoderma* **2009**, *153*, 347–352.

25. Correira, B.L.; Alleoni, L.R.F. Conteúdo de carbono e atributos químicos de Latossolo sob cana-de-açúcar colhida com e sem queima. *Pesq. Agropec. Bras.* **2011**, *46*, 944–952.

26. Christensen, B.T. Physical fractionation of soil and structural and functional complexity in organic matter turnover. *Eur. J. Soil Sci.* **2001**, *52*, 345–353.

27. Bayer, C.; Mielniczuk, J. Dinamica e funcao da material organica. In *Fundamentos da Materia Organica do Solo:Ecossistemas Tropicais e Subtropicais*, 2nd ed.; Santos, G.A., Silva, L.S., Canellas, L.P., Camargo, F.A.O., Eds.; Metrópole: Porto Alegre, Brazil, 2008; pp. 7–18.

28. Hennecke, A.M.; Faist, M.; Reinhardt, J.; Junquera, V.; Neeft, J.; Fehrenbach, H. Biofuel greenhouse gas calculations under the European Renewable Energy Directive—A comparison of the BioGrace tool *vs.* the tool of the Roundtable on Sustainable Biofuels. *Appl. Energy* **2013**, *102*, 55–62.

29. Huang, H.; Khanna, M.; Önal, H.; Chen, X. Stacking low carbon policies on the renewable fuels standard: Economic and greenhouse gas implications. *Energy Policy* **2013**, *56*, 5–15.

30. Soil Survey Staff, Natural Resources Conservation Service, United States Department of Agriculture. Official Soil Series Descriptions, 2013. Available online: https://soilseries.sc.egov. usda.gov/OSD_Docs/C/CONSUMO.html. (accessed on 8 June 2014).

31. Thomas, G.W. Soil pH and Soil Acidity. In *Methids of Soil Analysis: Chemical Methods. Part 3*; Sparks, D.L., Ed.; Soil Science Society of America: Madison, WI, USA, 1996.

32. Bouyoucos, G.J. Hydrometer method improved for making particle size analysis of soil. *Agronomy J.* **1962**, *54*, 464–465.

33. Warncke, D.; Brown, J.R. Potassium and other basic cations. In *Recommended Chemical Soil Test Procedures for the North Central Region*; NCR Publication No. 221; Missouri Agri Exp Station: Columbia, MO, USA, 1998; pp. 31–33.

34. Moiser, A.R.; Syers, J.K.; Freney, J.R. *Agriculture and the Nitrogen Cycle: Assessing the Impacts of Fertilizer use on Food Production and the Environment*; Island Press: Washington, DC, USA, 2004.

35. Hartemink, A.E. Sugarcane for bioethanol: Soil and environmental issues. *Adv. Agronomy* **2008**, *99*, 125–182.

36. Bahmani, O.; Nasab, S.B.; Behzad M.; Naseri, A.A. Assessment of Nitrogen Accumulation and Movement in Soil Profile under Different Irrigation and Fertigation Regime. *Asian J. Agric. Res.* **2009**, *3*, 38–46.

37. Rozeff, N. Sugarcane and Salinity—A review paper. *Sugarcane* **1995**, *5*, 8–19.

38. Kumar, S.; Naidu, K.M.; Sehtiya, H.L. Causes of growth reduction in elongating and expanding leaf tissue of sugarcane under saline conditions. *Aust. J. Plant Physiol.* **1988**, *21*, 79–83.

39. Dominquez, S.P. Salim and sodicity in new sugarcane regions of Ingenio Providencia. *Centro Investig. Cana Azúcar Columbia* **1993**, *l2*, 44–45.

40. Gomez, P.J.F.; Torres, A.J.S. Effect of salinity in the development and production of two varieties of sugarcane (*Saccharum* sp). *Centro de Investigacion de la Cana de Azucar de Colomba* **1993**, *12*, 35–36.

41. Watcharapirak, W. The Estimation of Carbon Storage in Various Growth Stages of Sugarcane in Sisatchanalai District, Sukhothai Province, Thailand. Master Thesis, Mahidol University, Bangkok, Thailand, 2009.

42. Dominy, C.S.; Haynes, R.J. Influence of agricultural land management on organic matter content, microbial activity and aggregate stability in the profiles of two Oxisols. *Biol. Fertil. Soils* **2002**, *36*, 298–305.

43. Orlando Filho, J.; Rosetto, R.; Muraok, T.; Zotelli, H.B. Efeitos do sistema de despalha (cana crua x cana queimada) sobre algumas propriedades do solo. *STAB* **1998**, *16*, 30–35.

44. Souza, R.A.; Telles, T.S.; Machado, W.; Hungria, M.; Tavares Filho, J.; Guimarães, M.F. Effects of sugarcane harvesting with burning on the chemical and microbiological properties of the soil. *Agric. Ecosyst. Environ.* **2012**, *155*, 1–6.

45. Ball-Coelho, B.; Sampaio, E.V.S.B.; Tiessen, H.; Stewart, J.W.B. Root dynamic in plant ratoon crops of sugar cane. *Plant Soil* **1992**, *142*, 297–305.

46. Cerri, C.C.; Galdos, M.V.; Maia, S.M.F.; Bernoux, M.; Feigl, B.J.; Powlson, D.; Cerri, C.E.P. Effect of sugarcane harvesting systems on soil carbon stocks in Brazil: an examination of existing data. *Eur. J. Soil Sci.* **2011**, *62*, 23–28.

47. Robertson, F.A.; Thorburn, P.J. Management of sugarcane harvest residues: Consequence for soil carbon and nitrogen. *Aust. J. Soil Res.* **2007**, *45*, 13–23.

48. Thorbourn, P.J.; Meier, E.A.; Collins, K.; Robertson, F.A. Changes in soil carbon sequestration, fractionation and soil fertility in response to sugarcane residue retention are site-specific. *Soil Tillage Res.* **2012**, *120*, 99–111.

49. Smernik, R.J.; Skjemstad, J.O.; Oades, J.M. Virtual fractionation of charcoal from soil organic matter using solid state13C NMR spectral editing. *Aust. J. Soil Res.* **2000**, *38*, 665–683, doi:10.1071/SR99115.

50. Canellas, L.P.; Velloso, A.C.X.; Marciano, C.R.; Ramalho, J.F.G.P.; Rumjanek, V.M.; Rezende, C.E.; Santos, G.A. Chemical soil properties of an Inceptisol under long-term sugarcane crops with vinasse application and without slash burning. *R. Bras. Ciênc. Solo* **2003**, *27*, 935–944.

51. Luca, E.F.; Feller, C.; Barthès, C.C.; Chaplot, B.V.; Campos, D.C.; Manechini, C. Avaliação de atributos físicos e estoques de carbono e nitrogênio em solos com queima e sem queima de canavial. *R. Bras. Ciênc. Solo* **2008**, *32*, 789–800.

52. Resende, A.S.; Xavier, R.P.; Oliveira, O.C.; Urquiaga, S.; Alves, R.J.; Boddey, R.M. Long-term effects of pre-harvest burning and nitrogen and vinasse application on yield of sugarcane and carbon and nitrogen stocks on a plantation in Pernambuco, N.E. Brazil. *Plant Soil* **2006**, *281*, 339–351.

53. Wiedenfeld, B.; Enciso, J. Sugarcane Responses to Irrigation and Nitrogen in Semiarid South Texas. *Agronomy J.* **2008**, *100*, 665–671.

54. Calcino, D.V. *Australian Sugarcane Nutrition Manual*; Sugar Research and Development Corporation/Bureau of Sugar Experiment Stations: Indooroopilly, Queensland, 1994.

55. Kwong, K.F.N.K; Deville, J. Application of 15N-labelled urea to sugarcane through a drip-irrigation system in Mauritius. *Fertil. Res.* **1994**, *39*, 223–228.

56. Takahashi, D.T. Effect of amount and timing on the fate of fertilizer nitrogen in lysimeter studies with N15. *Haw. Planters' Rec.* **1967**, *57*, 292–309.

57. Chang, J.H. The role of climatology in the Hawaiian sugar-cane industry: An example of applied agricultural climatology in the tropics. *Pac. Sci.* **1963**, *17*, 379–397.

58. Stanford, G.; Ayres, A.S.; Dol, M. Mineralizable soil nitrogen in relation to fertilizer needs of sugarcane in Hawaii. *Soil Sci.* **1965**, *99*, 132–137.

59. Wood, R.A. The effect of time of application on the utilization of fertilizer nitrogen by plant cane. *Proc. Int. Soc. Sugar Cane Technol.* **1974**, *15*, 618–629.

60. Thorburn, P.J.; Biggs, J.; Webster, A.; Biggs, I.M. An improved way to determine nitrogen fertiliser requirements of sugarcane crops to meet global environmental challenges. *Plant Soil.* **2011**, *339*, 51–67.

61. Mitchell, R.D.J.; Thorburn, P.J.; Larson, P. Quantifying the immediate loss of nutrients when sugarcane residues are burnt. *Proc. Aust. Soc. Sugar Cane Technol.* **2000**, *22*, 206–211.

62. Ghiberto, P.; Libardi, P.; Brito, A.; Trivelin, P. Leaching of nutrients from a sugarcane crop growing on an Ultisol in Brazil. *Agric. Water Manag.* **2009**, *96*, 1443–1448.

63. Di, H.J.; Cameron, K.C. Nitrate leaching in temperate agroecosystems: Sources, factors and mitigating strategies. *Nutr. Cycling Agroecosyst.* **2002**, *64*, 237–256.

64. Verloop, J.; Boumans, L.J.M.; van Keulen, H.; Oenema, J.; Hilhorst, G.J.; Aarts, H.F.M.; Sebek, L.B.J.. Reducing nitrate leaching to groundwater in an intensive dairy farming system. *Nutr. Cycling Agroecosyst.* **2006**, *57*, 67–73.

65. Pierzynski, G.M.; Sims, J.T.; Vance, G.F. *Soils and Environmental Quality*, 3rd ed.; Taylor and Francis: Boca Raton, FL, USA, 2005.

66. Van Antwerpen, R. Sugarcane root growth and relationships to above-ground biomass. *Proc. S. Afr. Sug. Technol. Ass.* **1999**, *73*, 89–94.

67. Wilhelm, W.W.; Johnson, J.M.F.; Hatfield, J.L.; Vorhees, W.B.; Linden, D.R. Crop and soil productivity response to corn residue removal: A literature review. *Agronomy J.* **2004**, *78*, 184–189.

68. Graham, M.H.; Haynes, R.J. Organic matter status and the size, activity and metabolic diversity of the soil microbial community in the row and inter-row of sugarcane under burning and trash retention. *Soil Biol. Biochem.* **2004**, *38*, 21–31.

69. Trumbore, S.E.; Davidson, E.A.; Barbosa, P.C.; Nepstadt, D.C.; Martinelle, L.A. Belowground cycling of carbon in forests and pastures of Eastern Amazonia. *Glob. Biogeochem. Cycles* **1995**, *9*, 515–528.

70. Canadell, J.; Jackson, R.B.; Ehleringer, J.R.; Mooney, H.A.; Sala, O.E.; Schulze, E.-D. Maximum root depth of vegetation types at the global scale. *Oecologia* **1996**, *108*, 583–595.

71. Brodie, H.W.; Yoshida, R.; Nickell, L.G. Effect of air and root temperatures on growth of four sugarcane clones. *Hawaii. Sugarcane Plant. Assoc.* **1965**, *58*, 21–52.

72. Rostron, H. Radiant energy interception, root growth, dry matter production and the apparent yield potential of two sugarcane varieties. *Proc. Int. Soc. Sugar Cane Technol.* **1974**, *15*, 1001–1010.

73. Smith, D.M.; Inman-Bamber, N.G.; Thorburn, P.J. Growth and function of the sugarcane root system. *Field Crops Res.* **2005**, *92*, 169–183.

This page left intentionally blank.

INDEX

This page left intentionally blank.